Landmark Essays

4 LB

Landmark Essays

on

Rhetoric of Science:
Case Studies

Edited by Randy Allen Harris

LEA Hermagoras Press
An Imprint of Lawrence Erlbaum Associates, Publishers

Landmark Essays Volume Eleven

Cover design by Kathi Zamminer

Lawrence Erlbaum Associates, Inc., Publishers
10 Industrial Avenue
Mahwah, New Jersey 07430

Library of Congress Cataloging-in-Publication Data

Landmark essays on the rhetoric of science : case studies / edited by
  Randy Allen Harris.
      p.   cm. -- (Landmark essays : v. 11)
    Includes bibliographical references and index.
    ISBN 1-880393-11-5 (pbk. : alk. paper)
    1. Communication in science--Philosophy. 2. Rhetoric--Philosophy.
  3. Persuasion (Rhetoric)   I. Harris, Randy Allen. II. Series.
  Q223.L26 1996
  808'.0665--dc20   96-28763
                                                        CIP

Books published by Lawrence Erlbaum Associates are printed
on acid-free paper, and their bindings are chosen
for strength and durability.

Printed in the United States of America

10 9 8 7 6 5 4 3 2

*This book is dedicated to
Pixie Parsons and Larry Naidoo
for Indira*

*and to
Barrie Parsons and Susan Naidoo*

But yet an union in partition.
—William Shakespeare

About the Editor

Randy Allen Harris is Associate Professor of Rhetoric and Professional Writing in the Department of English at the University of Waterloo. He publishes widely on rhetoric, linguistics, and professional writing.

Table of Contents

Introduction

Writing Science

Introduction

We scientists do not attend professional meetings in order to announce our findings ex cathedra, *but in order to argue.*

—John Polanyi

Scientists argue.[1] Nobody disputes that. Their standards of evidence are frequently more rigid than those of other arguers, and they are so good at arguing that they won't let each other get away with weak cases for very long, and the ultimate matter of their arguments is so concrete that we can stub our toes on it, or burn our fingers on it, or have our cars chased by it, and the overall level of agreement they achieve is amazing when compared to politics or religion or literary criticism. Science is a remarkable, remarkably successful, enterprise. But scientists argue, and their arguing is absolutely central to their success: science is rhetorical.

Still, the phrase "rhetoric of science" does not come trippingly off all tongues. The standards of evidence, and the self-policing rigour, and the toe-stubbingly hard subject matter, and the agreement, and the success, frequently lead people to suppose (scientists not least among them) that there is something so different about the arguing in science that a category shift of some profound sort happens, that we can't talk about the arguing of scientists in the same ways as we talk about the arguing of politicians and priests and literary critics and grocers,—that there is something so different about the arguing in science that "rhetoric of science" is blasphemy. This impulse stems from the immensely opposite connotations of the two words on either side of the preposition in that phrase, the tarring blame of *rhetoric* and the imperious praise of *science*.

These words, and the pursuits they denote, have had very different trajectories since the Enlightenment. Science has undergone an almost-literal apotheosis, taking over so many of the functions of religion that the primary meaning of *lay-person* has shifted from "non-cleric" to "non-scientist". The incorporation of the word *science* carries such prestige that entire communities graft it onto their concerns titularly: political science, computer science, scientology, . . . It brings such commanding authority to assertions that we regularly get headlines like "Science proves Shroud of Turin fake" and "Science close to cure for AIDS".

Meanwhile rhetoric underwent a slide deep into ignominy. Once a member of the seven sciences, rhetoric is now "rant", "bombast", and "twaddle", to take only a few of Roget's choicest synonyms. Outside a few cloistered academic communities, it is attached readily only to pursuits the attacher wants to demean as empty, or specious, or merely florid: "political rhetoric" is not the name of an academic discipline courting respectability, but another label for a politician's bafflegab; the default reading for "computer rhetoric" would be deliberately obfuscating technical talk; the likelihood of a religion with the label "rhetorology" recruiting Tom Cruise and John Travolta to its ranks is low in the extreme. Well, perhaps not, religion and Hollywood being what they are. But the appearance of *rhetoric* in the newspaper is virtually always derisive. We are a very long way from headlines like "Rhetoric proves Reagan lied" or "Rhetoric close to cure for unethical advertising".[2]

But let's back away from connotation. Rhetoric is the study and practice of suasion, per- or dis-, and suasion is the motive and the meat of all arguments. Science is the study of nature and the practice of making knowledge about nature (or, in some extremely mechanistic views, the practice of *finding* knowledge about nature).

Rhetoric of science is simply, then, the study of how scientists persuade and dissuade each other and the rest of us about nature,—the study of how scientists argue in the making of knowledge. In fragmented form, it goes back as long as the two fields have existed, and it makes various appearances throughout the history of each. In the ancient world, there are the mangled works of the Sophists, many of whom were both rhetoricians and scientists, and Aristotle, who was seminal in both rhetoric and science, occasionally speaks about them in the same breath (as when he says that dialectic and rhetoric are employed by the sciences, but are not sciences themselves; *Rhetoric* 1359b).[3] Francis Bacon's many writings on induction include a great deal of rhetorical and scientific commentary in concert, such as his discussion of invention in both. And Joseph Priestley's *Course of Lectures on Oratory and Criticism* contains a wide range of rhetorical advice directed explicitly at his fellow scientists. But the development of a specific genre of rhetorical analysis directed at scientific discourse is a very recent phenomenon, only now beginning to flower.[4]

The papers in this volume are those flowers. They apply rhetorical criticism to the arguments of scientists,—to the claims of giants like Newton and Darwin, to the controversies of archaeologists and biologists, to the public discourse of geneticists and virologists, to the evolution of experimental reports and the peer-negotiated journal publications that constitute and represent the knowledge of science.

The Rhetorical Turn

Rhetorical analysis will be politically more effective when rhetoric is taken more, and science less, seriously.

—Alan Gross

As an academic field, rhetoric has fallen far from the heights of the Middle Ages and the Rennaisance. It has never sunken to quite the depths the word occupies in ordinary language, but it no longer has the significant role it once had in the preparation of university graduates for participation in civic discourse. In North America especially, it found itself housed mostly in departments of public address and speech communication, with its research energies focused almost entirely on criticism, on the examination of particular cases of discourse for their suasive elements.

Until very recently, though, almost nobody felt that the subject matter of rhetorical criticism, those particular cases, could come from the sciences. Take the important and famous Wingspread conference, when a collection of rhetorical luminaries gathered in 1970 to fret and crow about "The Prospects of Rhetoric". They talked in fact a great deal about science, but only as method. They asked: Is rhetoric scientific? Should it be? To what extent can it borrow from the science of language (linguistics) or the science of mind (psychology) or the science of groups (sociology)? To what extent is rhetoric "menaced" by those fields? Rhetoricians were lured by the prestige of science and repelled by its anti-humanist reputation.

Despite the presence of scholars like Herbert Simons, later destined to wade into rhetoric of science up to his chin, and Robert L. Scott, the insistent modern voice that rhetoric should concern itself with the making of knowledge, and Henry W. Johnstone Jr., the editor who would publish some of the earliest rhetoric of science articles; despite a notion of rhetoric growing so voracious it was all-but-indistinguishable from 'communication'; despite the recurrent concern with opening new vistas of discourse to rhetorical investigation; despite, even, occasional proclamations like "the rhetorical dimension is unavoidable . . . in every scientific discussion which is not restricted to mere calculation"; despite, that is, an assembly of the right people with the right attitudes, virtually nothing was said at Wingspread about science as an object of interest or study.[5] Science was alien, the other, the opposite of rhetoric.

What changed? What changed to bring about the thriving subfield that this volume represents? What changed to bring us rhetoric of science?

In a word, *Kuhn*.

Kuhn and Aristotle's Walls

The late Thomas Samuel Kuhn's (1922-1996) massively influential *Structure of Scientific Revolutions*[6] was actually published well before Wingspread, in 1962, and provoked several howling rounds of debate in philosophy of science about its vision of scientific change, a vision that crucially involved persuasion. But it was with the second edition, published the same year as Wingspread, that Kuhn's notions really began to echo widely outside of the corridors of history and philosophy, and brought rhetoric to the absolute heart of science studies. "It might be carrying matters too far to describe [*Structure of Scientific Revolutions*] as a manual of rhetoric", Ernan McMullin says. "But," he adds, "there can be no doubt that rhetorical concerns manifest themselves on almost every page".[7]

It was this traffic with rhetoric that provoked the philosophical howling. Scientific change happens in two general ways according to Kuhn. In (what he calls) Normal Science, change occurs by way of additions to the data base, extensions of mechanisms, elaborations of principles, . . . an aggregate of minor growths and shiftings that follow from solving the daily puzzles of the lab; essentially, a steady accretion of knowledge. A gene is linked to a syndrome. A new behavior is attributed to a quark. A more precise calculation of cosmic dark matter is achieved. But some puzzles can't be solved within the accepted framework, within (what Kuhn calls) the paradigm of a field.[8] These puzzles, on rare occasions, lead to the second, and much bigger, type of change, to reconceptualizations so dramatic that the old framework must be discarded and a new one adopted. An earth-centered universe is replaced by a sun-centered universe. Absolute species are replaced by species related historically through evolution and natural selection. Stable continents are replaced with moving continents.

Kuhn said little new about the first type of change in *Structures*, except for his characterization of such changes as paradigm-defined puzzle solving. The big, revolutionary changes, however—where one paradigm comes to replace another—Kuhn says happen through a process antithetical to the philosophy of the time, through the defining notion of rhetoric: through persuasion. Kuhn also removed two very important conceptual barriers to rhetoric of science, barriers Aristotle had erected to keep science and rhetoric asunder,—the wall of certainty and the wall of expertise.

Rhetoric, since its very earliest formulations, has concerned itself with the probable and the contingent; science, with its fixation on certainty, was therefore out of bounds. As Thomas DeQuincey put it, "Whatever is certain, or matter of fixed science, can be no subject for the rhetorician"[9] But Kuhn showed very clearly that there is no such thing as a "fixed science". Positions and theories and data in science are not certain, he argued. They are contingent, resting on a "nest of commitments", a paradigm. They are invented, adopted, discarded, and adjudicated on the basis of likelihood.[10]

Bam! Down goes the wall of certainty.

The wall of expertise ran parallel. When discourse becomes too specialized, Aristotle said, it ceases to be rhetorical.[11] We can't be sure modern rhetoricians really believed that specialist vocabularies and appeals magically created some other, transcendent form of discourse, but they were sufficiently cowed by such vocabularies and appeals to let Aristotle's wall to stand uncontested in the case of science. Karl Wallace, for instance, allowed that some aspects of scientific discourse might be studied, particularly popularizations of science, but pointedly excluded the possibility of looking at the discourse of "experts speaking as experts to other experts".[12]

Kuhn is of a very different mind. "To understand why science develops as it does", he said,

> one must understand . . . the manner in which a particular set of shared values interacts with the particular experiences shared by a community of specialists to ensure that most members of the group will ultimately find one set of arguments rather than another decisive.

That process is persuasion.[13]

What looked to many rhetoricians like a qualitative difference between ordinary discourse and expert discourse was revealed as at-best a quantitative difference, a difference in degree, maybe a refinenement of vocabulary, maybe not even that.

Bam! Down goes the wall of expertise.

With these two walls in ruin, Kuhn's framework indisputably warrant the rhetorical investigation of science. His governing notion is that science proceeds in fits and starts, one model of reality, one paradigm, replacing another at key junctions—a heliocentric universe replacing a geocentric universe, moving continents replacing stable ones, probablisitic physics replacing deterministic physics—and at each of these key junctures, rhetoric is the engine of change. One scientific picture wins and one loses, and the winner is not objectively truer, Kuhn says, just more persuasive. It is a gross caricature of modern thought to say that everyone everywhere believed science proceeded in a linear, accretionary fashion, scientists piling more and more knowledge on their towering heaps each day before they left the lab, in some pristinely non-rhetorical fashion, until Kuhn stepped forward and said "Lo, all is suasion in the kingdom of science." But it seems that most rhetoricians thought along those lines before they encountered Kuhn, or encountered another rhetorican who had encountered Kuhn.

Other Influences

Still, *Kuhn* is, I confess, too glib an answer for the change in the 1970s that validated rhetorical investigations of science. A few more words are appropriate: *Polanyi, Toulmin, Ziman, Hanson, Habermas, Bronowski, Feyerabend.*[14] Like Kuhn, none of these scholars had much interest in speech communication, composition, suasion, or, indeed, any direct concern in the issues and themes that rhetoricians swear allegiance to. Their use of the term *rhetoric*, when they used it at all, "was at best colloquial and at worst pejorative".[15] But, like Kuhn, they contributed to a shift in the image of science so dramatically toward suasion that it now makes sense to speak, as Richard Rorty does, of "the rhetorical turn" in epistemological and science studies over the last three decades.[16] The turn comes down to a major shift, two claims that accompany the shift (in part, they are results of the shift, in part justifications), and an implication.

The shift is:

- from a concern with the products of science to the processes and practices of scientists.

The claims are:

- that science is not solely the province of individuals with beakers or telescopes or accelerator photographs, but of communities with conferences, journals, values, and goals; and

- that there is no single scientific method, but a (constrained) plurality of approaches, or styles, that differ from science to science, program to program, community to community,—all of them powerfully mediated by language.

The implication is:

- that the virtues of a scientific claim come not only from the way it is mapped against nature, but from the way it is mapped into the context of specific approaches and communities.[17]

The implication is often where the opposition to this turn comes from, because the most important property most people associate with a scientific claim is its truth (or falsity), which many, many scholars want to dissociate completely from rhetoric. And because some rhetoricians and others want to remove the *only* from its formulation.

Rorty coined the phrase, "rhetorical turn", on a perfectly reasonable analogy with the much vaunted "linguistic turn" that characterized most of the previous few decades' investigation of science and knowledge. The analogy is particularly strong in a way rarely noticed. The linguistic turn was not effected by linguists either.

An important difference in the two turns, though, and one that makes this collection possible, is that rhetoricians rapidly became interested in science; linguists, for the most part, never showed much interest in science during their time in the adjectival sun. The linguistic turn was pursued largely by philosophers, who usually had only the most casual acquaintance with linguistic theory; what they meant by *linguistic* was "semantic", in some more or less unspecified sense. The rhetorical turn, too, is being pursued by philosophers (and historians and sociologists) who usually have little acquaintance with rhetorical theory; what they mean by *rhetoric* is "nonrational", in some more or less unspecified sense.

But they have recently and eagerly been joined by a constituency with a far more intimate knowledge of the turn-defining term, and a much stronger allegiance to its traditions: rhetoricians. The reasons for this involvement are many, two of which stand especially prominent, one theoretical, one professional. The theoretical difference follows from a movement which takes its inspiration and its slogan from Robert Scott's germinal paper, "On viewing rhetoric as epistemic". The position—that rhetoric makes knowledge and truth, rather than (in the traditional Platonic-Aristotelian view) merely helping to sweeten and distribute it once some other enterprise (say, science) has uncovered it—has been hugely influential, sparking interest in relativism, rekindling study of the sophists, and permeating discussions of argument fields.

The professional difference follows from the rapid increase of programs in technical communication, and, more broadly, from the writing-across-the-curriculm movement. For most of this century, rhetoric had been in the exclusive, able custody of scholars housed in departments of speech communication, but the exponential growth of personal and small-business technologies in the 1970s and 1980s, with its need for specialists in writing about those technologies, drove the development of technical communication, a new scholarly discipline which took much of its inspiration from classical rhetoric. In roughly the same period, declining literacy among university students (or, in any case, the *perception* of declining literacy) brought composition teachers into contact with student writers outside of the humanities. These developments broadened the operating theatre of people who pledged allegiance to rhetoric and put many of them into daily contact with technical and scientific language. A substantial proportion of the rhetoricians studying science are, or have been, connected with such programs, and many of them began their rhetoric-of-science research in part to find solutions to the composition problems of their science and engineering students.[18]

There is one more difference between the linguistic and rhetorical turns that is at least as significant as the difference in levels of participation. The linguistic turn had virtually no influence on the valuation of science,—in particular, it had virtually no influence on the esteem in which science was universally held. The rhetorical turn, even without the contaminating presence of genuine, card-carrying rhetoricians, reduced that stature very substantially. In one complaint about Kuhn, this picture of science made science "irrational, a matter for mob psychology".[19]

Reciprocally, this reduced image of science was important to some, even inspirational, for the development of rhetoric of science. J. R. Ravetz, an early contributor to contemporary analyses of science, now sees their drift as distinctly unpleasant. "Working scientists have been enduring a bad press from the scholars who analyze their activity", he recently complained.[20] And worse has come to worst-possible for him with the advent of rhetorical analyses of science: "even more than a sociologist," he says, "a rhetorician has a professional motivation for reducing science to the same lowly status to which [rhetoric] has been assigned".[21] From the other side of the fence, this deflation of scientific prestige is not a debasement, not a matter "of debunking or downgrading the natural scientist", according to Rorty, looking over that fence, "but simply of ceasing to see him as a priest".[22]

Ravetz is at least partially right, however. Most rhetoricians have taken their critical tools to science because of its compelling success in achieving the ultimate goal of all rhetoric, agreement. But others, not especially to their credit, have exhibited exactly the sort of glee he fears. E. Claire Jerry, for instance, and Michael McGuire trade on the curious perspective that "science is not truth but a form of lie" to argue that rhetoric has an obligation to right the "balance in academia", to serve as "a counter-force to science's 'god-term' status".[23]

Rorty's nicely turned phrase, "the rhetorical turn", was coined to evoke and displace "the linguistic turn", but viewed from perspectives like Jerry's, or from the low, belly-dragging perspective of rhetoric in popular notions, it evokes another phrase altogether: "the worm turns".

The Rhetoricians Turn

We may well be entering a new golden era of development in rhetorical theory.

-Michael Osborn, 1983

Rhetoric of science, as a field involving rhetoricians, got under way in the mid 1970s, and expanded rapidly in the 1980s. The earliest article to string together our key terms in a defining way was Philip Wander's "The Rhetoric of Science", wherein the author notices that the increasing importance of science to contemporary public policy "obliges" rhetoricians to concern themselves with science:

> how it is used in debate; how it relates to other sources of information; what occurs when there is conflicting scientific evidence. [24]

But scientists also debate among themselves, Wander said, suggesting that

> there is yet another area within science amenable to rhetorical investigation. This has to do with the efforts made by scientists to persuade one another. Grant proposals, journal articles, and convention papers are designed to influence a professional audience (granting agencies, journal editors, and so on).[25]

Early Theory Papers

Wander's article was followed the next year by a flurry of related efforts, most of them in Henry W. Johnstone's journal, *Philosophy and Rhetoric*. Wander was concerned more with description and projection than with justification, and he had flashed only a single warrant (Ziman's discussion of the peer review process in science). Wander was also concerned much less with rhetorical theory than with rhetorical criticism. But the principal missions of this subsequent cluster of papers were (1) justifying rhetorical analyses of science, and (2) mapping out a theoretical framework for those analyses. All of them contain the word *toward* in their titles. Walter Weimer brought the news from Kuhn and Hanson and Feyerabend that science was a "rhetorical transaction". Michael Overington brought the news from Kuhn and Polanyi and Ziman that there was room for "a rhetorical analysis of science". Maurice Finocchiaro brought the news from his own investigations that "rhetoric is sometimes crucial in science".[26]

It is not at all clear how influential this cluster was. Certainly there is no direct line of descent from any of them to the far more fruitful applied work of the 1980s, and the *Philosophy and Rhetoric* papers are not especially inspiring. All of them are by non-rhetoricians (Weimer is a psychologist, Overington a sociologist, Finocchiaro a philosopher), with degrees of knowledge about rhetoric easily surpassed by their condescension toward it. Finocchiaro, for instance, says that rhetoric plays a role "in addition to argument",[27] suggesting that those two notions are independent, and Overington's rhetorical theory consists essentially of a substitution operation, *audience* for *community*. Weimer has an even more confused substitution, *persona* for *audience*.[28] And they came not so much to invest in rhetorical theory as to rehabilitate it: all of them serve up their personal frameworks as *the* structure whereby rhetoricians can now analyze science. Thanks, but no thanks, seems to have been the response from the field.

More warrant-serving papers followed over the next few years. Carolyn Miller pointed out that the sophists licensed the rhetorical investigation of science, and added a catalogue of other authorities a few years later. Tom Cook and Ron Seamon, and, in another appearance, Walter Weimer, wrote to say that Feyerabend's work was a model for rhetoric of science. Herbert Simons played the ubiquitous Kuhn backwards and forwards, concluding that "science is rhetorical through and through".[29]

Conspicuously absent in this period, though, were examples of what all this thought and talk was about: rhetorical analyses of science. There was an early forensic piece by Richard Weaver (included in this volume), the shining example of John Angus Campbell's work on Darwin (also in this volume), and a few other outliers.[30] But rhetorical investigations of science were vastly outnumbered by musings about rhetorical investigations of science. This balance shifted dramatically, and salubriously, in the mid 1980s. Finally, less than a decade ago, there was a field.

Again, it is not clear how much influence the justificatory and theorizing papers had on the explosion of case studies. They no doubt served the general purpose of adding a lot of chatter about science to rhetorical air waves which already included congenial bandwidths devoted to a resuscitation of the sophists, to the social construction of knowledge, and to rhetoric's epistemic consequences, all rapidly propagating at the time. But each case study is also personal, with a specific and individual set of motives behind it (Charles Bazerman's confessional first few pages of *Shaping Written Knowledge* is a good synechdoche for such sets[31]). Three general developments, though, have clear lines of influence on the rhetoric-of-science case-study movement: the rhetoric of inquiry; the concept of argument fields; and the sociology of scientific knowledge.

Rhetoric of Inquiry

Rhetoric of inquiry developed in the early 1980s at the University of Iowa, where Donald McCloskey, an economist more concerned than most about the discourse of his field, met Alan Nagel, an English professor with a deep interest in the Italian Renaissance, and, therefore, a deep awareness of the traditional theoretical framework for discourse studies: Nagel gave McCloskey that leading word for his interests, *rhetoric*, and together they formed a weekly seminar.[32] The seminar attracted, among others, Alan Megill, a historian, and John Nelson, a political scientist and the provider of that trailing qualifier, *of inquiry*. With their help, the movement explored ways in which rationality is a function of rhetoric, and the seminar quickly bloomed into the bracing 1984 University of Iowa Humanities Symposium on the Rhetoric of the Human Sciences, a symposium expressly called to

> explore an emerging field of interdisciplinary research on rhetoric of inquiry, a new field that stems from increased attention to language and argument in scholarship and public affairs . . . As the name implies, it differs from other postmodern accounts of science in appreciating the importance of *rhetoric*—the quality of speaking and writing, the interplay of media and messages, the judgment of evidence and arguments.[33]

The constitution of the founding group indicates, though, and the Symposium confirms, that the defining word of this passage is not the italicized flagship noun, *rhetoric*, so much the early unstressed adjective, *interdisciplinary*. The seminar enthusiastically crossed disciplinary boundaries, the Symposium even more so. The modus operandi was participant anthropology. The organizers reported on the discourse of their tribes—economists, historians, and political scientists—and invited others to do likewise. Legal scholars came to report on legal scholars, psychologists on psychologists, mathematicians on mathematicians. Anthropologists even came to report on anthropologists. Philosophers came for the sport.

The result was exhilaratingly uneven, with notions of what constituted *rhetoric* ranging from colloquial and even pejorative to expert and even laudatory. McCloskey, the guiding voice of the approach, uses *rhetoric* and *literary* pretty much as synonymous in his many writings, and Nelson has come under attack for voraciously vague "lists of undefined categories" splaying out from the rhetoric of inquiry.[34] Many of the Symposium contributions are more prescriptive than investigative, and some of the participants came to offer what they saw as an exposé of shoddy and manipulative language by the bad or ignorant members of their tribe, to lay bare the rhetoric in order to clear the way for true scientists. Listen to a pair of mathematicians tell us that the traditional division of math into pure and applied is only two thirds of the story. Pure mathematics is "what mathematicians do to please themselves". Applied mathematics is what they do when they carry out "the tasks set by the rest of society".

> Finally, there is rhetorical mathematics. What is that? It is what is neither pure nor applied. Not pure, because nothing of mathematical interest is done, no new mathematical ideas are brought forward, no mathematical difficulties are overcome; and not applied, because no real-world consequences are produced. No practical results issue from rhetorical mathematics—except publications, reports, and grant proposals. The word "rhetoric" means many things. One of its invidious meanings is empty

verbiage, or pretentious obfuscation. Mathematics can be rhetorical in this sense of the term. We call it rhetorical mathematics. (Davis and Hersh 1987:54-55)

But, glaring liabilities notwithstanding, the conference, and the rhetoric of inquiry movement generally, motivated a great deal of interest around the ways in which sciences are thick with texts and suasions. The guiding theme of the approach, that discourse not only propagates knowledge but makes it too, has been widely adopted, and the strong papers at the conference were very strong. Herbert Simons, for one, came away ebullient, suggesting it "may well be remembered as a watershed event in the history of rhetoric",[35] and it may well.

Argument Fields

At virtually the same time, and cross-pollinating regularly with rhetoric of inquiry scholars, two rhetorical societies were pursuing highly compatible research into a concept of Stephen Toulmin's. The members of the Speech Communication Association (SCA) and the American Forensic Association (AFA) were exploring Toulmin's notion of argument fields at their biennial joint conferences.[36] A number of itinerant rhetoricians in the early 1980s were becoming interested in how scientific discourse functioned, but it was here that the first "movement" sprouted of card-carrying rhetoricians investigating science. Rhetoricians had known about Toulmin long before they knew about Kuhn. His *Uses of Argument* was extremely influential.[37] It had fallen on the deaf, embarrassed, or puzzled ears of philosophers, to whom it was addressed, but, sparked by an important Ehninger and Brockreide article, rhetoricians adopted it eagerly.[38] Toulmin's terminology quickly percolated through the discipline, and versions of his argument schematics now populate most textbooks.

Particularly attractive in Toulmin's work was his notion of argument fields, which, effectively reinvented the rhetorical wheel of topoi: Toulmin noticed that some argument appeals are close to universal (the "field invariant" *koinoi topoi*) while others are quite specific (the "field dependent" *eide topoi*).[39] These observations kindled a new interest in argumentation, particularly among the community defined by the SCA and AFA, and a concomitant expansion of rhetorical focus from public address to other areas of discourse. Especially after Toulmin's explicit argument-field forays into science in his *Human Understanding*, the realization built among rhetoricians that science was within the province of this work, and specific explorations began. In 1983, at the third SCA/AFA Summer Conference, John Lyne looked at the growing sociobiology debate; John Angus Campbell and James T. Hayes looked separately at the scientific creationism debate; Jeanne Fahnestock and Marie Secor offered an early version of the notions behind their soon-to-be-classic scientific stases paper; Anne Holmquest proposed that "the 'rhetoric is epistemic' and the 'science is rhetorical' projects" were salt and pepper, ham and eggs, sauce for the goose and sauce for the gander. At the following conference, Holmquest was back, with work on Darwin, and Lyne was back, with the beginnings of his landmark investigation of punctuated equilibria.

The SCA/AFA argument-fields program was more modest than the rhetoric of inquiry project, and the participants were generally less schooled in the subject of their investigations, specific sciences, but it proceeded with a much better informed notion of rhetoric, and therefore moved much more quickly beyond platitudes like 'science uses literary devices' and 'science uses established forms of communication'. Here is Lyne, knocking down the wall of the experts:

> Even the most intramural discourses among specialists could be taken up under the 'hermeneutic' mode, since each writer must bridge to the otherness of colleagues in the field, and perhaps be the representative of an individual sub-specialty . . . one could apply the 'rhetorical' frame to track down the ways in which scientists act as propagators to their peers, . . . even *within* 'a small group of initiates,' rhetoric is ubiquitous.[40] (1983:406).

Here is Charles Arthur Willard on the historical roots of the public-policy dimensions of rhetoric of science:

> Debates among scientists and their nonscientific advocates in the late 19th and early 20th Centuries [concerned] whether scientists should have—and should actively seek—more influence upon public policy than they had hitherto been given. One conclusion of [these debates] was that scientists were obliged to ensure that their hard-won knowledge was not ignored and thus had a duty to seek influence in the political sphere. A picture gradually evolved which emphasized the ways scientific evidence gets distorted in the thrust and parry of daily decision making—the public sphere being a mudhole filled with unrecognizable facts. Scientific evidence can speak for itself, all things being equal, until contaminated by the cupidity, credulity, obstinacy, venality, stupidity, and sheer cussedness of political life.[41]

Here is Holmquest on the promise and the burden of rhetorical investigations of science:

> "Rhetoric is epistemic" may surely be a cliché in the summer of 1983, in the safe company of fellow rhetoricians. But in the vast and growing corpus of works about science, one is much more likely to encounter views that: "Rhetoric is figures of speech and amplification;" "Rhetoric is flourish and trickery;" and "Rhetoric is one thing, substance another." . . . Of course, by defining "rhetoric" as "manipulation" or "extravagant speech", other students of science can routinely rule rhetoric out of its substantive claims. Yet to us, this state of affairs should offer a challenge: to show that like the philosophy of science, history of science, sociology of science, and psychology of science there exists a legitimate rhetoric of science. And this rhetoric of science is a species of the genus meta-sci-

ence like the others, but differing in perspective and circumference . . .
By the year 1990, there may be as many people writing about science as
there are now writing about metaphor. The opportunity to create genuine
dialogue about rhetorical argument seems too good to pass up.[42]

Her prediction has proven true—at least the one about quantity of research
into rhetoric of science—in part because of the beneficial influence of the
rhetoric-of-inquiry and argument-fields work, but perhaps more strongly because
another recent tradition of analyzing science was influencing other rhetoricians,
and beginning to give the lie to Holmquest's facile characterization of external
hostility to rhetoric, and helping open the channels to genuine dialogue about
rhetorical argument: Sociology of Scientific Knowledge.[43]

Sociology of Scientific Knowledge (SSK)

This Kuhn-incited school of science criticism is inspirited, like many Kuhn-
incited analyses of science, with the ordinary language sense of *criticism*; Steve
Fuller calls it "a sociological dressing down of science",[44] and it is the school that
Ravetz was targetting directly a few pages back, in his plaint about bad press for
the working scientist. SSK is an ethno-sociological, case-study-focused approach
to science with a sharp eye on discourse and a centering concern with science's
amazing capacity to generate solid, applicable, bridge-supporting, missile-
launching, eclipse-predicting knowledge. It just happens to find that such knowledge grows mostly in scientists' negotiations with each other, rather than in their
negotiations with nature. The school flowered in the seventies and eighties, a
blossom or two ahead of most rhetoricians' interest in science, and has (so far)
contributed more widely to the rhetorical turn in science studies than has the field
of rhetoric.

The pioneering figures in the approach are Michael Mulkay, Barry Barnes,
and David Bloor, but a philosopher who has aligned himself with them (and they
with him) has perhaps been the most influential SSK exponent for rhetoricians,
Bruno Latour. Mulkay, Barnes, Bloor, and many of their colleagues in this school,
like others we have seen in this introduction, are only casually aware of the themes
and rhemes of rhetoric. Latour embraces them, making regular pronouncements
that would look at home in the *Quarterly Journal of Speech* or *Rhetorica*; to wit,

> Rhetoric used to be despised because it mobilized *external allies* in favor
> of an argument, such as passion, style, emotions, interest, lawyer's tricks,
> and so on. It has been hated since Aristotle's time because the regular
> path of reason was unfairly distorted or reversed by any passing sophist
> who invoked passion and style . . . [The wrong way to interpret my
> analyses is to say that they study] the 'rhetorical aspects' of technical
> literature, as if the other aspects could be left to reason, logic and
> technical details. My contention is that on the contrary we must eventu-

ally come to call scientific the rhetoric able to mobilize on one spot more resources than older ones.[45]

OK, "hated since Aristotle's time" would get scrubbed by an editor or reviewer versed in the history of rhetoric, who might remember a few patches in the ensuing millennia where rhetoric was admired, often *for* its external allies. But Latour has certainly got the nub of the ancient antipathy and *rhetoric* is not only a widely embracing term in his work, it is widely approbatory,—virtually isomorphic with *successful*. Latour is therefore—particularly his *Laboratory Life* with Steve Woolgar—very frequently endorsed by rhetoricians looking at science, and at knowledge-making generally.

Greg Myers is the rhetorician who pays the most explicit and recurrent homage to Sociology of Scientific Knowledge; in fact, he often seems more comfortable traveling with sociologists than with rhetoricians. But Charles Bazerman clearly shows their influence among his many divergent sources, Lawrence Prelli has borrowed from their case studies for the extended examples in his *A Rhetoric of Science*, and Alan Gross included one of their number as a guest non-rhetorician in his special issue of *Technical Communication Quarterly* on rhetoric of science.[46] Fruitful collaborations between rhetoricians and sociologists on matters scientific are, one hopes, imminent, and would mark the real beginnings of genuine dialogue about rhetorical argument.[47]

Rhetoric of Science and Its Older Cousins

Prelli's rereading of Kuhn is instructive, and characteristic of [rhetoric of science] as a whole. It shows that Kuhn's work is implicitly giving a rhetorical account of science, but, alas, does not go far enough, and so a bona fide *rhetorician must come in to do his/her appointed duty.*
—Dilip Parameshwar Gaonkar

How, then, is one to tell them apart? Where do the boundaries lie among history, philosophy, sociology, and rhetoric of science, particularly when there is so much current overlap not just in subject (which is inevitable) but in approach, warrants, and results (which is not)?

The Pervasiveness of Rhetoric
in Contemporary Science Studies

Let's take the toughest notion, the one defining the turn in which they are all presently participating,—the notion (not the field) of rhetoric. It is deeply, deeply pervasive. Historians, like the ones who contributed to the collection of essays in *The Figural and the Literal*, whose introduction is subtitled "Rhetoric and Writing in Early Modern Philosophy and Science", are bringing it to the center of their field.[48] Philosophers put the earliest contemporary shovels into the rhetorical turn, like Feyerabend, with his preoccupation over the suasive processes he labels "propaganda", and statements that, coming from a philosopher, are enough to make most rhetoricians swoon:

> Learning to argue is part of learning how to get along with people. One does not study the 'nature of man'; one studies individual people and learns the many ways of living with them, arguments included. One learns how to adapt one's persuasion to the idiosyncrasies of the person one confronts rather than to an abstract creature 'rational man'. There is no distinction between logic and rhetoric.[49]

More recently, philosophers have taken to collecting essays in books like *Persuading Science: The Art of Scientific Rhetoric* (Pera and Shea). And sociologists are currently the most active and astute, if not the most rhetorically informed, participants in the analysis of scientific discourse. Rhetoric is everywhere in studies of science.

Knowing Rhetoric

So, back to the question: How are we to distinguish rhetoric of science from philosophy, sociology, and history of science, when they are all so darn rhetorical, and when rhetoricians have cheerfully plundered those fields for the warrants (and, in some cases, the material) to get started in the first place?

Under a broad construal, we don't even need to worry: there is simply no reason to distinguish them: rhetoric of science is not a truly interdisciplinary project, since little serious collaboration is underway, but it is a thoroughly multidisciplinary enterprise, since it is being pursued by such diverse scholars. If we use casual, general definitions—*rhetoric* as the analysis and practice of suasion, *science* as the making of knowledge about the natural world—then *rhetoric of science* is a wide and welcoming umbrella. Anyone with a more than passing interest in the field is obliged to read widely in all the contemporary disciplines that investigate the flow of discursive influence in the sciences.

But (even pretending we had the luxury of sampling broadly in this collection, an essay or two from each relevant subfield), we still need a tidier definition for the purposes of this book, and the umbrella does in fact have a center pole, identified by the guiding term, *rhetoric*. That *rhetoric* is the critical term (not *sociology*, for instance, or *philosophy*, or *history*) suggests one immediate criterion, knowledge. People who call themselves rhetoricians of science usually know the doctrines and traditions associated with the word *rhetoric*, and it shows in their work. Dilip Gaonkar has jumped on Prelli for saturating his writings with the word *rhetoric* and its variants. "It is as if Prelli wishes [to] make his analysis rhetorical", he complains, "by simply repeating those words again and again".[50] Gaonkar, though, would do well to look at publications by contemporary philosophers, sociologists, and historians of science, where *rhetoric* rarely functions as more than a hand waving at some vague discursive or literary or postmodern concept. That's where he will find scholars trying to make their work rhetorical by incantation. What Gross grouches about one philosophical and historical collection—that its papers frequently display "an embarrassing ignorance of rhetorical theory and criticism"[51]—is true in general for the vast majority of analyses of science outside the traditional field of rhetoric.

Where Prelli's work departs, and departs very sharply, from research in history or philosophy or sociology of science which tosses around some of the same terminology is, not putting too fine a point on it, that he *knows* rhetoric. Richard Westfall, a superb historian but a rhetorical neophyte, is uncommonly frank about his superficial knowledge of rhetorical theory, opening his paper on Galileo and Newton at a rhetoric-of-science conference with an apology for having "never formally studied the discipline of rhetoric",[52]—but his confession would be well at home in scores of other papers and books, in which his brethren proceed as if the ignorance is irrelevant, as if using the word *rhetoric* or *persuasion* recurrently is enough to qualify their analyses as rhetorical criticism. Prelli's *A Rhetoric of Science*, on the rhetorical hand, is as much about topoi and stases and symbolic inducement as it is about memory transfer experiments and the double helix and scientific heresy. It is, in short, as much about rhetoric as it is about science, and while the proportions may differ in other rhetoricians looking at science, comparable interests and conceptual backgrounds are still at work.

Still, just knowing the principles and traditions of rhetoric will not really do as a criterion, since there are honorable exceptions among scholars in other fields. Philip Kitcher and Marcello Pera are philosophers who do considerably more than wave at rhetoric. Peter Dear is a historian of the same cloth, Steve Woolgar a sociologist, Donald McCloskey an economist, Michael Billig a psychologist.[53] And, alas, there are also dishonorable exceptions in rhetoric, though we can do without names here.

One might, as some have been tempted, simply use the same label, *rhetoric of science*, for all investigations of science which are knowledgeable about rhetoric. We might call Kitcher and Dear and Woolgar and McCloskey and Billig "rhetoricians of science". The dishonorable exceptions in our own field we could find some other label for. But, truth to tell, Kitcher, Pera, Dear, Woolgar, and a few comparably knowledgable others are exactly that. They are "others" who happen to deploy their informed borrowings from rhetoric very ably,—just as Gross is a rhetorician who deploys his philosophical borrowings ably, Campbell his historical borrowings, Myers his sociological borrowings.

Defining Allegiances

We need a more tightly defining term than *knowledge*, then. That term is *allegiance*. I mean allegiance to rhetoric, certainly, but that only pushes the definitional question down one semantic level. The next question is the notorious but necessary "What do you mean by *rhetoric*?" and each scholar will have a somewhat personal constellation of allegiances. Scholarly disciplines are akin to languages, and complications can arise when two or more disciplines use the same word in different ways.

Take English and Russian. In English, the word *dog* refers to any hairy, barking, domestic quadruped,—any example of the species *Canus familiaris*. In Russian that word (that sound) refers to only one specific breed of hairy, barking, domestic quadruped,—the example of *Canus familiaris* that English speakers call a *Great Dane*. Being related European languages, English and Russian have phonological and semantic overlaps. Speakers of each can even point to the same quadruped, utter the same noise, and be making highly similar, true statements about the world. They won't, however, mean the same thing; their linguistic (and maybe ontological) allegiances are quite distinct.

Sociologists speak another, closely related language to ours in which *rhetoric* is a vaguer, shallower, though decidedly similar, term. In contrast to the historical tradition of rhetoricians, sociologists view rhetoric "as a late-twentieth-century invention, conceptually born of French theorists and methodologically realized by British sociologists".[54] Speakers of the language in which rhetoric means a late twentieth-century Western European notion can point to the same object (say, Darwin's *Origin*) and say "lo, rhetoric", but they don't mean precisely the same thing by that utterance as speakers of the language in which *rhetoric* means an ancient Mediteranean notion. Philosophers have their own usage, and historians, and there is plenty of individual variation.

The defining allegiances of the long rhetorical tradition that begins in the ancient Mediterranean—the notions evoked when we rhetoricians make the noise, or type the marks, *rhetoric*—are to suasion, ethos, pathos, logos, motive, action, identification, argumentation. That's what we mean when we use the word. These are our *primary* allegiances. Sociologists have a primary allegiance to communities, philosophers to reasonableness and conduct, historians to temporal and cultural circumstances. As it happens, there are also scholars who do investigations we could label literary criticism of science, psychology of science, linguistics of science, anthropology of science. Not surprisingly, their primary allegiances are to notions central in literary criticism, psychology, linguistics, anthropology.[55] When scholars with these other primary allegiances admit an organizing role for rhetoric in their work, the best definitional route is not nominal, but adjectival. We can call Pera's work rhetorical philosophy of science; Woolgar's, rhetorical sociology of science; Dear's, rhetorical history of science. (We might also call Gross's studies the philosophical rhetoric of science; Myers's the sociological rhetoric of science; Campbell's the historical rhetoric of science; . . .[56])

Or, at any rate, that is the best definitional route for this introduction, which needs to tidy up a welter of overlapping disciplinary concerns. The move, like every attempt to nail down words, is artificial. And the move comes down to this: rhetoric of science is the analysis of scientific discourse by scholars whose primary allegiances are to the guiding notions of rhetorical theory, and who place their work in the tradition of others with those allegiances, some of whom invented those allegiances (a cast prominently including such folks as Protagoras, Aristotle, Cicero, Vico, Burke, and Perelman).

Gaonkar berates Prelli for more than his frequent use of *rhetoric*. He also goes after Prelli for (in this section's epigram, which Gaonkar means to sting) reading Kuhn as a nascent and dumb rhetorician of science, someone who didn't have the courage or the tools to follow the implications of his work all the way, from whom we rhetoricians have had to wrest the torch and carry it to the promised land. Gaonkar takes Prelli, correctly if snidely, to stand for his entire subfield here, rhetoric of science.

But strip this reading of its hooting at Kuhn, which neither Prelli nor any other rhetorician of science manifests, and it is perfectly legitimate; from another prevalent contemporary perspective, Kuhn is a nascent sociologist of science who just didn't know enough, or care enough, about sociology to do the job fully. Such readings are problematic or illegitimate only if they hold Kuhn's rhetorical or sociological shortfalls—which are genuine—to be failings. But if, as Prelli and others do, in rhetoric and in sociology, we simply thank Kuhn for opening up the space, for netting us fish we can fry more ably than he, these readings are just a standard operating procedure of scholarship; quantum theorists, for instance, read Einstein in an exactly parallel way. Besides, the invitation for this reading *comes* from Kuhn, and it is an invitation that calls more loudly to rhetoric, the study of suasion, than to any other field: "Just because it is asked about techniques of persuasion, or about argument and counterargument in a situation in which there can be no proof", Kuhn said, his guiding idea "is a new one, demanding a sort of study that has not previously been undertaken".[57]

Kuhn provided us with a new ground on which our allegiances can play.

The Landmarks

Rhetoric in all its applications is focused on the particular.
                                                                                —Richard McKeon

Which brings us to our last slate of rhetorical questions: *Landmarks*? In *Rhetoric of Science*? Have rhetoricians shouldered the burden that Holmquest set out for them barely a decade ago,—"to show that like the philosophy of science, history of science, sociology of science, and psychology of science there exists a legitimate rhetoric of science"?[58]

Not entirely. *Landmarks* in connection with *rhetoric of science* does construe as doubly premature: in contrast to the rhetorical investigation of public address, or stylistics, or invention, the rhetorical investigation of science is still shaking the albumen from its wings; in contrast to the philosophical, sociological, or historical investigation of science, it is barely peering over the edge of the nest. Or maybe it is just a bit older than that: The field is very young and the quality disparate, but it exhibits what Myers has called, in another context, "the messy adolescent phase of inquiry".[59]   But that's not so bad. Adolescence is messy, no question, but it is also a period when defining physical traits make their appearance, when defining anatomical landmarks arise, or drop, and it is a period where acts of startling maturity occur, all the more startling for the background of immaturity. In the rhetoric of science, this gives us Campbell, and Bazerman, and Fahnestock, and Myers, against the backdrop of less developed research.

"Philosophical problems", Ludwig Wittgestein once argued, characteristically show up "when language *goes on holiday*",[60] providing the perfect counter-point to rhetorical criticism generally, and rhetoric of science specifically. As with many of his cool-sounding enigmaphorisms, one can't be entirely sure what Wittgenstein means here, though it has something to do with the barrels of philosophical ink spilled over pseudo-problems that arise when people get careless with language. But, whatever his precise meaning, Wittgenstein's encapsulation is starboard to rhetoric's port: problems which reward rhetorical analysis characteristically show up when language *goes to work*, and language nowhere works harder than it does in science, where its job is to make the most robust knowledge of our culture. And when language leaves the house with its lunchpail, it goes, just like its employers, to a specific job. Language doesn't just float in an abstract Realm of Words; it links and evokes and causes; it connects ideas and points minds in new directions and gets the salt down the table. Rhetorical criticism, like science, has always been a strongly empirical field. Its eye is always fast on the particular case. It investigates the symbolic evocations and causes and connections of specific jobs. The essays in this volume, the strength of rhetoric of science, epitomize that concern. They are case studies.[61]

The studies in this volume are exemplars for rhetoric of science. They chart the field. They exhibit the governing themes of rhetorical criticism when its eye turns to science: suasive greatness, paradigmatic debates, public policy concerns, and composition issues. They do so by starting at the top. They take as their main courses the two disciplines highest in the scientific food chain, physics and biology, with side orders of archaeology and experimental psychology. They employ a methodological tool-set largely inherited from Aristotle, but also draw pluralistically on related enterprises, such as pragmatics, ethnology, and literary criticism. And they engage the ruling theoretical issues of the field. They are landmarks. They define the field.

Giants in Science

This book is not a book about theory, but one theoretical issue is unavoidable. It has already raised its head several times in this introduction,—the issue of how deeply rhetoric runs in science. Recall the principal implication that follows the rhetorical turn in science studies:

the virtues of a scientific claim come not only from the way it is mapped against nature, but from the way it is mapped into the context of specific approaches and communities.

"How big is that *not only*?" is the inevitable question for rhetoric of science. Does rhetoric add a little salt and pepper to the stew of science, which would be just as nourishing, if not so tasty, without it? Or is rhetoric the meat and potatoes and rutabagas of the stew, the very substance of science? Since the job of science is to make knowledge, these questions are necessarilly epistemological ones, and all of the papers in this volume sit on an epistemic continuum that runs from a Platonic salt-and-pepper, rhetoric-just-propagates-knowledge to a Sophistic meat-and-potatoes, knowledge-is-rhetorical-through-and-through position. [62] The first two papers in the collection mark the opposite ends of that continuum. The first essay, by Campbell, typifies the epistemologically conservative version of rhetoric of science; it might be fairly characterized as agnostic. The second essay, by Gross, is unequivocally radical, seeking to banish the *only* altogether. It exemplifies Gross's thesis that science is rhetoric, wholly rhetoric, and nothing but rhetoric ("without remainder" as he puts it elsewhere[63]).

John Angus Campbell's essay is the first case study in this volume. The lead-off spot is appropriate, since Campbell's work on *The Origin of Species* provided the strongest case for many people that there could be such a thing as rhetoric of science. It began when the phrase "rhetoric of science" was unspoken, the pursuit unheard of. But quietly, unimpeachably, Campbell demonstrated the value of rhetoric for investigating science. His work has grown into an impressive, many-peaked series of articles on Darwin. The essay included here, "Charles Darwin: Rhetorician of Science", is perhaps the slightest of that series—there was little new in it when it appeared, and it rose quite late, a decade and a half into the deep shadow of his first Darwin paper—but it is precisely this lateness that makes it valuable for our collection.[64]

Campbell wrote the paper for the 1984 Iowa Symposium, when other people were finally beginning to embark on the kinds of investigations he had been exploring since the late 1960s, and it brilliantly summarizes his earlier work for that audience. Despite its participation in that Symposium, though, the paper is not much concerned with the rhetoric of inquiry, with discursive practices that constitute knowledge. Rather, it investigates a "rhetoric of dissemination" characteristic of all Campbell's early essays, belonging clearly at the just-propagating end of the epistemological continuum. [65] "To claim that Darwin was a rhetorician", Campbell offers cautiously, "is not to dismiss his science, but to draw attention to his accommodation of his message to the professional and lay audiences whose support was necessary for its acceptance".[66] *Accommodation* is the watchword for this paper,—a thorough exploration of the ways in which Darwin marshaled his evidence, managed his ethos, and molded his argument to the scientific and religious expectations of the day, successfully mobilizing his suasions about what quickly became one of the most influential packages of ideas in Western thought.

Darwin has his knowledge already, Campbell suggests; he made it or he found it elsewhere; *Origin of Species* deploys it to best effect.

The contrast with **Alan Gross**, whose "On the Shoulders of Giants" follows Campbell's paper in this collection, could not be starker. Gross's essay examines Newton's magnificently successful campaign to refashion the physics of light, concluding radically that the success was solely rhetorical. Newton's *Opticks* succeeded dramatically through, in Gross's analysis, its careful consonance with previous work, through its geometrical arrangement, through its copious experimental appeals, and through a series of field-guiding rhetorical questions. His triumph stands in sharp relief against his failed earlier attempt to influence the physics of light, in a brash, confrontational paper, that snubbed earlier work, that argued much more informally, that hangs on only one, sketchily described experiment, and that makes no serious effort to guide future research.

The comparison of these two documents—which had very different impacts despite precisely the same core implications for the nature of light—forms the basis of Gross's argument that Newton's phenomenal success with the *Opticks* (and his failure with the 1672 paper) came "solely by means of its rhetoric".[67] More radically yet, Gross says that "science is rhetorically constituted, a network of persuasive structures, patterns that extend upward through style and arrangement to invention itself, to science itself." His case study is therefore made to carry very heavy freight: Newton's physics stand for all sciences in the paper, for science itself. If the work of *Newton*, a hugely important scientist, in *physics*, the science with the solidest reputation, is rhetorical stem-to-stern, then the entire scientific enterprise is rhetorical, without remainder.

Gross's and Campbell's essays highlight another way in which *landmark* is appropriate for the works in this collection. Many of them chart pinnacles of science, as Campbell charts Darwin's *Origin* and Gross Newton's *Opticks*.

Michael Halloran's contribution to this volume, "The Birth of Molecular Biology", is another such paper. It examines the ethos of an extremely prominent twentieth-century document, one which virtually invented a science. This document, the modestly entitled "A Structure for Deoxyribose Nucleic Acid", has received a goodly amount of rhetorical attention. "If there is a canonical text in this still-early period of the rhetorical criticism of science", Carolyn Miller notes, "it is the 1953 paper in which James D. Watson and Francis H. Crick proposed the double-helix structure for DNA".[68] Halloran's essay was the first rhetorical analysis of the paper, and it remains the best. A large part of what consitutes a scientific paradigm, Halloran notes, is the ethos of its practioners,—the characteristic attitude with which they attack problems and propogate claims. He finds that the oddly brash-but-genteel ethos he explicates in Watson and Crick's paper set the rhetorical agenda for their new science in almost exactly the way that their model of DNA set the research agenda.

Together, these three papers—Gross's, Campbell's, and Halloran's—represent one of the major themes in rhetoric of science, the investigation of the suasive greatness of scientific giants, individuals whose accomplishments have stamped their respective sciences in epoch-defining ways. A second major strain is the investigation of periods of conflict.

Conflict in Science

The rhetorical study of scientific conflicts grew naturally from the Kuhnian roots of rhetoric of science, with its emphasis on upheaval and revolution; indeed, under some readings of Kuhn, it is only in times of revolution that scientists let slip the dogs of rhetoric. But, Kuhnian roots notwithstanding, none of the scientific debates investigated rhetorically (in this volume or elsewhere) are truly revolutionary. Some rhetoricians have brushed up against revolutions—Campbell's work on the natural selection controversy, for instance, Gross's and Bazerman's on the corpuscular light controversy, Gross's on the heliocentric revolution—but these studies are invariably sympathetic with only one side, and always with the winning side at that. The study of revolutions is very ripe for rhetoric of science, and the fine studies of smaller clashes in this collection can serve as templates.[69]

Jeanne Fahnestock's superb examination of a dispute in contemporary archaeology is an especially fine example of this strain. The question at issue in the dispute is when humans began to populate North America,—whether we came over roughly ten to twelve thousand years ago, as one camp holds, or much earlier than that, perhaps as much as thirty thousand years ago, as their opponents contend. The article tracks a number of arguments from both sides, addressing a number of different audiences—hence, Fahnestock's title, "Arguing in Different Forums"—exploring an important rhetoric of science *leit motif*, how positions shift as a function of audience. Her essay also illustrates a fascinating Kuhnian wrinkle in late-twentieth century scientific debates (as, in fact, do both of the other papers in this section): how often Kuhn is invoked by disputants defending their position. Fahnestock's essay is particularly good at depicting the way both sides paint themselves as the embattled, outnumbered revolutionaries, their opponents as the plodding, dogmatic majority of normal scientists.

John Lyne and **Henry Howe's** (Howe being the only interloper in this book of rhetoricians; a, gasp, scientist) study of the controversy surrounding a recent subordinate clause of the neoDarwinian synthesis also examines these two themes,—the way arguments change as they move from audience to audience, and the invocation of Kuhn as Patron Saint of (alledgedly) embattled scientists. They trace the reception of (shifting versions of) the punctuated-equilibria theory of evolutionary change. Darwin's picture of evolution is essentially gradualist, species changing very slowly as their environment blindly rewards random mutations, but Stephen J. Gould and Niles Eldredge suggest that some changes are much more rapid, that the steady equilibrium of species variation is occasionally punctuated by quick jumps. Lyne and Howe follow this position—how it is framed and how it is received—as it moves among audiences of paleontologists, biologists, and creationists, returning a valuable analysis of rhetorical dynamics based on ethos, audience, and situation.

Lawrence Prelli's "The Rhetorical Construction of Scientific Ethos", the last entry in the scientific conflicts section, is more singleminded than either Fahnnestock's or Lyne and Howe's essay. In the most systematic analysis of ethos as a tool of scientific legitimation in this volume, Prelli examines an exchange over the ability of apes to acquire language, in a single forum, the court of public opinion. He explores the claims of primate-researcher Francine Patterson that she taught a gorilla to use American Sign Language and the attack on those claims by Thomas Sebeok, a prominent linguist. Prelli frames his study in terms of the expected (Mertonian) norms of scientific argumentation, and their unexpected (Mitroffian) counter-norms, finding that scientists on either side of a debate will often exploit mutually incommensurable appeals.

While focussing on the norms of scientific ethos, Prelli also broaches another significant theme in rhetoric of science, the quest for legitimation and authority that science often pursues in public, evoking what has been one of the driving concerns in rhetoric of science since its beginning, public science policy. This recurrent focus on matters of science policy is a wonted one. The prototypical domain of rhetoric, from the moment the word has had denotation, is civic life: it is natural, if not inevitable, that a third major strain in the rhetorical criticism of science is the investigation of scientific matters that impinge directly upon the business of the polity.[70] In fact, this area may be the one in which rhetoricians can make their most distinctive and enlightening contributions to science studies.

Public Science

Richard Weaver recognized the opportunity and obligation of studying public science long before there was anything one could even call rhetoric of science, arguing that the public understanding of science was "one of the serious problems of our age".[71] Weaver, of course, is a rhetorical landmark in his own right, though very little discussion of his work notices that he had an (atypical for rhetoricians of his generation) interest in science. His work comments regularly on the "strongly rhetorical character" of science, and he authored several investigations of appeals to science.[72] His contribution to this volume, "Dialectic and Rhetoric at Dayton, Tennessee", sorts through the melange of issues—epideictic and forensic appeals, education and ignorance, church and state, science and anti-science, dialectic and rhetoric—that constituted the famous Scopes "Monkey trial" in the 1920s bible-belt America.

Weaver saw the rise of science and the decline of the fortunes of rhetoric as inextricably linked modern phenomena, keyed, in part, to a growing distrust of emotion:

> However curious it may appear, [the] notion gained that man should live
> down his humanity and make himself a more efficient source of those

logical inferences upon which a scientifically accurate understanding of
the world depends. As the impulse spread, it was the emotional and
subjective components of his being that chiefly came under criticism . . .
Emotion and logic or science do not consort; the latter must be objective,
faithful to what is out there in the public domain and conformable to the
processes of reason. Whenever emotion is allowed to put in an oar, it gets
the boat off true course.[73]

The charge that emotion can only corrupt science is a familiar one, captured
most succinctly in Aristotle's analogy that infecting reason with emotion "is the
same as if someone made a straight-edge crooked before using it".[74] The other
two essays in the public science section join with Weaver's in putting the lie to
the belief that emotion and science do not consort. But each paper does so in
completely inverse and complementary ways to the other.

Craig Waddell's "The Role of Pathos in the Decision Making Process: A
Study in the Rhetoric of Science Policy" turns Aristotle on his head, showing that
emotion can sometimes *be* the straight-edge, guiding reason aright. Waddell
closely investigates the Cambridge rDNA hearings of the mid 1970s, when
genetic engineering was first rearing its head in public and there was widespread
fears that it might result in deadly new organisms which might escape the lab.
Waddell interviewed the members of the board that conducted and adjudicated
those hearings, finding that while they disparaged emotional argumentation
(principally as they associated it with the fear tactics of one testifier) their decision
had a heavy emotional basis (principally in its dependence on the appeals of
another testifier to the potential benefits rDNA research could bring to sick and
dying children).

Carol Reeves's "Owning a Virus: The Rhetoric of Scientific Discovery
Accounts", on the dramatic other hand, is virtually a proof of Aristotle's
bending-the-ruler claim. Like Waddell, Reeves lifts the covers of science to
show a matress of emotion just under the bedsheets of logic, but she gives a
startling example of how emotion can indeed distort judgment, at least the
emotion of greed. Her essay is not just about science done in the public eye, but
about how all the personal trappings that go with public science—fame, prizes,
filthy lucre—can twist science, even an applied science waving the banner of
Hippocrates, away from the public good. She examines Robert Gallo's fierce
campaign for recognition as the discoverer of the AIDS virus, a campaign which
played itself out in a variety of increasingly public forums.

Both Waddell and Reeves show that it is delusional and self-defeating to
pretend emotion is not a significant factor in decisions (about science or anything
else), and that, just like in the rest of life, whether the consequences of that factor
are for the good or for the bad depends on the constitution and the context of
the pathos. Emotion is not a single oar, as Weaver termed it, but many; some get
the boat of science off course, some help to true the course.

Writing Science

Rhetoric in the modern academic era began almost exclusively as the study of public speaking, but it has subsequently come to be allied closely with the study of writing as well.

Charles Bazerman is the most dedicated champion of this alliance as it plays out in the rhetoric of science. His landmark book, *Shaping Written Knowledge* begins with the perfectly titled section, "Writing Matters". Bazerman's contribution to this collection, which leads off the final section, is "Reporting the Experiment", an excitingly representative sample of that book.[75] It is a breezy, scholarly, thoroughly rhetorical look at the evolution of experimentation in early modern science through the texts it generated. "Experimental reports tell a special kind of story," he says, the story "of an event created so that it might be told".[76] And the challenge of telling such stories is great: not only must the event itself be created and managed in a way that reaches appropriately into the traditions of knowledge in the field, and embraces appropriately the allegiances and goals of the present, and projects appropriately into the future; not only, that is, must the event be meaningful in very narrow ways, the language that frames it must make the reaching and embracing and projecting seem virtually inevitable. The burden is too much for anyone to discharge individually, so the genre of experimental reports developed to constrain the stories. The shape of the experimental report is a rhetorical achievement, the product of three hundred years of argument and accommodation. Bazerman's paper charts the first century and a bit of this achievement, showing how it grew into a form preoccupied with (and tailored beautifully for) representing the rigid and narrow test of flexible and general beliefs.

Where Bazerman looks broadly at hundreds of articles over more than a century, the final offering in this volume, Greg Myers's "Texts as Knowledge Claims: The Social Construction of Two Biology Articles", looks deeply at only two, over the periods between their conceptions and their publications.

Starting this collection with John Angus Campbell is appropriate because of his early exemplary work in the field; ending with **Greg Myers** is appropriate for the converse reason, for how far he takes the field. Several of the scholars collected here have important books: Prelli has *A Rhetoric of Science*, contrasting nicely with Gross's *The Rhetoric of Science*, Bazerman has *Shaping Written Knowledge*; and Myers has *Writing Biology*. Each book has its virtues, but *Writing Biology* is the best of the field. It is the most patient, most consistent, most contextually deferential, and most text-driven book in rhetoric of science. His contribution here, represents it well. "Texts as Knowledge Claims" uses the evolution of two quite different biology articles to plumb the text-and-knowledge construction practices inherent in the negotiative processes of scientific peer review.

With access to an unusually rich body of material (both papers were reviewed four times, and rewritten as often; Myers worked from all the referee reports, all the correspondence surrounding those reports, all the submitted manuscripts, and several intermediate drafts), Myers unravels a tale of competing drives in the growth of communal texts. The authors strive to achieve recognition for the originality of their work, while the referees and editors strive to fit that work into the paradigmatic knowledge of the field. The results are published papers barely recognizable as the descendants of the initial submissions, and the change in epistemic status is equally dramatic. Publication makes communally sanctioned knowledge out of individual assertions.

Rhetoric of science is too capacious and too pluralistic for a monolithic *Principia*. On one side of the preposition, the traditions, themes, and rhemes of rhetoric are as broad as discourse. Some (Moss, for instance, and Gaonkar[77]) see this omnivorousness as a failing. Others (including your humble narrator) see it as no different in principle or in practice from the omnivorousness of philosophy, which wants to understand everything, or of education, which tries to teach everything, or of any language, which expresses everything, or of any enterprise whose goals are both fundamental and expansive. There are many such enterprises. But, will you, nill you, praise it or blame it, it's a fact: rhetoric has an exceptionally big appetite, and a digestive tract to match. Moving to the other side of the preposition, the appetite and the gut of science are even greater; it is physics, after all, which relentlessly hunts for the Theory of Everything. The traditions, themes, and rhemes of science are as broad as knowledge. Science is overwhelmingly more successful than rhetoric, or anything else, at achieving consensus, and in generating concrete celebrations of its consensuses, success which is largely a product of narrow standards of evidence, rigorously applied. That is why it has a *Principia* (in fact, several) and we don't. But science as a conglomerate is no more amenable to tidy, necessary, sufficient definitions than rhetoric. Indeed, even the preposition that squeezes between *rhetoric* and *science* can maddeningly point both ways (in "hat of Noam" *of* points forward and the hat belongs to Noam, but in "collection of hats", *of* points backwards and the hats belong to the collection). A single book by a single scholar will not give you rhetoric of science.

John Lyne, however, tells us what will, and the collection of articles you're holding has taken his advice. Ten years ago he ended a discussion on rhetoric of science with the claim that what will matter most to the field "is having good exemplars of rhetorical scholarship, works of such merit that others will want to use them and emulate them"[78]. He was right. They follow.

A Note on the Texts and Citations

The essays in this volume come form a wide range of books and journals, and therefore have a wide range of citation styles. I have regularized these to some degree, but a few differences remain. In particular, the papers by Bazerman, Fahnestock, Reeves, and Waddell have in-text citations with works-cited lists, while the remainder use reference notes. Also, the reader may note that the chapter by Lyne and Howe, which principally employs the endnote citation style, departs inexplicably from this style for a few quotations in the middle of the text, only to return to the endnote style toward the end. The editor and publisher would merely like to state that for technical reasons the article is republished here exactly as it appeared in the *Quarterly Journal of Speech* with respect to these citations. Additionally, I have lightly edited some of the essays, removing typos, supplying missing information, and adding or updating a few references. (Alas, new errors and omissions very likely crept in with our scanning and typesetting procedures as well.)

Notes

1. I would like to acknowledge, very gratefully, the general contributions to this introduction and this collection by the following teachers, students, and colleagues (though, as many of them would eagerly attest, *contribution* does not mean *endorsement*): Charles Bazerman, Dorina Belu, John Campbell, Val Clark, Kate Conway, Keri Downs, Marianne Evans, Alan Gross, Michael Halloran, Russ Hirst, Elaine Hudson, Stewart Lindsay, John Lyne, Donna MacKinnon, Carolyn Miller, Shirley Moore, Greg Myers, Linda Sanderson, Judy Segal, Michael Strickland, Ron Struck, Anne Thomas, Kathleen Venema, Craig Waddell, and Jim Zappen. I would also like to acknowledge, with increasing gratitude, the more direct contributions of the following commentators (repeat above disclaimer): Hugh Chamberlain, Matthew Chaput, Evan Jones, Andrea Kelso, Gerald Holton, Greg McMehen, James Murphy, Lawrence Prelli, Charles Bazerman (again), Greg Myers (again), Jim Zappen (again), and an anonymous reviewer for Hermagoras Press. I would also like to acknowledge, with even more gratitude, the support and inspiration of Indira and Galen. And there's a coffee joint near the University of Waterloo that has been very helpful, too, but the acknowledgementometer is starting to red-line.

2. See, for instance, Richard Weaver, "Language is Sermonic*", Language is Sermonic,* ed., Richard L. Johannesen, Rennard Strickland, and Ralph T. Eubanks, (Baton Rouge: Louisiana State University Press, 1970): 201ff, Karl Wallace "The Fundamentals of Rhetoric", *The Prospects of Rhetoric,* ed., Lloyd Bitzer and Edwin Black (Englewood Cliffs: Prentice-Hall, 1971): 9-10, and/or Lawrence Rosenfield, "An Autopsy of the Rhetorical Tradition", *Prospects of Rhetoric,* 65ff, for some discussion of the respective waxing and waning of science and rhetoric. *Science,* aside from sporadic humanist sniping, has been an almost universal term of approbation as long as it has been used in English. Certainly it is one of those terms in the twentieth century which causes knees to jerk most quickly in approval. It is one of Weaver's original god-terms,—"Ultimate Terms in Contemporary Rhetoric", *The Ethics of Rhetoric* (Davis, CA: Hermagoras, 1985): 215ff. John Searle calls it "an honorific", *Minds, Brains and Science* (Cambridge, MA: Harvard University Press, 1984): 11; Willard Van Orman Quine and J. S. Ullian, *The Web of Belief* (New York: Random House, 1978): 6, "a thumbs-up word"; Richard Rorty, "Science as Solidarity", *The Rhetoric of the Human Sciences,* ed., John S. Nelson, Allan Megill, and Donald N. McCloskey (Madison, WI: University of Wisconsin Press, 1984): 38, calls it a label whose principal value is to help secure any field "a place at the trough".

3. It was much less clear in Greek antiquity (especially before Plato) who was a rhetorician, who a scientist, and why one would want to wall those job descriptions apart. As some of the older names reveal, science is primarily rhetorical. The familiar *-logy* is "discoursing": biology is discourse about life, zoology is discourse about animals, geology is discourse about earth . . . In any case, there should be no dispute that presocratic schools like the Milesians, the Eleatics, and the Pythagoreans were equally concerned with what we now call rhetoric and science, and certainly there is no dispute that the sophists were rhetoricians. But that the sophists were also scientists is less well known now than it was before the word *sophist* became a pejorative, when the word applied easily to people like Thales, Anaxagoras, and Pythagoras. But even if we stick to the canonical sophists in the history of rhetoric: Protagoras was an anthropologist, mathematician, and linguist, whose treatises included *On Science*; Gorgias was an astronomer and physicist who wrote *On Nature*; Prodicus was a physicist, medical scientist, and linguist; Hippias was a linguist and a student of astronomy. See G. B. Kerferd, *The Sophistic Movement* (Cambridge: Cambridge University Press, 1981) : 38-40 for a general discussion of the sophists as scientists. See Diogenes Laertius, *Lives* IX.6, for Protagoras; G. B. Kerferd, "Gorgias and Empedocles", *Sicvorvm gymnasivm* 38 (1985): 595-605, for Gorgias; Kerferd, *Sophistic,* 39-40 on Prodicus as physicist and medical scientist; Everett Lee Hunt, "Plato and Aristotle on Rhetoric and Rhetoricians", *Historical Studies of Rhetoric and Rhetoricians,* ed., R. J. Hower (Ithaca: Cornell University Press, 1961): 24 , on Prodicus and Hippias as

linguists; Plato, *Hippias Minor*, 368[B] on Hippias as astronomer. Kerferd, in fact, takes an interest in science to be so characteristic of the sophists that he uses it as a criterion to argue that Socrates should be included in the sophistic camp (*Sophistic*, 56).

4. For Bacon, see especially his *The Advancement of Learning*, William Aldis Wright, ed. (Oxford: Clarendon Press, 1900); for Priestley, *A Course of Lectures on Oratory and Criticism* (Menston, England: The Scolar Press, 1968; facsimile of the 1777 edition), especially Lecture IX, on the analytic method. See Charles Bazerman and David Russell's introduction to *Landmark Essays on Writing Across the Curriculum* (1994), especially xxvii-xxxv. The intermingling history of rhetoric and science is yet to be written, though there is much scattered information towards such a history. In particular, most historically based rhetorical studies of science discuss their interrelations in the period at hand. For instance, Jean Dietz Moss (although she works with a narrower—in fact, historically more consonant—definition of rhetoric than the one in this introduction) is especially thorough on their linkages in the important Aristotelian infused age of Copernicus and Galileo,—see her *Novelties in the Heavens: Rhetoric and Science in the Copernican Controversy* (Chicago: University of Chicago Press, 1993). James Zappen is the scholar who has most systematically pursued the twining histories of rhetoric and science,—see his "Historical Perspectives on the Philosophy and the Rhetoric of Science, *PRE/TEXT* 6 (1985): 9-29,"Historical Studies in the Rhetoric of Science and Technology, *The Technical Writing Teacher 14* (1987): 285-99, "Francis Bacon and the Historiography of Scientific Rhetoric", *Rhetoric Review 8* (1989): 74-90, and "Scientific Rhetoric in the Nineteenth and Early Twentieth Centuries: Herbert Spencer, Thomas H. Huxley, and John Dewey", *Textual Dynamics of the Professions: Historical and Contemporary Studies of Writing in Professional Communities*, Charles Bazerman and James Paradis, ed., (Madison: University of Wisconsin Press, 1991):145-69.

5. See Bitzer and Black's *The Prospect of Rhetoric* for the papers and reports from the Wingspread conference. The most explicit discussions of the methodological links between science and rhetoric are in the addresses by Brockreide and Bryant; "menaced" is from Rosenfield (64); "the rhetorical dimension . . ." is from Perelman (119). Simons's most notable contributions to rhetoric of science are his early paper, "Are Scientists Rhetors in Disguise?" *Rhetoric in Transition*, ed. E. E. White (University Park, PA: The Pennsylvania University Press, 1980): 115-30, and his two edited volumes, *Rhetoric in the Human Sciences* (London: Sage, 1989), and *The Rhetorical Turn* (Chicago: University of Chicago Press. 1990). Scott's most relevant papers are "On Viewing Rhetoric as Epistemic", *Central States Speech Journal 18* (1967): 9-17, and "On Viewing Rhetoric as Epistemic: Ten Years Later", *Central States Speech Journal 27* (1976): 258-66. Johnstone, editor of *Philosophy and Rhetoric*, published Paul Newell Campbell's "Poetic-rhetorical, Philosophical, and Scientific Discourse" 6 (1973): 1-29, James Stephens's "Rhetorical Problems in Renaissance Science" 8 (1975): 213-29, and a 1977 volume (10) containing early theoretical papers by Walter B. Weimer ("Science as a Rhetorical Transaction: Toward a Nonjustificational Conception of Rhetoric", 1-29), Maurice A. Finocchiaro ("Logic and Rhetoric in Lavoisier's Sealed Note: Toward a Rhetoric of Science", 111-22), and Michael A. Overington ("The Scientific Community as Audience: Toward a Rhetorical Analysis of Science", 143-64).

6. Chicago: University of Chicago Press. The second edition came out in 1970; all references in this introduction are to the second edition.

7. "Rhetoric and Theory Choice in Science", *Persuading Science*, ed., Marcello Pera and William R. Shea (Canton, MA: Science History Publications, 1991): 59.

8. *Paradigm* is a somewhat controversial term, which Kuhn in fact repudiated under assault from philosophical hair-splitters (cf. *Structures*, 174ff); the most concerted attack was by Margaret Masterman, "The Nature of Paradigm", *Criticism and the Growth of Knowledge*, I. Lakatos and A. Musgrave, ed. (Cambridge: Cambridge University Press, 1970): 59-90, where she charts 29 different uses for the term in *Structures*). But the term Kuhn substitutes, *disciplinary matrix*, is also unsatisfactory, and (however loose its designation may be) *paradigm* has become well accepted in all fields studying science, and many that don't. I will use it, as do several of the essays in this volume, without further comment.

9. "Rhetoric", *Selected Essays on Rhetoric by Thomas De Quincey*, ed., Frederick Burwick (Carbondale, IL: Southern Illinois University Press, 1967): 90.

10. The wall of certainty, of course, had long since crumbled in science, and Kuhn was merely holding up some of the rubble. Lobachevski had showed that Euclidian geometry was not certain. Einstein had showed that Newtonian physics was not certain. Heisenberg had showed that certainty was impossible in principle about some subatomic phenomena. Bohr's quantum model was inescapably probablistic. Kuhn's most thorough discussion of uncertainty and probability in science is in Chapter XII (144-159); "nest of

commitments" (41). For further discussion, see my "Assent, Dissent, and Rhetoric in Science", *Rhetoric Society Quarterly 20* (1990): 22-6.

11. E.g., *Rhetoric* 1358a, 1359b.

12. See his germinal paper, "The Substance of Rhetoric: Good Reasons", *Quarterly Journal of Speech 49* (1963), especially pages 241-242 for the availability of some scientific discourse to rhetorical criticism; the quotation above is from his address in Bitzer and Black's *Prospects* (3). In the modern period, this position dates at least to Herbert A. Wichelns's 1925 "The Literary Criticism of Oratory", reprinted most recently in *Landmark Essays on American Public Address*, Martin J. Medhurst, ed. (Davis, CA: Hermagoras Press, 1993): that rhetoric "is to be thought of as the art of popularization. Its practitioners are the Huxleys, not the Darwins of science; the Jeffersons, not the Lockes and Rousseaus, of politics" (31). For a more recent version of this position, from another influential theorist of rhetorical criticism (who, however, has the goal posts in a slightly different place), see Edwin Black, *Rhetorical Questions* (Chicago: University of Chicago Press, 1992):187n.3.

13. *Structure of Scientific Revolutions*, 200.

14. See, as representative works, Michael Polanyi, *Personal Knowledge* (Chicago, University of Chicago Press, 1962), Stephen Toulmin, *Human Understanding* (Oxford: Oxford University Press, 1972), John Ziman, *Public Knowledge* (Cambridge: Cambridge University Press, 1968), Norbert Hanson, *Patterns of Discovery* (Cambridge: Cambridge University Press, 1958), Jürgen Habermas, *Communication and the Evolution of Society*, trans., J. J. Shapiro (Boston: Beacon Press, 1971), Jacob Bronowski, *Science and Human Values* (New York: Harper and Row, 1956), Paul Feyerabend, *Against Method* (London: Verso, 1975). See Prelli's "Editor's Introduction" to his special rhetoric-of-science issue of *Argumentation*, 8.1 (1994): 1-8, for a similar, though greatly abbreviated, account of the development of rhetoric of science to the one in this section. (Perhaps because of abbreviation, however, Prelli misleadingly suggests that one can watch in Kuhn, Feyearbend, Polanyi, and the others Prelli and I treat as their confreres, as "the language of rhetorical theory increasingly gained currency" (2). In fact, the language of rhetorical theory must be read *into* Kuhn and the others; their use of words like *persuasion* and *rhetorical* comes from ordinary language, and owes almost nothing to theories of rhetoric.)

15. This is Sonja K. Foss, Daren A. Foss, and Robert Trapp, *Contemporary Perspectives on Rhetoric* (Prospect Heights, IL: Waveland Press, 1985) :77-78, on Toulmin, but it applies equally of the others in this list (and, of course, many outside the list).

16. Rorty coined the phrase, "rhetorical turn", at a conference to be discussed shortly (the 1984 Iowa Symposium on the Human Sciences); see Herbert W. Simons, *The Rhetorical Turn* (vii).

17. I adopted this list, rather liberally, from Marga Vicedo's review article, "Scientific Styles: Toward some Common Ground in the History, Philosophy, and Sociology of Science", *Perspectives on Science 3* (1995): 249-251. Vicedo does not mention rhetoric in his article, and might well be aghast that I have invoked his name in this context. But his discussion of styles, and those of the people whose work he reviews (Ian Hacking, Jonathan Harwood, and Jane Maienschein), is effectively a discussion of suasive contexts. Another article in the same volume—Miriam Solomon, "The Pragmatic Turn in Naturalistic Philosophy of Science", *Perspective on Science 3* (1995): 206–30—which also fails to mention rhetoric, provides another label for the focus on the negotiative processes of scientists, despite the fact that the leading contemporary pragmatist, Richard Rorty, coined the phrase "rhetorical turn". Solomon is attempting to distinguish the work of some turners she approves of from that of others, whom she disapproves of (labeled the "epistemological anarchists"). In this connection, too, see Marcello Pera's *The Discourses of Science*, Clarissa Botsford, trans. (Chicago: University of Chicago Press, 1994), who goes to very great lengths to detatch himself from the word *rhetoric* in favour of *dialectics*, while articulating the shift, findings, and implication above. Philosophers apparently like to close their eyes and think of some alledgedly harder edged notion, like pragmatism or dialectics, when they lie down with rhetoric.

18. For technical communication, see Dwight W. Stevenson's "Toward a Rhetoric of Scientific and Technical Discourse", *The Technical Writing Teacher 5* (1977): 4-10, S. Michael Halloran's "Technical Writing and the Rhetoric of Science", *Journal of Technical Writing and Communication 8* (1978): 77-88, Victoria Winkler's "The Role of Models in Technical and Scientific Writing", *New Essays in Technical and Scientific Communication*, ed. Paul V. Anderson, R. John Brockman, and Carolyn R. Miller (Farmingdale, NY: Baywood, 1983): 111-22, James P. Zappen's "A Rhetoric for Research in Science and Technologies", *New Essays in Technical and Scientific Communication*,123-38, and Alan G. Gross's "The Form of an Experimental Paper", *Journal of Technical Writing and Communication 15* (1985): 15-26, for some of the connections between rhetoric of science and technical communication. The link has remained strong: *The Journal of Technical Writing and Communication* regularly publishes rhetoric of science articles, and

Gross recently guest-edited an issue of *Technical Communication Quarterly* dedicated to rhetoric of science (3, 1994); the Association of Teachers of Technical Writing also maintains a yearly bibliography of research in rhetoric of science (in the final number of its *Technical Communication Quarterly*); see also Paul M. Dombrowski's section, "Rhetoric of Science", in his edition, *Humanistic Aspects of Technical Communication* (Amityville, NY: Baywood, 1994): 15-79. The link to rhetoric of science from writing across the curriculum is less easy to provide a paper trail for. Institutionally, it is most closely associated with the National Council of Teachers of English's annual Conference on College Composition and Communication, though there is little publication presence by this group in the related organ (*College Composition and Communication*). The people most influential in this development include Charles Bazerman and Greg Myers.

19. *The Methodology of Scientific Research Programmes* (Cambridge: Cambridge University Press, 1978):91.
20. "Just Words", *Nature* 350 (7 March, 1991):30. For his early contributions, see *Scientific Knowledge and its Social Problems* (Oxford, Clarendon Press, 1971).
21. *Ibid.*
22. Richard Rorty, "Science as Solidarity", 39.
23. "Form of a lie" (Michael McGuire, "The Ethics of Rhetoric", *Southern Speech Communications Journal* 45 [1980]: 147); "counterforce" (E. Claire Jerry, "Rhetoric as Epistemic: Implications of a Theoretical Position", *Visions of Rhetoric*, ed., Charles W. Kneupper [Arlington: Rhetoric Society of America, 1987]: 126). In 1975, Douglas Ehninger, "Science, Philosophy—and Rhetoric: A Look Toward the Future", *The Rhetoric of Western Thought*, eds. James L. Golden, Goodwin F. Berquist, and William E. Coleman, fourth edition (Dubuque, IA: Kendall/Hunt, 1989): 631, commenting on the disparity between the respect paid to the sciences and to rhetoric, and casting the fortune of rhetoric of science, said that we need either to tear down the former or build up the latter and that "the first alternative, I think you will agree, is quite unacceptable". Many people, apparently, did not agree.
24. *Western Speech Communication* 50 (1976): 227. There were earlier uses of the phrase "rhetoric of science", but they played on the decorative or dishonest connotations of *rhetoric*. William Powell Jones's book, *The Rhetoric of Science* (Berkeley, CA: University of California Press, 1966), for instance, investigated the use of science as a source of figures and tropes for 18th century poetry, and Jack Douglas's "The Rhetoric of Science and the Origins of Statistical Social Thought", *The Phenomena of Sociology*, ed., Edward A. Tiryakian (New York: Appleton-Century-Crofts, 1971), is largely an assault on sociology that apes the natural sciences by adopting their rhetoric (a usage of the phrase which might, more properly, be replaced by Malcolm Sillars's earlier label, "the rhetoric of scientism", in his "Rhetoric as Act", *Quarterly Journal of Speech* 50 (1964): 282). Sociologist Joseph Gusfield's "The Literary Rhetoric of Science", *American Sociological Review* 41 (1976):16-34, which came out a little earlier in the year than Wander's paper, and has had some influence on the field, is somewhat closer to Wander's usage of the phrase than either of the others. But Gusfield's conception of rhetoric, as the title indicates, owes more to literary criticism than to the traditions of rhetoric; see Carolyn Miller's "Public Knowledge in Science and Society", *Pre/Text 3* (1982): 31-49 for a rhetorical commentary on Gusfield's research, Michael A. Overington's "A Critical Celebration of Gusfield's 'The Literary Rhetoric of Science'", *American Sociological Review* 62 (1977):170-3, for a critique by a like-minded sociologist, and Gusfield's response to the latter in the same issue (173-4). The focus Wander introduces on how scientific evidence is used in debates, of course, did not concern how scientists use evidence in debate with each other, but how they use it with the laity, and how it is used in arguments among laity alone; Wander here is still operating on the far side of Aristotle's wall of expertise,—under Wicheln's notion that Huxleys, not Darwins, are the practitioners of rhetoric and consequently the former not the latter are fit subjects for rhetoricians. While much of the work in this volume gives the lie to that position, popular science is still one of the most prominent themes in rhetoric of science; a large part of what Fahnestock means by "arguing in different forums", for instance, is identified by her sub-head, "The Rhetoric of Popularization". For a more direct discussion of the issues of popularization, see her "Accommodating science", *Written Communication* 3 (1986): 275-96. For other rhetorical studies of scientific popularization, see John Lyne, "Ways of Going Public"; Ray Lynn Anderson, "Rhetoric and Science Journalism"; Diane Dowdy, "Rhetorical Techniques of Audience Adaptation in Popular Science Writing", *Journal of Technical Writing and Communication* 17 (1987): 275-85; Thomas M. Lessl, "Science and the Sacred Cosmos", *The Quarterly Journal of Speech* 71 (1985): 175-87; Greg Myers, "Nineteenth Century Popularizations of Thermodynamics and the Rhetoric of Social Prophecy", *Victorian Studies* 29 (1985): 35-66, "Science for Women and Children", *Nature Transfigured: Literature and Science 1700-1800*, ed. Sally Shuttleworth and J. R. R. Christie (Manchester; Manchester University Press, 1989), and (especially) *Writing Biology,* 141-92; and Kather-

ine Rowan "Moving Beyond the What to the Why: Differences in Professional and Popular Science Writing," *Journal of Technical Writing and Communication* 19 (1989): 161-79. (Two other fine articles by Rowan are also of note—"Strategies for Explaining Complex Science News", *Journalism Educator* 45 (1990): 25-31, and "When Simple Language Fails: Presenting Difficult Science to the Public," *Journal of Technical Writing and Communication* 21 (1991): 369-82—though they tend to be more prescriptive than critical.) Additionally, popularization is often a motif in essays more directly interested in other subjects,—as in Campbell's work on *The Origin*, which (contra Wichelns) was written for a general audience, and Lyne and Howe's discussion of creationist rhetoric, which is virtually by-definition for non-technical audiences.

25. *Ibid, 227.*

26. The Weimer and Overington quotations are from their titles; Finocchiaro's is from page 112. Two earlier *Philosophy and Rhetoric* papers are worth noting in this regard,—Paul Newell Campbell's "Poetic-rhetorical", a small slice of which is concerned with justifying rhetorical analyses of science (21-2), and James Stephens's "Rhetorical Problems", which is a genuine *application* of rhetoric to scientific discourse. Another early paper, P. N. Campbell's "Personae", after a little bit of justification, gets down to discussing a notion (persona) not previously applied to scientific discourse.

27. "Logic and Rhetoric", 112.

28. "Science as a Rhetorical Transaction", 19 - 22. Dilip Parameshwar Gaonkar, "The Idea of Rhetoric in the Rhetoric of Science", *The Southern Communication Journal* 58 (1993): 271, calls Weimer's paper "perhaps the most influential philosophical brief for [rhetoric of science] in the 70s", but I see no evidence at all in the literature to support that claim.

29. "Are Scientists Rhetors in Disguise?" (127). On the sophists, Carolyn Miller, "Technology as a Form of Consciousness", *Central States Speech Journal* 29 (1978): 228-36; her catalogue, "Public Knowledge", 40-1. The Feyerabend articles are both in *PreText* 1 (1980),—Cook and Seamon, "Ein Feyerabenteur", 124-60; Weimer, "For and Against Method", 161-203.Other papers also populate this period, arguing for or about the rhetoric of science: Campbell's "Personae"; Stevenson's "Toward a Rhetoric"; Halloran's "Technical Communication"; J. A. Kelso's "Science and the Rhetoric of Reality", *Central States Speech Journal* 31 (1980): 17-29; Fred Carlisle's "Literature, Science, and Language", *PreText* 1 (1980): 39-72.

30. Notably, Charles Bazerman's "What Written Knowledge Does: Three Examples of Academic Discourse, *Philosophy of the Social Sciences* 11 (1981): 361-88. James Stephens's work on Renaissance science and Wilda Anderson's "The Rhetoric of Scientific Language", *Modern Language Notes* 96 (1981): 746-70, on Lavoisier. Too, Halloran's "Technical Writing" foreshadows his treatment of Watson and Crick in "The Birth of Molecular Biology", *Rhetoric Review* 3 (1984): 70-83 (and this volume), and Ray Lynn Anderson's "Rhetoric and Science Journalism", *Quarterly Journal of Speech* 56 (1970): 358-68, has partial membership in this group, particularly to the theme that developed in rhetoric of science work following popularizations of science. Weaver's forensic essay, "Dialectic and Rhetoric at Dayton, Tennessee", is from his 1953 book, *Ethics of Rhetoric*, 27-54. It is included in this collection. Two other early papers by Weaver also fit loosely into the outlier category, since they deal with social science, "The Rhetoric of Social Science" [1953], *Ethics*, 186-210, and "The Concealed Rhetoric in Scientistic Sociology" [1959], *Language*, 139-59. Campbell's work on Darwin began appearing with his 1970 paper, "Darwin and *The Origin of Species*", *Speech Monographs* 37: 1-14. The essay in this collection, "Charles Darwin: Rhetorician of Science", summarizes the findings of that paper and some of his other earlier work.

31. Pages 1-9.

32. James Hikins, in his review of *The Rhetorical Turn, Quarterly Journal of Speech* 79 (1993): 366, has the relations between rhetoric of science and rhetoric of inquiry a little backward, suggesting that the former evolved into the higher life form of the latter. Others (Simons, *Rhetoric in the Humans Sciences*, 1, for instance) see the phrases effectively as synonyms. Neither of these characterizations is accurate, and, if anything, whatever evolutionary connections the two fields have is close to the opposite of Hikins's story,—rhetoric of science growing out of the rhetoric of inquiry. Many people (certainly not all, but including me) would see rhetoric of science as a subset of rhetoric of inquiry (which also includes the study of many practices not generally considered sciences, such as history and literary criticism), and the institutional apparatus around rhetoric of inquiry (especially the conferences and the University of Wisconsin Press series) has certainly been very valuable to the growth of rhetoric of science. But there is no necessary connection between the two, and the intertwined evolution of both is much more complicated than one begetting the other.

33. Nelson and others 1987:ix.

34. By Greg Myers, in his review of the Symposium proceedings, "Persuasion, Power, and the Conversational Model", *Economy and Society* 18 (1989): 235-6.

35. This is from his "Chronicle and Critique of the Conference", *Quarterly Journal of Speech* 71 (1985):52; Michael Osborn says much the same thing in response to papers by Nelson, Megill, and McCloskey at a conference the year before the symposium,—"Rhetoric in the Human Sciences", *Argument in Transition: Proceedings of the Third Summer SCA/AFA Conference on Argumentation*, eds. Jack Rhodes and others (Annandale, VA: Speech Communication Association, 1983): 234. See also John Lyne's "Rhetorics of Inquiry", *Quarterly Journal of Speech* 71 (1985):65-71, and the preface to Nelson, Megill, and McCloskey's collection, *The Rhetoric of the Human Sciences*, ix-xiii, for other accounts of the symposium. See Greg Myers's "Persuasion, Power, and the Conversational Model", Carolyn Miller's "Some Perspectives on Rhetoric, Science, and History", *Rhetorica* 7 (1989): 101-14, and Malcolm O. Sillars "When Science comes to Rhetoric's House", *Text and Performance Quarterly* 9 (1989): 229-42 for review articles of the book it produced. Herbert W. Simons' collections, *Rhetoric in the Human Sciences* and *Rhetorical Turn*, gather later work conducted under the rhetoric of inquiry banner, both of which contain (as does the Nelson, Megill, and McCloskey volume) commentaries on the rhetoric of inquiry, a very reflexive enterprise.

36. Toulmin was a strong influence on both the rhetoric of inquiry and the argument fields movements—physically sitting on the planning committee for the 1984 rhetoric of inquiry Symposium, and intellectually sponsoring the SCA and AFA research into argument fields,—which indicates some of the sources of the convergence of the two schools. McCloskey, Nelson, and Megill all participated in the 1983 SCA/AFA conference on Argumentation and Social Practice, while Michael McGee and John Lyne returned the visit, participating in the 1984 Iowa Symposium. The role of the SCA in the general development rhetoric of science has remained steady, and has recently reached a more formal stage. At the 1991 conference, Alan Gross and John Lyne organized the seminar, "The Rhetoric of Science: New Directions for the Nineties", and the 1993 conference saw the emergence of a promising organization, The American Association for the Rhetoric of Science and Technology.

37. Cambridge: Cambridge University Press, 1958.

38. "Toulmin on Argument" *Quarterly Journal of Speech* 46 (1960): 44-53.

39. See, in particular, Toulmin's "Logic and the Criticism of Arguments, *The Rhetoric of Western Thought*, especially page 80, where he confesses that he rediscovered the topoi "sleepwalkingly", and Carolyn Miller's "Fields of Argument and Special Topoi" *Argument in Transition*, 147-58, for discussion of the connections.

40. "Ways of Going Public: The Projection of Expertise in the Sociobiology Controversy" *Argument in Transition*, 406.

41. "The Science of Values and the Value of Science", *Argument and Social Practice: Proceedings of the Fourth SCA/AFA Conference on Argumentation*, ed., Robert J. Cox, Malcolm O. Sillars, and Greg B. Walker (Annandale, VA: Speech Communication Association, 1985): 438.

42. "Rhetorical Argument in Science", *Argument in Transition*, 258.

43. *Sociology of Scientific Knowledge*, often abbreviated as *SSK*, is also called in the literature simply *Sociology of Science*, as well as the *Edinburgh School [of Sociology of Science]*, and the *Strong Programme [of Sociology of Science]*. But the first of these alternate label fails to distinguish this qualitative, epistemically directed sociology from more quantitative and epistemically agnostic variants of sociology examining science (particularly as influenced by Robert Merton's research); it is not the only flavour sociology of science comes in. And the other two labels make that distinction at the cost of clarity and breadth. See Charles Bazerman's "Scientific Writing as a Social Act", *New Essays in Technical and Scientific Communication*, 156-81, and Greg Myers' "Writing Research and the Sociology of Scientific Knowledge", *College English* 48 (1986): 595-610, for reviews of this work by the rhetoricians it has most heavily influenced. Merton's work, too, has been influential in rhetoric of science, but to a lesser degree; see, in particular, Bazerman's *Shaping Written Knowledge* (Madison, WI: University of Wisconsin Press, 1988): *passim*, and Lawrence J. Prelli's *A Rhetoric of Science* (Columbia, SC: University of South Carolina Press, 1989): *passim*. I don't mean to imply that there was no overlap between rhetoricians interested in SSK and rhetoricians involved in the SCA/AFA investigation of argument fields—Miller's "Argument Fields", for instance, demonstrates such an overlap—but it was far more muted than the mutual influences of the rhetoric of inquiry and argument field research.

44. *Philosophy, Rhetoric, and the End of Knowledge* (Madison: University of Wisconsin Press, 1992): 10.

45. *Science in Action* (Cambridge, MA: Harvard University Press 1987): 61.

46. Trevor Pinch, "Cold Fusion and the Sociology of Scientific Knowledge" *Technical Communication Quarterly* 3 (1994): 85-100.

47. See Steve Woolgar's appeal for such collaboration, "What is the Analysis of Scientific Rhetoric for?", *Science, Technology, and Human Values* 14 (1989):47-49.

48. Andrew E. Benjamin, Geoffrey N. Cantor, and John R. R. Christie, eds. (Manchester: Manchester University Press, 1987). See also Gillian Beer and Herminio Martins's special issue of *The History of the Human Sciences* (3, 1990), and Peter Dear's collection, *The Literary Structure of Scientific Argument* (Philadelphia: University of Pennsylvania Press, 1991).

49. *Philosophical Papers 1: Realism, Rationality, and Scientific Method* (Cambridge: Cambridge University Press, 1981): 6.

50. "The Idea of Rhetoric in the Rhetoric of Science", 289.

51. "Rhetorical Imperialism in Science", *College English* 55 (1993): 85.

52. "Galileo and Newton: Different Rhetorical Strategies." *Persuading Science*, 107.

53. For representative work of Latour and of Woolgar, see their *Laboratory Life*, 2nd ed (Princeton: Princeton University Press, 1986); for Kitcher, see "Persuasion", *Persuading Science*, 1-28; for Pera, see *The Discourses of Science*; for Dear, see *"Totius in verba"*, *Isis* 76 (1985): 145-61; for McCloskey, see *The Rhetoric of Economics* (Madison, WI: Wisconsin University Press, 1985); for Billig, see *Arguing and Thinking* (Cambridge: Cambridge University Press, 1987).

54. This is Charles Bazerman's take—"Introduction: Rhetoricians on the Rhetoric of Science", *Science, Technology, and Human Values* 14 (1989): 3—and it is a reliable take. Bazerman is thoroughly bilingual. He speaks the language of rhetoricians and the language of sociology.

55. For literary criticism of science, see Wilda Anderson's *Between the Library and the Laboratory* (Baltimore, MD: The Johns Hopkins University Press, 1984), L. J. Jordanova, ed., *Languages of Nature*, (New Brunswick, NJ: Rutgers University Press, 1986), Paul Hernadi, "Literary Interpretation and the Rhetoric of the Human Sciences", *The Rhetoric of the Human Sciences*, 263-75. For psychology of science, see Abraham Harold Maslow, *The Psychology of Science* (New York: Harper and Row, 1966), Barry Gholson, *et al.*, ed., *Psychology of Science* (Cambridge: Cambridge University Press, 1989). For linguistics of science, see Zellig Harris, *The Form of Information in Science* (Boston: Kluwer Academic, 1989), John Swales, *Genre Analysis* (Cambridge: Cambridge University Press, 1990), Greg Myers, "The Pragmatics of Politeness in Scientific Articles", *Applied Linguistics* 10 (1989): 1-35, Jay L. Lemke, *Talking Science: Language, Learning, and Values* (Norwood, NJ: Ablex Pub. Corp., 1990) and M. A. K. Halliday and J. R. Martin, *Writing Science* (Pittsburgh: University of Pittsburgh Press, 1994). For anthropology of science, see Sharon Traweek, *Beamtimes and Lifetimes: The World of High Energy Physicists* (Cambridge, MA: Harvard University Press, 1988). Notably, some non-anthropologists pledge allegiance to anthropology,—Paul Feyerabend, *Against Method*, 249ff, Latour and Woolgar, *Laboratory Life*, 27ff. There are also significant overlaps between some of these other subfields and rhetoric of science, most notably with literary criticism of science. There are enough smearings of categories in this introduction already, so I won't pursue these areas further, but my remarks about rhetoric of science and the three main subfield comparisons extrapolate to these others. For instance, the primary allegiance of literary criticism is to the text; of linguistics, to the code; of psychology, to the mind; of anthropology, to the culture.

56. Please do not take this passage, or this section generally, as a plea for 'keeping rhetoric pure', or an insistence on the sanctity of disciplinary borders, or anything of the like. Rhetoric is, and should be, a field for generalists. I am only pursuing a descriptive course here, not a prescriptive one, and to the extent that I am being a bit hard-headed about knowledge and tradition and allegiance, it is not to make rhetoricians feel smug, but to suggest to scholars of science in other fields who can be somewhat cavalier about the knowledge and traditions of rhetoric that we might be worthwhile comrades.

57. *Structure*, 152.

58. "Rhetorical Argument in Science", 258; see note 39 above.

59. *Writing Biology* (Madison, WI: University of Wisconsin Press, 1990): 195.

60. "Die philosophischen Probleme enstehen, wenn die Sprache *feiert.*" *Philosophische Untersuchungen / Philosophical Investigations*, trans., G. E. M. Anscombe (Oxford: Basil Blackwell, 1958): 19/19e (38.31).

61. Several of them are quite self-conscious about their particularity. They are explicit attempts to draw the energy of field toward case studies and away from underdetermined speculation. Rhetoricians like Michael Halloran and John Lyne noticed in the mid 80s that there was talk aplenty in journals about rhetorical criticism of science, with very few rhetoricians putting their critical money where their theoretical mouths were. "While a number of scholars have been arguing theoretically that science is rhetorical", Halloran said, before launching his case study, "very little attention has been paid to particular cases of scientific

rhetoric" (This volume, p. 39). And, so the lesson isn't lost, once he has docked that study, Halloran calls for more work, for " a body of critical discourse on particular cases of scientific discourse" (This volume, p. 48). Lyne, collaborating with Henry Howe, begins with the identical warrant: "one way to begin constructing a rhetoric of science . . . is to follow what happens when specific scientific arguments go before different audiences. We will trace the dynamics of one such rhetorical process" (This volume, p. 69). And they conclude with the identical clarion: "case histories of scientific controversy need to be analyzed in order to begin identifying the ways scientific discourses and audiences condition one another" (This volume, p. 82).

62. The terms shift regularly, but a continuum of this sort is an obsessive theoretical concern in rhetoric of science. Herbert Simons frames the continuum in a way that reaches from a circumstantial role, rhetoric showing up only in deviant or fringe activities, to an inherent role, rhetoric constituting science ("Are Scientists?"). Carolyn Miller, loading her terms, places the span between a "weak" rhetoric of science, in which there is no knowledge-building role for rhetoric, and a "strong" rhetoric of science, in which rhetoric makes knowledge ("Public Knowledge", 34). For Alan Gross the continuum is shorter and the meat-and-potatoes end is more extreme. For him, the difference is between the possibility of a hard, recalcitrant core of rhetorically uncontaminated scientific knowledge at one end of the continuum, and, at the other end, the possibility that science, knowledge and all, is rhetoric and only rhetoric (most tidily, in his "Rhetoric of Science Without Constraints", *Rhetorica* 9 (1991): 283-99; see J. E. McGuire and Trevor Melia, "The Rhetoric of the Radical Rhetoric of Science" for a response to that position, and Gross's response to the response, in the same issue, 301-16).

63. *The Rhetoric of Science* (Cambridge: Harvard University Press, 1990): 33.

64. Besides, Thomas Benson beat me to the punch by including Campbell's first Darwin paper, quite rightly, in his *Landmark Essays* volume on *Rhetorical Criticism* (Davis, CA: Hermagoras, 1993).

65. The phrase is Carolyn Miller's, about this essay,—"Some Perspectives", 107. Gaonkar, "The Idea of Rhetoric", 276-82, divides Campbell's investigations of Darwin into two stages, focussing on Darwin respectively as rhetorical tactician and as cultural conduit: the first, he calls "intentionalist"; the second he calls "intertextual", though another term he uses, "constitutive" is perhaps more appropriate; Campbell's "Scientific Discovery and Rhetorical Invention", *The Rhetorical Turn*, 58-91, in particular, begins to admit a constitutive role for rhetoric in Darwin's ideas. See Campbell's "Reply to Gaonkar and Fuller", *The Southern Communication Journal* 58 (1993): 312-18, for his comments on this reading, and or other aspects of Gaonkar's position. "Charles Darwin: Rhetorician of Science" succinctly captures—almost wraps up—the first stage. For other rhetorical treatments of Darwin, who has been subject to many, none with the perception of Campbell's, see Avon Crismore and Rodney Farnsworth's "Mr. Darwin and His Readers: Exploring Interpersonal Metadiscourse as a Dimension of Ethos", *Rhetoric Review* 8 (1989): 91-112; their "Scientific Rhetoric, Metadiscourse, and Power: Darwin's *Origin*", *Rhetoric and Ideology*, ed Charles W. Kneuper (Arlington, TX: Rhetoric Society of America): 174-88; their (authorship reversed this time) "On the Reefs: The Verbal and Visual Rhetoric of Darwin's Other Big Theory", *Rhetoric Society Quarterly* 21 (1991): 11-25; Anne Holmquest's "Rhetoric and Semiotic in Scientific Argumentation", *Argument and Social Practice*, 376-402; Barbara Warnick's "A Rhetorical Analysis of Episteme Shift: Darwin's *Origin of Species*", *The Southern Speech Communication Journal* 49 (1983): 26-42; Phillip Sipiora, "Ethical Argumentation in Darwin's *Origin of Species*", *Ethos: New Essays in Rhetorical and Critical Theory*, ed., James S. Bauman and Tita French Bauman (Dallas: Southern Methodist University Press, 1994): 265-292, and Marcello Pera, *Discourses of Science*, 71-88, *et passim*.

66. This volume, p. 3.

67. This volume, p.33. For a less compact and less contentious, but deeper, rhetorical treatment of Newton's *Opticks*, see Bazerman's chapter in *Shaping Written Knowledge* (Madison, WI: University of Wisconsin Press, 1989): 80-127.

68. "Kairos in the Rhetoric of Science", *A Rhetoric of Doing*, ed., Stephen P. Witte, Neil Nakadate, and Roger D. Cherry (Carbondale: Southern Illinois University Press, 1992): 310. Miller is right that Watson and Crick's paper is a key text for rhetoric of science: see Bazerman, *Shaping Written Knowledge*, 18-48, Gross, *The Rhetoric of Science*, 54-65, Lawrence Prelli, *A Rhetoric of Science*, 236-56, Walter R. Fisher, "Narrative Rationality and the Logic of Scientific Discourse", *Argumentation* 8 (1994): 21-32. Her implication, however, that this paper is the clearest candidate for canonization is a little strong; *The Origin of Species* has it beat substantially, in both quantity and quality of rhetorical analysis.

69. Two books are also worth noting. My *Linguistics Wars* (New York: Oxford University Press, 1993) treats an important dispute in the modern history of linguistics at some length, and attends carefully to both

sides, but much of the rhetorical analysis is superficial. And Jean Dietz Moss's *Novelties in the Heavens* follows the geo/heliocentric controversies from Copernicus to Galileo.

70. Wander, "The Rhetoric of Science" (226-7), identified two principal areas of the rhetorical investigation of science, one concerning the efforts of scientists to persuade one another, the other concerning how science intersects with deliberative rhetoric. Cook and Seamon, "Ein Feyerabenteur" (146), do likewise,—with their "internal" and "external" categories for rhetoric of science. My own effort at taxonomy, "Rhetoric of Science", cut things somewhat differently, but also identified a "rhetoric of public science policy" as a major area of the field.

71. "The Rhetoric of Social Science", *Ethics*, 186.

72. See "The Rhetoric of Social Science", *Ethics of Rhetoric*, 186-210, "Concealed Rhetoric in Scientistic Sociology", *Language*, 138-58. As the terms of the latter paper make clear, though, this work belongs more fully in the category Sillars calls "rhetoric of scientism" (see note 24 above). Weaver's clearest statement of the connections between science and rhetoric, which locates him at the Platonic just-propagating end of the epistemic continuum, is in "Concealed Rhetoric" (140-2). "Strongly rhetorical"is from "Ultimate Terms in Contemporary Rhetoric", *Ethics of Rhetoric*, 215.

73. "Language is Sermonic", *Language*, 205.

74. *Rhetoric* 1354a. The translation is by George A. Kennedy, *On Rhetoric: A Theory of Civic Discourse* (New York: Oxford University Press, 1991): 30. Aristotle's analogy directly concerns forensic rhetoric, but I don't think I have bent it unduly for this context. See Barbara G. Cos and Charles G. Roland, "How Rhetoric Confuses Scientific Issues", *I.E.E.E Transactions on Professional Communication* PC-16, No. 3 (September, 1973): 140-3, for a typical rant against emotion in scientific rhetoric.

75. In addition to more work by Bazerman, mostly in *Shaping Written Knowledge*, see Mary B. Coney, "Terministic Screens: A Burkean Reading of the Experimental Article", *Journal of Technical Writing and Communication* 22 (1992): 149-59; Alan G. Gross, "The Form of an Experimental Paper", *Journal of Technical Writing and Communication* 15 (1985): 15-26, and "Experiment as Text", *Rhetoric Review* 11 (1993): 290-300, for other work on the rhetorical nature of experimental reports.

76. This volume, p.169.

77. For Moss, see *Novelties in the Heavens*, 21-23, 330-332; for Gaonkar, "The Idea of Rhetoric", *passim*.

78. "Rhetorics of Inquiry", 72-3.

Giants in Science

Charles Darwin:
Rhetorician of Science

by John Angus Campbell

To claim that Charles Darwin was a "rhetorician" may seem to confuse the provinces of rhetoric and science. Their juxtaposition, however, is not only warranted; it is also inescapable. Even scientific discourse must be persuasive to rescue insight from indifference, misunderstanding, contempt, or rejection. Aristarchus was not believed when he argued that the earth moved around the sun, and although Mendel discovered the laws of inheritance, he failed to convince his scientific peers.[1] To claim that Darwin was a rhetorician, therefore, is not to dismiss his science, but to draw attention to his accommodation of his message to the professional and lay audiences whose support was necessary for its acceptance. Commonly overlooked in studies of Darwin is that he persuaded his peers and the wider community by using plain English words and plain English thoughts.[2]

Prior to Darwin, no evolutionist, whether popularizer or professional scientist, enjoyed both a popular and a professional following.[3] (Some enjoyed neither.) To understand why Darwin was persuasive with the reading public as well as with a key minority of his professional peers requires an examination of Darwin as a rhetorician of science.

I

That *The Origin of Species* was a popular book should hardly be surprising. *The Origin* is rhetorical from the ground up. The brevity of Darwin's classic work—indeed, its appearance as an "abstract"—is evidence of its rhetorical character. That *The Origin* made its appearance as a single compact volume, accessible to a general audience, was the result of a remarkable circumstance. In June 1858 Darwin was in the second year of writing *Natural Selection*, a book on transmutation which he had been planning since 1837. On the sixteenth of that month Darwin was startled to receive from the young naturalist Alfred Russell Wallace the sketch of a theory virtually identical to his own. In the wake of the Wallace letter, Darwin put aside his mammoth text, then two-thirds complete, and in nine months produced the work on which his fame rests.[4] Darwin received Wallace's letter on June eighteen, 1858.[5] He began *The Origin* on July 20, and by March 22 the book was written. *The Origin* went on sale on November 24, 1859.[6]

The ethos of its author is further proof that *The Origin* is rhetorical. Darwin directly appeals to the reader's sympathy: "my health is far from strong. . . . This Abstract . . . must necessarily be imperfect. I cannot here give references and authorities for my several statements; and I must trust to the reader reposing some confidence in my accuracy."[7] As Darwin's son Francis observed, "The reader feels like a friend who is being talked to by a courteous gentleman, not like a pupil being lectured by a professor. The tone of . . . *The Origin* is charming, and almost pathetic."[8]

The rhetorical character of *The Origin* is further established by its everyday language. Darwin's very title, *On the Origin of Species by Means of Natural Selection, or, The Preservation of Favoured Races in the Struggle for Life*, is colloquial. The themes of "origin," "selection," "preservation," "race," "struggle," and "life" underscore the intimacy, not the distance, between the author and the everyday world. Further, Darwin's exposition is as down to earth as his title. C. C. Gillespie's list of Darwin's commonplaces could be easily duplicated by any reader:

> So ordinary is the language that it almost seems as if we could be in the midst of a lay sermon on self-help in nature. All the proverbs on profit and loss are there, from pulpit and from counting house—On many a mickle making a muckle: 'Natural selection acts only by the preservation and accumulation of small inherited modifications, each profitable to the preserved being'; On the race being to the swift: 'The less fleet ones would be rigidly destroyed'; On progress through competition: 'Rejecting those that are bad, preserving and adding up all that are good; silently and insensibly working, whenever and wherever opportunity offers, at the improvement of each organic being'; . . . . On saving time: 'I could give many examples of how anxious bees are to save time'; . . . On the compensation that all is, nevertheless, for the best: 'When we reflect on this struggle, we may console ourselves with the full belief, that the war of nature is not incessant, that no fear is felt, the death is generally prompt, and that the healthy and the happy survive and multiply.'[9]

Further evidence that *The Origin* is rhetorical is seen in Darwin's deference to English natural theology. Everyone knows that theological objections were raised against *The Origin*. What might surprise the modern reader is the theological defense within it. In the first edition, Darwin's flyleaf contained two citations from works in the tradition of English natural theology, one from William Whewell's *Bridgewater Treatise* and one from Francis Bacon's *Advancement of Learning*. In the second edition, the first two citations were reinforced by a third from Bishop Butler's *Analogy of Revealed Religion*. In the first edition, the famous final line, which begins, "There is grandeur in this view of life," continues, "with its several powers, having been originally breathed into a few forms or into one." Starting in the second edition, the line has been changed to read "breathed by the Creator into a few forms or into one." In the fourth edition of the work, the reader finds the following postscript at the end of the table of contents: "An admirable . . . Review of this work including an able discussion on the Theological bearing of the belief in the descent of species, has now been . . . published by Professor Asa Gray, M.D., Fisher Professor of Natural History in Harvard University."[10] The reader of *The Origin* would not know that Darwin himself was responsible for financing the publication of Gray's essays in pamphlet form (originally they appeared as unsigned essays in the *Atlantic Monthly*), and until 1867 he would have no way of knowing that Darwin did not believe in the argument they contained.[11] Although Darwin privately expressed his difference with Gray in a letter in the fall of 1860, it was not until 1867 that he publicly rejected Gray's argument in the conclusion to his two-volume *Variation in Plants and Animals Under Domestication*.[12] No mention of this refutation was ever made in the subsequent two editions of *The Origin* (1869, 1872). Indeed, throughout the body of his book, whether the reader examines Darwin's case for the common ancestry of the horse, hemionus, quagga, and zebra or his account of how natural selection could have formed the eye, Darwin urges his views as more in keeping with proper respect to the ways of Providence than the views of his opponents.[13]

The rhetorical character of *The Origin* is also seen in Darwin's appeal to common sense. In language reminiscent of Scottish Commonsense Philosophy, Darwin urged that we can trust a theory which explains so many large classes of facts because this "is a method used in judging in the common events of life."[14]

II

In light of the manifest rhetorical features which would have recommended *The Origin* to a general audience, an obvious question suggests itself. Why did the clearly popular character of Darwin's writing not impede the reception of his ideas among his scientific peers? One reason Darwin's literary language did not pose the kind of obstacle to professional acceptance it would today is, as Susan Gliserman has noted, that all science was so plainly literary in Darwin's day: "I have considered the literary structure of the science writers as no difference from that of Tennyson's poems."[15] Yet, as Darwin's imagistic language was an issue, even by the standards of his own time, something more than Darwin's conformity with accepted literary conventions seems to have been involved in his generating both professional support and popular appeal.

A reputation for eloquence can be a dangerous thing. Although the art of rhetoric may make a speech or book striking, if its artistry is detected, that very fact may be advanced as reason for rejecting it. If it seems unlikely that anyone in real life could claim, "I am no orator as Brutus is," and then deliver an eloquent address without the audience's getting suspicious, it is well to recall the example of Thomas Henry Huxley. It was the no-nonsense Huxley who coined the term "agnosticism" and who characterized Comte's religion of humanity as "Catholicism *minus* Christianity."[16] Both Darwin and Huxley enjoyed solid reputations as scientists, both were unusually gifted writers, yet neither man's literary skills ever compromised his reputation for fact and dusty sobriety. Like Huxley, Darwin minimized his literary gifts. He also minimized his formidable theorectical power. Darwin's dismissal of his own colorful language and deemphasizing of the hard, sustained theoretical work behind his theory are connected.

Darwin introduced the major theoretical work of modern biology by minimizing the importance of his own speculative powers; he used provocative images throughout his exposition, yet he explained away his originality by insisting that his ideas were the result of "facts" and his metaphors mere expressions of convenience.

The thesis I am arguing is that Darwin was able to make his rhetoric seem unimportant or at best incidental to his scientific point and to persuade his professional peers because his narrative was governed by the conventions of Baconian induction and quasi-positivist standards of proof. Examination of the discrepancies between Darwin's public and private attitudes toward his method, language, and achievement offers a rare glimpse of a process which, in successful science at least, is infrequently observed: the production of the Mark Anthony effect, in which rhetoric is freely employed and effectively masked.

One of the most striking discrepancies between Darwin's public and private attitudes toward the conventions of proper scientific theory is the contrast between his declared and his actual path to discovery. In the opening paragraph of *The Origin* we read the following account:

> When on board H.M.S. 'Beagle,' as naturalist, I was much struck with certain facts in the distribution of the inhabitants of South America, and in the geological relations of the present to the past inhabitants of that continent. These facts seemed to me to throw some light on the origin of species—that mystery of mysteries as it has been called by one of our greatest philosophers. On my return home, it occurred to me, in 1837, that something might perhaps be made out on this question by patiently accumulating and reflecting on all sorts of facts which could possibly have any bearing on it. After five years work I allowed myself to speculate on the subject, and drew up some short notes; these I enlarged in 1844 into a sketch of the conclusions, which then seemed to be probable: from that period to the present day I have steadily pursued the same object. I hope that I may be excused for entering on these personal details, as I give them to show that I have not been hasty in coming to a decision.[17]

In his *Autobiography* Darwin similarly affirms: "I worked on true Baconian principles, and without any theory collected facts on a wholesale scale." Of his famous insight on reading Malthus, Darwin records: "Here, then, I had at last got a theory by which to work."[18]

What one finds when one examines Darwin's private notebooks, however, is irreconcilable with Darwin's public statements about his research method. One of the closest students of these notebooks, Howard Gruber, says of Darwin's public comments on method: "Insofar as he said anything publicly on the subject of method, Darwin presented himself in ways that are not supported by the evidence of the notebooks." In response to Darwin's granddaughter, Nora Barlow, who affirmed that in the earlier days there was a closer fit between her grandfather's theorizing and observations, Gruber observed that "it seems to me that even in these early notebooks, . . . he delighted in far-ranging speculations and saw himself as creating ideas of the same grandeur and cosmic scale as the 'early astronomers' to whom he likened himself."[19] Of the specific citations we have noted from *The Origin* and the *Autobiography*, Gruber comments:

Taken together, these statements give an extremely misleading picture. Darwin certainly began the notebooks with a definite theory, and when he gave it up it was for what he thought was a better theory. True, when he gave up his second theory he remained in a theoretical limbo for some months. But even then he was always trying to solve theoretical problems. . . . he almost *never* collected facts without some theoretical end in view. It was not simply from observations but from hard theoretical work that he was so well prepared to grasp the significance of Malthus' essay.[20]

Occasionally in his correspondence, Darwin would similarly present himself as a firm inductionist. In a letter to Herbert Spencer's American disciple John Fiske, Darwin diplomatically avoided discussing Fiske's books by affirming: "my mind is so fixed by the inductive method, that I cannot appreciate deductive reasoning. I must begin with a good body of facts and not from principle (in which I always suspect a fallacy), and then as much deduction as you please."[21] But in letters to his associates, Darwin expressed himself quite differently. In a letter written in June 1860 to his long-time friend Charles Lyell, Darwin bemoaned a paper by Hopkins, who would not accept the argument of *The Origin* on the ground that the mere explanatory value of a theory did not prove its correctness: "on his standard of proof, natural science would never progress, for without the making of theories, I am convinced there would be no observations."[22] In a letter written in 1861 to his colleague Henry Fawcett, Darwin criticized strict inductionists in these words: "About 30 years ago there was much talk that geologists ought only to observe and not theorise; and I well remember some one saying that at this rate a man might as well go into a gravel pit and count the pebbles and describe the colours. How odd it is that anyone should not see that all observation must be for or against some view if it is to be of any service!"[23]

Given that Darwin not only understood the importance of theory, but began his own research with a conclusion that transmutation had occurred, and held to that conclusion even when he could not factually support it, how are we to account for the discrepancy between Darwin's private and public statements on method? The discrepancy, I believe, is explained by the view that Darwin was using a methodological convention important to his colleagues, though irrelevant to his science, to give a traditional warrant to a controversial thesis and hence make it persuasive.

That Darwin's public account of his method was rhetorically motivated is supported by the esteem in which Baconian induction was held by all English philosophers of science in the mid-nineteenth century. John Herschel, William Whewell, and John Stuart Mill disagreed about many particulars, but on one thing they were resolved—true science was inductive. In analyzing the place of induction in mid-century philosophy or science, David Hull makes the wry observation: "It would be nice to be able to set out at this point the meaning which the disputants attached to this word, but I cannot. Everyone meant something different by it, and in the works of a single man, one is likely to find many different uses of the word."[24] In short, by Darwin's time "Baconian Induction" had become what Bacon would have called an "Idol of the Theatre."

As Charles Bazerman points out in his paper on the history of the American Psychological Association's stylesheet, professional conventions dictate the form of scientific discourse.[25] In Darwin's time, no less than in our own, data certified by the appropriate method are far more likely to be accepted than argument about fundamentals. Even M. T. Ghiselin, who along with Gavin DeBeer holds that Darwin was true to the canons of the hypothetical-deductive method, describes Darwin's introductory paragraph to *The Origin* as a "dialectical maneuver" and observes that "Darwin, like other scientists of his day, gave much lip service to 'induction,' and such hypocrisy has long been the real nature of scientific discovery." Ghiselin's way of avoiding misunderstanding Darwin is "to abandon the study of words and to derive our understanding from concepts." In Ghiselin's view, "The structure of Darwin's systems explains his success and failure alike. When the process through which his discovery was generated has been understood, there is no reason whatever to treat his perfectly ingenuous accounts of the discovery as mistaken, contradictory, or hypocritical."[26]

I concur with Ghiselin's assessment of the importance of understanding "Darwin's systems." I reluctantly differ with his judgment that once this is done Darwin's statements on method emerge as "perfectly ingenuous." The testimony of Darwin's notebooks argues strongly that Darwin thought long and hard, not only about nature, but about persuasion, and that he went to great lengths, including not developing his views on the evolution of man, to minimize the shock of novelty *The Origin* would occasion.[27] No one serious about making a revolution can lightly ignore accepted professional standards. How far one goes in deferring to standards irrelevant or hostile to one's actual procedures determines the personal dimension in science. Some writers, like René Descartes or Noam Chomsky, may storm the citadel of convention directly. The fact is, however, that frontal assault was not Darwin's style, and thus a certain disingenuousness was necessary for Darwin to be persuasive. Edward Manier puts the issue of Darwin's rhetorical strategy succinctly: "the early drafts of the theory do not conform to the 'hypothetico-deductive model' of scientific explanation, although they indicate Darwin's intent to represent his views as *if* they did conform to that model."[28]

To appreciate how much rhetorical ingenuity went into the composition of *The Origin*, one has only to contrast the reassuring inductivist style of *The Origin* with the rapid sequence of topics, inferences, and reflections on strategies of persuasion one finds in Darwin's notebooks. There is ample science in Darwin's notebooks and much of it is outstanding science. But the story-line is not the same as that in *The Origin*. In the notebooks, we see the young Darwin, even before he solved the technical problem of speciation, thinking of ways to solve the problem of persuasive exposition. In the "C" notebook, Darwin reminds himself to point out to his audience the moral responsibility of the scientist as epochal truth-bearer:

> Mention persecution of early Astronomers,—then add chief good of individual scientific men is to push their science a few years in advance only of their age . . . must remember that if they *believe* & not openly avow their belief they do as much to retard as those whose opinion they believe have endeavored to advance the cause of truth.[29]

The same notebook illustrates the intermingling of his scientific insight with his theological and strategic reflections:

> Study Bell on Expression & the Zoonomia, for if the former shows that a man grinning is to expose his canine teeth ((this may be made a capital argument. if man does move muscles for uncovering canines)) no doubt a habit gained by formerly being a baboon with great canine teeth.—((Blend this argument with his having canine teeth at all.—)) . . . Hensleigh says the love of the deity & thought of him / or eternity / only difference between mind of man & animals.—yet how faint in a Fuegian or Australian! Why not gradation.—no greater difficulty for Deity to choose. when perfect enough for Heaven or bad enough for Hell.—(Glimpses bursting on mind & giving rise to the wildest imagination & superstition.—York Minster story of storm of snow after his brother's murder.—good anecdote.[30]

In the "M" notebook, Darwin's awareness of the rhetorical dimension of his task is registered in his reflection on how best to make his underlying philosophy: "To avoid stating how far I believe in materialism, say only that emotions, instincts, degrees of talent, which are hereditary are so because brain of child resembles parent stock."[31] The "M" notebook also makes clear that from the first Darwin speculated freely on both science and philosophy and did not begin by amassing facts and postponing thought: "Origin of Man now proved.—Metaphysics must flourish.—He who understands baboon would do more toward metaphysics than Locke."[32] When one contrasts the breadth and exuberance of Darwin's early reflections, which freely move backward and forward through philosophy, theology, rhetoric, psychology, and numerous branches of natural science, and encompass ethics and aesthetics, with the chastened tone and narrow range of topics addressed in *The Origin*, one is little short of awed by the massive restraint and carefully premeditated adaptation of his public argument.

Darwin's care to redescribe his path to discovery so that it appeared to conform with conventional standards of Baconian inductionism is not the only way in which he adapted his ideas to his scientific peers. Darwin was rhetorical both in his concern with persuasion and in the heavily metaphorical character of his thought. His images lent his ideas popular appeal, but since they drew attention to themselves as images, explaining them away posed a distinct rhetorical challenge. As of his method, so of his metaphors. Darwin argued that his language conformed to accepted professional standards.

The highly imagistic character of Darwin's language was a center of controversy from the very first. Ghiselin's recommendation that Darwin's language simply be set aside indicates that the problem of how to interpret it is still an open question. C. C. Gillispie has long held that Darwin expressed himself in a needlessly misleading manner, and even Howard Gruber, who does not appear to share Ghiselin's view of the cogency of Darwin's approach to method, cautions that making too much of the social roots of Darwin's language is "unDarwinian."[33] It is at least curious that so many distinguished interpreters of Darwin, who do not necessarily agree on other points, concur in deemphasizing the importance of his language for an understanding of his achievement. The thesis is worth considering that Darwin used metaphorical language to make his scientific point and that the very connotations we are warned not to take seriously were instrumental in his ability to persuade both his professional peers and the general public. To determine what importance to attach to Darwin's language, let us contrast his public statements concerning language with the testimony of his private papers.

Starting with the third edition of *The Origin*, Darwin responded to the criticism of his imagistic language by pointing out that certain of his metaphors were in fact metaphors:

> In the literal sense of the word, no doubt, natural selection is a misnomer; but who ever objected to chemists speaking of the elective affinities of the various elements?—and yet an acid cannot strictly be said to elect the base with which it will in preference combine.

> It has been said that I speak of natural selection as an active power or Deity; but who objects to an author speaking of the attraction of gravity as ruling the movements of the planets? Everyone knows what is meant

and is implied by such metaphorical expressions; and they are almost necessary for brevity. So again it is difficult to avoid personifying the word Nature; but I mean by Nature, only the aggregate action and product of many natural laws, and by laws the sequence of events as ascertained by us. With a little familiarity such superficial objections will be forgotten.[34]

Darwin's public account of his metaphors creates the impression that his images could be replaced by literal statements if time were not a factor. But Darwin's philosophy of language, as well as his use of language generally, shows that rhetoric is essential, not incidental, to his case.

Darwin's minimizing of metaphor manifests his seeming deference to the linguistic standards of Comtean positivism. In August 1838 Darwin read David Brewster's review of the first two volumes of Comte's *Philisophie Positive*.[35] Brewster's review convinced Darwin of Comte's thesis that, like humankind in general, each science goes through the stages of myth and metaphysics before reaching a final positive stage. After reading Brewster's review, Darwin took as his own the mission of bringing biology out of the metaphysical stage.[36] A significant point of difference between Darwin and Comte, revealed by Darwin's notebooks and underscored by his published writing, however, concerns the language proper to science. Comte's philosophy of language was thoroughly nominalist. Both in his theory of language and in his use of language, Darwin was a realist. Comte, for example, would ban from chemistry such expressions as "elective affinities" and ban "attraction" from the language of astronomy. In the above quotation from the *The Origin*, although he retains the offensive terms from astronomy and chemistry, Darwin's definition of "nature" and "natural law" are solidly in line with Comtean linguistic standards.[37]

Darwin's difference with Comte on the language proper to science in fact was radical. First, in keeping with the realism of Scottish Commonsense Philosophy, Darwin saw nature itself as expressive. Human language, in the Scottish Commonsense view, was a continuation of the natural expressiveness of all sentient life.[38] Having accepted this position, Darwin did not have the horror of anthropomorphism that was endemic to positivism, with its demand for a language appropriate to a Cartesian billiard-ball universe. Second, and as a corollary, Darwin saw the aim of scientific language as persuasive communication and not conceptual precision.[39]

Darwin's philosophy of language is as important to his *scientific* achievement as to his popular success because it, rather than the inductivist-positivist theory of language to which he publicly deferred, helps explain his success in establishing a novel research paradigm. We can see the distance between Darwin's public quasi-positivist account of his metaphors and the actual use he made of figurative language by examining his key terms, "natural selection" and "struggle for existence." In a crucial section of his chapter on "Natural Selection," Darwin dramatically contrasts man's puny powers with the powers of nature: "Man can act only on external and visible characters: nature cares nothing for appearances. . . . She can act on every internal organ, on every shade of constitutional difference, on the whole machinery of life." In the next paragraph, Darwin says, "It may metaphorically be said that natural selection is daily and hourly scrutinising, throughout the world, every variation, even the slightest; rejecting that which is bad, preserving and adding up all that is good; silently and insensibly working, whenever and wherever opportunity offers."[40]

There is a marked discrepancy in these passages between Darwin's claim that he is merely adopting a way of speaking and his inability to speak any other way. Since in Darwin's own terms nature's selection is invisible and insensible, his metaphor is a matter of necessity and not of convenience. In this passage, Darwin uses rhetorical language simultaneously to propose a new paradigm for science and to create a new popular understanding of humanity's relation to nature. The key element is the tension between Darwin's image of the human selector (the breeder), whose operations are known to the audience, and the operation of nature, whose ways are unknown. The image of the selector is persuasive precisely because it brilliantly exploits a technological symbol and thus competes with the idea of miracle in a concretely believable way. Miracles were more credible to Darwin's contemporaries than the obvious a fortiori argument, popularly advanced by Robert Chambers and Herbert Spencer, that since natural law governed every other department of science, it *must* govern biology as well. A common expression in the science of Darwin's time was "the laws of creation."[41] Darwin, we may surmise, was persuasive because he took the "confused notion" of "creation" by "law" and gave it a decisive naturalistic turn.[42] In comparison with Milton's "The grassie Clods now calv'd, now half appeer'd the Tawnie Lion, pawing to get free His hinder parts," Darwin's "natural selection" was equally concrete yet provided a more believable illusion because the reader knew how domestic varieties came into being.[43] The image of nature forming species, much as the cattle-breeder or the pigeon-fancier formed varieties, is a naturalistic image that for scientist and general reader is truer to experience than is miracle. For the scientist in particular, "natural selection" heuristically embodies a richer research program than the one embodied in the notion of "laws of creation." When we appreciate that Darwin had originally hoped to explain variation, and could not, we begin to understand why it is the rhetorical tradition of the Scottish Commonsense Philosophers, and not the positivist tradition of August Comte, that accounts for his language.

"Natural selection" does not explain how an imperceptible variation internal to the organism could be selected. Nor does it allow us to predict the kind of internal variations we would expect to find in organisms in a particular environment. What natural selection does is clear a semantic space that a natural law might fill. Indeed, Edward Manier describes the semantic-rhetorical function of natural selection precisely when he describes it as a "place-holding allusion."[44] Natural selection is not incompatible with any known law, and it is not supernaturalistic, because although Darwin magnifies nature's powers, his concept of "nature," like the breeder, acts only on variations when they happen to occur.

The nonpositivist character of Darwin's term is underscored when we consider its ancestry. In his sketch of 1842, and again in his draft of 1844, Darwin had asked the reader to imagine "a being more sagacious than man, (not an omniscient creator)."[45] Although "natural selection" is less anthropomorphic than the "being more sagacious than man," the function of the image remains identical. Rather than asking the reader to imagine "a being," Darwin simply has the reader project what is known of the operations of the domestic breeder onto nature. Although Darwin's image does not explain variation, or even how imperceptible unspecified internal variations could be of use to the organism, it does provide science with a heuristically rich "as if " to guide research.

What we have seen as true of "natural selection" holds equally for Darwin's other centrally important term, "struggle for existence." In both cases, the affective connotations of the terms seem to have been at least as important as their literal meanings. When we see the variety of terms Darwin considered, the self-consciously rhetorical character of Darwin's choice of "struggle for existence" becomes clear. In *Natural Selection*, the book Darwin abandoned when he received the Wallace letter, the section which corresponds to chapter 3 of *The Origin*, "The Struggle for Existence," had once been entitled "War of Nature." An early topic sentence had read, "The elder De Candolle in an eloquent passage has declared that all nature is at war."[46] Manier notes that Darwin at one time considered using Lyell's expression "equilibrium in the number of species." Indeed, Darwin affirmed that Lyell's expression was "more correct" than his own. Significantly, however, Darwin rejected Lyell's expression on the ground that it conveyed "far too much quiescence." By Darwin's own account, accuracy was not his criterion. He chose "struggle for existence" because it occupied a desirable semantic space mid-way between "war" and "equilibrium."[47] In *The Origin*, Darwin distinguished three uses of the term "struggle." He indicated that we could speak of organisms as "truly" engaged in struggle where two animals were in competition with one another for the same scarce resource and if one obtained more of the resource, that animal would increase its life expectancy or prospects of leaving progeny, while its adversary would not. Second, the "less proper" meaning of struggle would describe a situation in which an organism confronted a limited environment, as in the case of a plant in time of drought. Darwin recognized that it would be "more proper" in such an instance to say that the plant was "dependent" upon moisture than to say that it was struggling to survive. Finally, Darwin used "far fetched" to characterize a struggle in which a parasite so increased in power that it threatened its host's existence and , ultimately, its own. Darwin noted that the three meanings "pass" or "graduate" into each other.[48] As Manier observes of these three meanings:

it is necessary to consider the possibility that each meaning influenced his understanding of the other two. The domain of events referred to by the terms 'war' or 'conflict,' for example, may be significantly redescribed if the same term ('struggle') is used to designate it and two other domains (those more commonly designated by 'dependence' and 'chance') as well. The result is not the expression 'too much quiescence' but rather an elaborate qualification of the 'strict meaning' of 'struggle' within the context of Darwin's theory. Darwin's use of this metaphor may have been poetic as well as scientific. He was willing to risk the ambiguity resulting from the inter-connection of a variety of related but distinct meanings in a single, compressed metaphoric representation.[49]

Darwin's invocation of quasi-positivistic disclaimers for his use of metaphoric terms can be reconciled neither with his adherence to Scottish Commonsense linguistic philosophy nor with the functions of his key terms "natural selection" and "struggle for existence" in *The Origin*. His public insistence that his images were for convenience was an apparent attempt to defer to scientific conventions too professionally entrenched to challenge. Darwin's distinct genius for giving old terms new meanings in order to present persuasively a novel vision of nature was central to his scientific and popular success, even though the linguistic-rhetorical theory which informed his choice of language could not have been made explicit without damage to his credibility.

A final aspect of Darwin's adaptation to his professional audience concerns his endeavor to convince his peers that in natural selection he had identified the specific mechanism by which evolution occurs. Clearly, since Erasmus Darwin, Jean Baptiste Lamarck, Etienne Geoffrey Saint-Hilaire, Robert Chambers, and Herbert Spencer all had argued the general case for evolution, Darwin's unique scientific contribution was his theory of natural selection. In presenting his theory, he was careful to use language that would communicate to his peers the unique explanatory power he believed natural selection possessed. Following the theoretical language popularized by John Herschel, Darwin spoke of natural selection as the *vera causa* of organic change.[50]

As has been often remarked, the irony of Darwin's achievement is that he succeeded in popularizing all forms of evolutionism but his own.[51] Even Huxley, wholehearted as he was in advancing science through championship of Darwin, did not believe natural selection to be the sole cause of evolution. Whereas Darwin insisted that "natura non facit saltum," Huxley was willing to allow an occasional leap, particularly when Lord Kelvin insisted that data from physics denied that the earth was as old as Darwin needed it to be. Although personally, as Michael Ruse puts it, Darwin "miserably dug in his heels and refused to defer to the physicists," he concluded in private and in a letter to the *Athenaeum* that the specific theory one adopted was less important than the choice between evolution of whatever kind and special creation.[52]

From a rhetorical standpoint, the irony of Darwin's achievement is only partial. However scientifically important natural selection is for contemporary science, in historical and rhetorical perspective, Darwin's discovery was only an incident in the development of a general argument he already believed in on other grounds. In the spring of 1838, before he had read Malthus, Darwin observed in his notebook that there was scarcely any novelty in his theory of transmutation and that the whole object of his prospective book was proof. Dov Ospovat concluded his examination of the pre-and post-Malthus Darwin by underscoring the early emphasis on the general argument: "Darwin was fond of the theory of natural selection, but his greatest concern was to establish the doctrine of descent."[53] Darwin's use of the theoretically fashionable expression *vera causa* was not entirely lost on his professional peers and no doubt made his ideas seem all the more impressive to the general reader. His brilliant evolutionary reinterpretation of the known facts and theories of mid-century science persuaded a significant number of his peers and no doubt many of his lay contemporaries that some naturalistic *vera causa* could account for organic change.[54] We err when we think that Darwin's underlying intent was to offer an original scientific theory. Darwin's initial intent was to make evolutionism persuasive.

III

Charles Darwin was a brilliant scientist, but neither an iconoclast nor a martyr. Shortly after his return to England, probably as a result of reviewing the data on geographical distribution from the *Beagle* voyage, Darwin became a convinced transmutationist. As his was a bold and original mind, Darwin at once proceeded to draw out the various implications of his discovery. As Darwin's notebooks demonstrate, theological, aesthetic, and moral theorizing, as well as sustained reflection on how best to persuade, were integral to his thought from the first. After formulating and abandoning two theories of transmutation, Darwin at last recognized in Malthus the principle long implicit in his own thought. His personal identification with the professional scientific community of his time made him anxious that advocacy of evolutionism not damage his scientific reputation. Darwin found in the language of Baconian inductivism and positivism the protective coloration he needed for his unorthodox conclusions. Indeed, Darwin was so persuasive in redescribing his path to discovery and his philosophy of language that he even convinced himself. Various letters and the statement in his *Autobiography* that only with Malthus did he at last have a theory all indicate that as he grew older Darwin began to remember his path to discovery not as it had been, but as Baconian and positivist method held that it should have been.[55]

Darwin's disavowal of his own rhetoric was not without cost. Consequent to his denial of his philosophy of language, Darwin lost his ability to delight in what he beheld. In later life he complained, "My mind seems to have become a kind of machine for grinding general laws out of large collections of facts, but why this should have caused the atrophy of that part of the brain alone, on which the higher tastes depend, I cannot conceive."[56]

At the beginning of this essay I affirmed the propriety of juxtaposing rhetoric and science. For Darwin, the consequence of denying his own rhetoric was poignant. For us, affirming Darwin as a rhetorician of science underscores rhetoric as the bridge uniting science with culture and, far from denying the integrity of Darwin's vision, restores the motive which gave it life.

Notes

1. Giorgio de Santillana, *The Origins of Scientific Thought* (New York: Mentor Books,1961), pp. 248–50; Loren Eiseley, *Darwin's Century* (New York: Anchor Books,1961), pp. 205–7.
2. Paul N. Campbell, "Poetic-Rhetorical, Philosophic, and Scientific Discourse," *Philosophy and Rhetoric* 6 (1973);1–3. Dov Ospovat observes in *The Development of Darwin's Theory* (Cambridge: Cambridge University Press, 1981), p. 229, "that the formation and transformation of Darwin's theory represent not so much the results of an interaction between the creative scientist and nature as between the scientist and socially constructed conceptions of nature."
3. Michael Ruse, *The Darwinian Revolution* (Chicago: University of Chicago Press, 1979), pp. 94–131.
4. R. C. Stauffer, ed., *Charles Darwin's Natural Selection* (Cambridge: Cambridge University Press, 1975), pp. 8–10.
5. For a critical discussion of the date on which Darwin received the Wallace letter, see John Landon Brooks, *Just Before the Origin* (New York: Columbia University Press, 1975), pp. 229–57.
6. November 24 is the traditional date. For discussion of November 26 as the true date see Morse Peckham, *The Origin of Species by Charles Darwin:A Variorum Text* (Philadelphia: University of Pennsylvania Press, 1959), p. 18.
7. Charles Darwin, *On the Origin of Species: A Facsimile of the First Edition with an Introduciton by Ernst Mayr* (New York: Athenum, 1967), pp. 1–2.
8. Francis Darwin, ed., *Charles Darwin's Autobiography*, with introductory essay by Gaylord Simpson (New York: Collier Books, 1950), p. 115.
9. Charles Coulston Gillispie, *The Edge of Objectivity* (Princeton: Princeton University Press, 1960), pp. 303–4.
10. Peckham, *Origin of Species*, p. [iii], 759,57.
11. A. Hunter DuPree, *Asa Gray 1810–1888* (Cambridge: Harvard University Press, 1959), pp. 298–301.
12. Sir Francis Darwin, ed., *More Letters of Charles Darwin* vol. 1 (London: John Murray, 1903), p. 146; see also pp. 190–94; Charles Darwin, *Variation in Plants and Animals Under Domestication*, vol. 2 (New York: D. Appleton & Co., 1896), pp. 248–49.
13. Darwin, *The Origin*, 1st ed., pp. 167,188–89. For other examples of the same kind, see, for instance, pp. 243–44,484, and Peckham, *Origin of Species*, pp. 748,753.
14. Peckham, *Origin of Species*, p. 748, 6th ed.
15. Susan Gliserman, "Early Victorian Writers and Tennyson's 'In Memoriam': A Study in Cultural Exchange," Pt. 2, *Victorian Studies* 18 (1975):456.
16. Cited in William Irvine, *Apes, Angels and Victorians* (New York: Meridian Books, 1964), pp. 249–50.
17. Darwin, *The Origin*, 1st ed., p. 1.
18. Nora Barlow, ed., *The Autobiography of Charles Darwin: 1809–1882* (London: Collins, St. James Place, 1958), p. 120.
19. Howard Gruber, *Darwin on Man: A Psychological Study of Scientific Creativity: Together with Darwin's Early and Unpublished Notebooks*, transcribed and annotated by Paul Barrett, foreword by Jean Piaget (New York: E. P. Dutton, 1994), p. 123.
20. Ibid., p. 173.
21. Francis Darwin, ed., *Life and Letters of Charles Darwin*, vol. 2 (New York: D. Appleton, 1896), p. 371.
22. Francis Darwin, ed., *More Letters of Charles Darwin*, vol. 1 (New York: D. Appleton, 1903), p. 195.
23. *Ibid.*, p. 173.
24. David Hull, *Darwin and His Critics* (Cambridge: Harvard University Press, 1973), p. 4.
25. Charles Bazerman, "Codifying the Social Scientific Style: The APA *Publication Manual* as a Behaviorist Rhetoric," Nelson, John S., Allan Megill, and Donald S. McCloskey, eds. The Rhetoric of the Human Sciences (Madison: University of Wisconsin Press, 1987), 125–44.
26. Michael T. Ghiselin, *The Triumph of Darwinian Method* (Berkeley and Los Angeles: University of California Press, 1969), pp. 35, 75.

27. Gruber, *Darwin on Man*, pp. 23–24.

28. Edward Manier, *The Young Darwin and His Cultural Circle* (Boston: D. Reidel, 1978), p. 195.

29. Gruber, *Darwin on Man*, p. 450 (C123).

30. Ibid., p. 454 (C243).

31. Ibid., p. 276 (M57).

32. Ibid., p. 281 (M84).

33. Gillispie as cited in Manier, *The Young Darwin*, p. 19; Gruber, *Darwin on Man*, p. 12.

34. Peckham, *Origin of Species*, p. 165, 3d ed.

35. Manier, *The Young Darwin*, p. 40.

36. Gruber, *Darwin on Man*, p. 278 (M69–M73).

37. Manier, *The Young Darwin*, p. 39–40.

38. Ibid., p. 199.

39. Ibid., pp. 61–64, 149, 150, 154–56, 158, 161.

40. Peckham, *Origin of Species*, p. 168–69, 2d ed.

41. Ruse, *Darwinian Revolution*, pp. 152–57, 99, 100; Charles Coulston Gillispie, *Genesis and Geology* (New York: Harper & Row, 1951), pp. 146–50.

42. Chaim Perelman, and Lucie Olbrechts-Tyteca, *The New Rhetoric: A Treatise on Argumentation* (Notre Dame, Ind.: University of Notre Dame Press, 1969), pp. 79, 135–35, 450.

43. John Milton, *The Complete Poetry and Selected Prose*, vol. 7: *Paradise Lost*, introduction by Cleanth Brooks (New York: Modern Library, 1950), p. 260.

44. Manier, *The Young Darwin*, p. 174.

45. Ibid., 174–75; Charles Darwin and Alfred Russel Wallace, *Evolution by Natural Selection*, with a foreword by Sir Gavin DeBeer (Cambridge: Cambridge University Press, 1958), pp. 45, 114, 115.

46. Stauffer, *Darwin's Natural Selection*, pp. 175, 569. Darwin attributes the phrase "All nature . . . is at war" to the elder De Candolle rather than to Hobbes as Manier affirms. See Manier, *The Young Darwin*, p. 181.

47. Manier, *The Young Darwin*, p. 181.

48. *The Origin*, pp. 62–63, 1st ed.

49. Manier, *The Young Darwin*, p. 13.

50. Darwin, *The Origin*, e.g., pp. 159, 482, 1st ed. For the role of *vera causa* in the dispute, see Hull, *Darwin and His Critics*, pp. 27, 45, 109, 115, 163, 180, 355.

51. Ruse, *Darwinian Revolution*, pp. 205–6.

52. Ibid., pp. 222–25; Ospovat, *Development of Darwin's Theory*, p. 89.

53. Ospovat, *Development of Darwin's Theory*, p. 87, 88.

54. Ibid., pp. 210–35.

55. See nn. 18, 21, 22.

56. *Autobiography*, p. 139.

On the Shoulders of Giants: Seventeenth-Century Optics as an Argument Field

by Alan G. Gross

In science, there are two sorts of rhetorical masterpieces: those powerful enough to provoke revolution, and those ingenious enough to avoid it. *Dialogue Concerning the Two Chief World Systems* and *On the Origin of Species* are examples of masterpieces of the first sort. Initially, Galileo and Darwin caused more debate than assent, more turmoil than change. Descartes's optical works and Newton's *Opticks*, on the other hand, are masterpieces of the second sort: each in its own way successfully persuaded; each dominated optical research for nearly a century. In these works, each man argued for change less as revolution than as continuity: the extension of the best of the past. Despite the similar persuasive aims of these masterpieces, however, their notions of science could not be more opposed.

It is this opposition that is my subject, a contrast that reaches to the rhetorical heart of seventeenth-century optics. Published in 1637 as an integral, though now nearly forgotten, appendix to his *Discourse on Method*, Descartes's optical works have been quite properly called a first step in a new direction. Although Descartes did not violate traditional presuppositions, he created from these an original physics of light. Still, the newness of his program never seriously forestalled the conviction of his intended audience; he shared with traditional science two central views, one epistemological, the other explanatory. He believed that rational intuition, not experiment, was epistemologically prior: reason, not experience, was the bedrock, the touchstone, of knowledge. In addition, he was as convinced as Aristotle that a complete scientific explanation must include the traditional three causes: the formal, efficient, and the material. These shared convictions allowed Descartes to be rhetorically transparent and convincing at the same time: he could use his prose to clarify, indeed to highlight, his views.

In 1672, a time when Descartes's optics was already firmly entrenched, Newton published his first paper on that science. Its views were new in a new way. Unlike Descartes, Newton challenged a traditional tenet concerning the nature of light. For the first time in the history of optics, white light was revealed as a compound of all of the lights of the visible spectrum. Moreover, the ground Newton offered for certainty in optics differed radically from that of his predecessor, and of his predecessor's predecessors: by giving epistemological priority to experiment over rational intuition, Newton overturned a central presupposition of traditional science. In addition, Newton's explanation of light did not give a full account of its origin, an account that included the operation of its material cause. A startling claim, a new method, a different, more restrictive, style of explanation—seemingly, Newton needed to discharge a strong burden of proof. But in this early paper on light and color, in a rhetoric as transparent as that of Descartes, Newton did not discharge this burden; instead, he emphasized his conflict with traditional views and methods.

In retrospect, it is hardly surprising that Newton failed to convince such contemporaries as Hooke and Huygens. After a flurry of inconclusive debate, and a second early paper, he withheld his optical researches from the public for nearly three decades. In 1704, he published his *Opticks*, his second attempt at persuasion. In this work, Newton discarded the transparent rhetoric that made the epistemological and explanatory novelty of his early papers clear; in its stead, he substituted a rhetoric that invented an essential continuity between his work and the optical and scientific past. The rhetoric of the *Opticks* concealed his radical intent; it was designed to convince, even at the expense of perfect frankness. In his final masterpiece, Newton transformed optics, and experimental science, by allowing his fellow physicists to believe that an adherence to the new did not entail a fundamental rejection of the old. This rhetorical strategy was successful: throughout the eighteenth century, in England and on the Continent, the physics of light was Newton's physics.[1]

Rhetoric of Science as a Discipline

From an Aristotelian point of view, the rhetorical categories—style, arrangement, and invention—are not appropriate to the analysis of science, a method of inquiry designed to deduce from certain first principles necessary conclusions about the natural world.[2] In contrast, this paper presupposes what most now believe, and what Aristotle could never accept: the unavailability of certain knowledge. There are no principles for whose certainty we can authentically vouch; even if there were, there would be no foolproof method of deducing from them necessary conclusions about the natural world. Since such certainty is unavailable, rhetoric and dialectic, the arts of likelihood, become plausible candidates for the analysis of scientific texts.

My reasons for preferring rhetoric to dialectic as a framework for analysis are essentially those of Perelman and Olbrechts-Tyteca.[3] The texts scrutinized in this paper are all extended discourses aimed, not at individuals, but at communities, at audiences. Moreover, their appeal is not exclusively rational; rather their total effect also depends on factors not usually connected with the domain of pure intellection in which dialectic purportedly holds sway. Finally, as this paper will demonstrate, scientific texts respond well to an analysis of their style, arrangement, and invention. For me, therefore, rhetoric of science makes sense as a branch of inquiry that has as its goal "to find out in each case the existing means of persuasion,"[4] its proper task being the reconstruction of the means by which scientists convince themselves and others that their claims are true of the world.

In saying that the rhetoric of science must analyze the natural sciences from the point of view of the humanities, I in no way equate the two. The natural sciences differ from the humanities in a fundamental way. As Habermas says: "they do not have first to gain access to their object domain through hermeneutic means."[5] Nothing, however, prevents us from viewing the enterprise of science hermeneutically as a stream of texts exhibiting generally an epistemology based on understanding. It is by recourse to an insight as old at least as Vico that we can view the texts of natural science as a product of human interaction, a phenomenon to be understood in the way human beings understand each other.

From a rhetoric of science so conceived, no feature of scientific texts is exempt from rhetorical explanation. In this paper, I consider rhetorical Newton's "invention" of discontinuity, of scientific revolution, in his first paper on light and colors, and his "invention" of continuity, of scientific evolution, in his *Opticks*. I also call rhetorical Newton's attempt to make experimental evidence criterial for scientific claims, and formal and efficient causes sufficient for a scientific explanation. To me, these are all tasks of persuasion, jobs of changing minds.

In principle, then, rhetoric of science is a discipline as complete as history, philosophy, or sociology of science. And as incomplete, For Gadamer's point about physics applies to any discipline that insists on ultimate coherence:

> The world of physics cannot seek to be the whole of what exists. For even a world formula that contained everything, so that the observer of the system would also be included in the latter's equations, would still assume the existence of a physicist who, as the calculator, would not be the object calculated.[6]

It is in this sense that the scientific work of Descartes and Newton is completely analyzable from a rhetorical point of view.

My view of optics as a total rhetorical construct derives directly from Willard on argument theory: the central assumptions and agreed-upon methods that form traditional optics are an "argument field," the description of which "is largely a matter of describing the things its actors take for granted, their self-evident truths."[7] As a general rule, these truths are field specific. In constitutional theory, for example, we assume that all men are created equal; in quantum physics, that in a sub-atomic universe individual events can never be predicted with accuracy.

From such "self-evident," field-specific truths, conclusions are drawn by means of agreed-upon procedures. To secure a claim rationally is "[to use] a rule of inference vouched for by agreeable authorities" and "[to violate] no taken-for-granted assumptions."[8] From field to field, rules of inference differ in pattern and rigor: in number theory and in rhetorical theory, for example, such rules will probably continue to differ in these important ways. Rules can also vary within the same field over time: for instance, the seventeenth-century development of the calculus depended importantly on a relaxation in the rigor demanded of mathematical proofs. Finally, it is a mistake to feel that all of these inferential procedures can be reduced to a single set: although mathematics and logic may share an equal rigor, the former cannot be reduced to the latter.

Cartesian Optics and Tradition

A scientific argument field is a set of field-specific assumptions and their inference rules. For example, the central assumptions and agreed-upon methods of traditional optics constitute an argument field in science. All are present in Descartes. Descartes agreed that white light was basic, and color derivative, an alteration in white light. Descartes also concurred with tradition in representing light as straight lines that can cross without interfering. This characteristic permitted him to use geometry to solve problems in the physics of light: the axioms and theorems of Euclidean geometry are central to the methodology of both traditional and Cartesian optics.

To illustrate the two central geometrical properties of sunlight—the rectilinear propagation and non-interference of its rays—Descartes compares light's action to the pressure on a vat of half-pressed grapes. This pressure transmits itself throughout the vat, squeezing the juice equally through two widely separated holes in the bottom:

> And in the same way considering that it is not so much the movement as the action of luminous bodies that must be taken for their light, you must judge that the rays of this light are nothing else but the lines along which this action tends. So that there is an infinity of such rays which come from all points of luminous bodies, toward all points of those that they illuminate, in such a manner that you can imagine an infinity of straight lines, along which the actions coming from all points of the surface of the wine . . . tend toward [one hole], and another infinity, along which the actions coming from these same points tend also toward [the other hole], without either impeding the other.[9]

When geometric methods are applied, light is found to exhibit two regularities: reflection and refraction. When a ray of light is reflected, its angle of incidence, the angle at which it strikes the reflecting surface, is always equal to the angle of reflection, the angle at which the ray leaves the reflecting surface (Fig. 1).[10] When light penetrates a transparent body at an angle other than right, it refracts; it bends toward or away from an imaginary line, called the normal, a line perpendicular to the refracting surface. When the light crosses into a medium of higher optical density, for example from air to water, it bends toward the normal; when it crosses into a medium of lower optical density, it bends away (Fig. 2).[11]

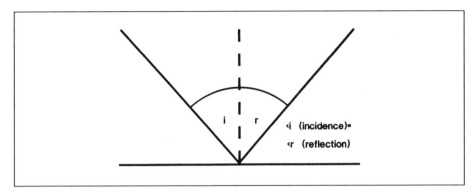

FIGURE 1
Reflection

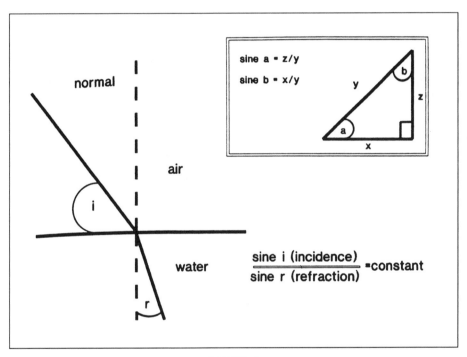

FIGURE 2
Refraction

In reflection, the angles of incidence and reflection are equal, a quantitative relationship known since antiquity. An analogous quantitative relationship for refraction eluded all investigators until Snell and Descartes. In refraction, as both discovered, rays of light conform to a definite geometrical relationship, that of sines. In a triangle one of whose angles is right, the sine is the ratio of the side opposite either acute angle and the side opposite the right angle (Fig. 2-insert). A ray refracting in most transparent substances bends so that the ratio of the sine of its angle of incidence to the sine of its angle of refraction equals a specific number. For each regularity-abiding refracting substance, this number never changes:

$$\frac{\text{sine i}}{\text{sine r}} = \text{constant}$$

Although the sine relationship was a new discovery, it was an empirical regularity compatible with traditional assumptions and methods: there was no question of its immediate absorption into the body of traditional optics.

The New Cartesian Science

But so uncomplicated an absorption would not satisfy Descartes. Although he practiced traditional geometric optics, he did so within the framework of the new science he created: Cartesian optics transforms traditional optics into a new argument field with a radically altered view of what constitutes a law of science. Scientists working within the argument field of traditional optics might easily be persuaded that the refractive regularity of Snell and Descartes was not merely a useful empirical discovery, but a law of science; wrongly persuaded, in Descartes's view. To Descartes, an empirical regularity becomes a scientific law only when it is understood as an integral part of a true and coherent physics, his physics. The refractive regularity of Snel and Descartes can be promoted to a law, in this case about sines, only when we understand it as correct according to the central and absolutely certain principles of Cartesian physics.

Although Descartes shared with tradition his conviction that the goal of his science is absolute certainty, rather than moral certainty (what we would call very high probability), the source of his certainty represents a new interpretation of the primacy of reason in epistemology. The intuitions of Descartes's unaided reason, provided they are clear and distinct, form the incorrigible foundation of his new science.[12] According to this reason, the material world consists only of extension and its laws of motion. The universe is a plenum, completely full: space itself is material, because extended. Local motion, the relative repositioning of extended substances, can occur only through contact. Everything we experience, even light, is the result of local motion originating in contact.

In Cartesian physics, the sun consists of small, rapidly moving particles of the first element which exert a pressure on the second, the element that fills interplanetary spaces. This pressure is transmitted instantaneously in straight lines in all directions. Experienced on the retina, this pressure creates the sensation called light. Light *is* this pressure; it is not, as it was for Aristotle, a qualitative alteration in a medium, the air; it is not, as it will be for Newton, the effect on the retina of a stream of tiny particles traveling from a luminous source.

At the end of his *Principles of Philosophy* Descartes sums up the full scientific implications of his philosophical position:

These reasonings of ours will perhaps be included among the number of these absolutely certain things by those who consider how they have been deduced in a continuous series from the first and simplest principles of human knowledge. Especially if they sufficiently understand that we can feel no external objects unless some local movement is excited by them in our nerves; and that such movement cannot be excited by the fixed stars, very far distant from here, unless some movement also occurs in these and the whole intermediate heaven; for once these things have been accepted, it will scarcely seem possible for all the rest, at least the more general things which I have written about the World and the Earth, to be understood otherwise than as I have explained them.[13]

In Descartes's physics, then, reason bears the burden of sole guarantor of the certainty of general principles of which the specific truths of this science are the consequences. In a typical comment on the relationship between reason and experiment, Descartes asserts: "And the demonstrations of this [law of impact] are so certain that, even if experience were to appear to show us the opposite, we would nevertheless be obliged to place more trust in our reason than in our senses."[14] It is our reason alone that assures us: no explanation of light, no explanation in physics, can be persuasive if it is not also mechanical, the result, direct or indirect, of matter in motion: "Whatever I concluded to be possible from the principles of my philosophy actually happens whenever the appropriate agents are applied to the appropriate matter."[15]

Like light, color must be explained in terms of matter in motion. In Descartes's view, we recall, the particles of the second element, those that fill all the space between the sun and the earth, transmit the sun's first-element pressure. In refraction, these second-element particles take on differing rotational speeds, speeds whose pressures are communicated directly to the retina. It is from these differing pressures that the sensations of color arise: "the nature of the colors appearing at [the lower end of the spectrum] consists only in the fact that the particles of the fine substance that transmits the action of the light have a stronger tendency to rotate than to move in a straight line; so that those which have a much stronger tendency to rotate cause the color red."[16] The sensation of white light is the result of the simple pressure of particles of the second element on the retina; color is the result of the action of these same particles to which a rotating motion has been imparted.[17]

Although reason is central to Cartesian physics, experience cannot be ignored; indeed, experimentation has an important role.[18] Though everything flows from general principles, everything cannot be deduced from them. For example, the constant of optical density, which accounts for the degree of refraction, differs for differing substances, and cannot be deduced from the law of refraction. To obtain this constant in each particular case, "we must appeal to experience."[19] Furthermore, the law of refraction itself, though it conforms to the laws of motion, cannot be deduced from them. Light always acts in accordance with the laws of motion, which are general principles, but it does not always act in accordance with the law of refraction. Experience is the sole desideratum for empirical regularities. The sine law, for example, must square with experience in the case of each transparent substance: "For, seeing that these parts could have been regulated by God in an infinity of diverse ways; experience alone should teach us which of all these ways He chose."[20]

From an examination of Descartes's philosophy, it is true, we may discover a role for experiment even more nearly central. In that philosophy, reason tells us that the real qualities of material objects with which science is concerned consist "clearly and distinctly" of "magnitude or extension in length, breadth, or depth." These real qualities also include "[1] figure which results from a termination of this extension, [2] the situation which bodies of different figure preserve in relation to one another, and [3] movement or change of situation; to which we may also add [4] substance, duration and number."[21]

Such a view seems to entail a central role in physics for both mathematics and measurement. But neither mathematics nor measurement appears regularly in Descartes's published science, pages filled with accounts no experiment could ever confirm.[22] Despite his statements to the contrary, Descartes's physics seems essentialist and qualitative.[23]

In all of Descartes, there is one exception to this subordinate epistemological role for measurement, an exception that proves this rule. In his discussion of the rainbow, by measurement alone, Descartes signally advances the cause, not only of his, but of our, physics; by measurement alone, he deduces the width and height of both rainbow arcs: "the radius of the interior arc must not be greater than 42°, nor that of the exterior one smaller than 51°."[24] *Must be!* From his discourse on the rainbow, which Newton knew well,[25] one could learn a whole new way of doing science, a way in which experimentation and measurement combined to form an epistemological vanguard in search of new, equally durable regularities.

But from this analysis, which he regarded as an authentic example of his method,[26] Descartes learned nothing new about the fundamental role of measurement: he gives equal credence to his theory of color, for which he offers, in addition to a general conformity with experience, only the certainty of his unique, reason-derived mechanics.[27] Concerning this latter certainty, Descartes sounds a characteristic note when he asserts "how little faith we must have in observations which are not accompanied by true reason."[28]

To sum up. In fundamental ways, the argument field of Cartesian optics is continuous with that of traditional optics. It is these threads of continuity, these shared ontological and epistemological commitments, that ease his persuasive task. Descartes's interpretation of the incorrigibility of reason is his own, but the general supremacy of reason over experiment and measurement is part of the Aristotelian inheritance. Descartes's promotion of mathematization in physics, though not new, breaks with the qualitative bias of traditional science; moreover, it has clear implications for the importance of measurement in science. But Descartes's practice is generally one in which neither mathematics nor measurement has a central place. Even his few permanent contributions to science do not depart from traditional presuppositions: light propagates rectilinearly, white light is simple in nature.

Newton's First Paper

Newton's paper of 1672 starts as follows:

> in the beginning of the year 1666 . . . I procured me a Triangular glass-Prisme, to try therewith the celebrated *Phaenomena of Colours*. And in order thereto having darkened my chamber, and made a small hole in my window-shuts, to let in an convenient quantity of the Suns light, I placed my Prisme at his entrance, that it might be thereby refracted to the opposite wall. It was at first a very pleasing divertisement, to view the vivid and intense colours produced thereby; but after a while applying my self to consider them more circumspectly, I became surprised to see them in an *oblong* form; which, according to the received laws of Refraction, I expected should have been *circular*.[29]

If light behaved as Descartes said, the spectrum would disperse equally in all directions, like spray from a hose: the anomaly of the oblong spectrum demands an explanation. Like the plague of Thebes, this explanation has fatal consequences. At first, Newton attempts innocent explanations, those without any, or without serious, theoretical import: perhaps it was the thickness of the prism-glass, perhaps its unevenness; perhaps the rays curve. All of these "suspicions" being removed, Newton performed the *"Experimentum Crucis,"* the decisive experiment; the final searching question is put to nature.[30]

Sure enough, the optical tradition exhibits a fatal flaw, a deep-seated but hitherto unsuspected incoherence. Unless we abandon the law of sines—an unthinkable prospect—we must abandon the fundamental assumption that white light is basic, color derivative. Indeed, it is color that is basic, while white light is derivative: "[white] *Light* consists of *Rays differently refrangible* [bendable]."[31] When these rays are separated, they form the radiant colors of the spectrum; when combined again, in the proper proportions, they re-create white light.

In thirteen propositions that follow the *Experimentum Crucis*, Newton shows, point by point, how this property, the differing refrangibilities of light, explains "the celebrated *Phaenomena of Colours*."[32] In opposition to Descartes's views, sunlight is "ever compounded, and to its composition are requisite all the aforesaid primary Colours, mixed in a due proportion."[33] It follows from the combinatory nature of white light that "Colours are not *Qualifications of Light*, derived from Refractions, or Reflections of natural Bodies [as in Descartes] . . . but *Original* and *connate properties*, which in diverse Rays are divers."[34] This difformity of light explains, not only radiant colors (those produced by a prism), but "the Colours of all natural Bodies [which] have no other origin than this, that they are variously qualified to reflect one sort of light in greater plenty then another."[35]

It is a deliberate irony that the spearhead of Newton's attack on Descartes is the certainty of an experimental anomaly: it will not go away; it cannot be explained away. It is equally ironic that only a wholly quantitative point of view can reveal the anomaly: unlike Descartes, Newton projects his spectrum, not a few inches, where the anomaly is not apparent, but twenty-two feet, where it is unmistakable.[36] Like Kepler's famous eight minutes of arc, these twenty-two feet have revolutionary implications in Newton's hands and, in his early paper, he makes the most of them.

In this early paper, following in the footsteps of Boyle and Hooke, Newton uses experiment systematically as the primary epistemological instrument. But Newton's relentless use represents a new departure, as a comparison with *Experiments and Considerations Concerning Colors* and with *Micrographia* will confirm. In theory and in practice, Newton clearly and uncompromisingly reverses the traditional and Cartesian roles of reason and experiment: "if the Experiments, which I urge, be defective, it cannot be difficult to show the defects; but if valid, then by proving the Theory they must render all Objections invalid."[37] In Newton's science, experiment bears the full weight both of discovery and of theory.

In a justly famous, early, Baconian[38] pronouncement on method, Newton affirms experimentation as the driving force behind theory, and the prime source of certainty. In the same passage, he discloses his perception of the central defect of Cartesian physics, indeed of any physics that overemphasizes the hypothetical—the fatal ease with which explanations may be invented:

> the best and safest way of philosophizing [doing science] seems to be this: first search carefully for the properties of things, establishing them by experiments, and then more warily to assert any explanatory hypotheses. For hypotheses should be fitted to the properties which call for explanation, and not be made use of for determining them, except in so far as they can furnish experiments. And if anyone makes a guess at the truth of things by starting from the mere possibility of hypotheses, [I] do not see how to determine any certainty in any science; if indeed it be permissible to think up more and more hypotheses, which will be seen to raise new difficulties.[39]

Newton's notion of a crucial experiment is central to his first paper. It is derived from Bacon, and designed to be as conclusive as a Euclidean QED, but is seriously flawed as a persuasive device. Part of the difficulty is remediable. The persuasiveness of the crucial experiment depends on its replicability; but the crucial experiment in this first paper is accompanied by neither diagram nor clear directions.

More significant, because far less remediable, the persuasive effect even of replicatible results depends absolutely on their univocal interpretation. At one point, Newton becomes exasperated with some rival experimental results: "it is not the number of Experiments," Newton cries, "but weight to be regarded; and where one will do, what need many?"[40] True; but it was folly to expect such unanimity of interpretation in so hotly disputed a seventeenth-century subject as the nature and action of light. In his first paper, Newton draws what for him is the obvious ontological inference: "since Colours are the *qualities* of Light, having its Rays for their intire and immediate subject, how can we think those Rays *qualities* also, unless one quality may be the subject of and sustain another; which in effect is to call it *Substance*."[41] But Hooke, and Pardies, had different, though plausible, explanations of Newton's experimental result, explanations founded on the rival ontology of wave theory.

Critics objected not only to the unwarranted univocality of Newton's interpretations, but to the incompleteness of his explanations. Huygens animadverted that "he hath not taught us, what it is wherein consists the nature and difference of Colours, but only this accident . . . of their *different Refrangibility*."[42] Indeed, Newton had avoided any mechanical explanation of refraction or color formation: "I shall not mingle conjectures with certainties," he grandly asserted.[43] But to Huygens, a scientific explanation that insisted on the corpuscularity of light must include a specific mechanism for the production of the sensations of light and color by means of matter in motion.

Newton's 1672 paper must be counted as a failure of persuasion—and Newton counted it as such. It succeeded in establishing neither his method of doing science nor his beliefs about the nature and action of light. In his *Opticks* of 1704, Newton created a second opportunity to persuade the scientific community of the efficacy of his method and the truth of his optical beliefs, the same method and, essentially, the same beliefs contained in his optical papers of thirty years before. It is to the *Opticks*, Newton's final masterpiece, that I now turn.

The Opticks, a Rhetorical Masterpiece

The rhetoric of Newton's first paper—riveting to this day—was that of youth: brash and brilliant, relying for its persuasive effect on a clash of principles, a decisive confrontation, clearly and unequivocally presented. The debate to which this paper led was noisy and inconclusive. Disputes of this sort resolved nothing, Newton was to assert in the preface to the *Opticks*, an opinion he held to the end of his life.[44] In contrast to his early papers, the rhetoric of the *Opticks* was that of late middle age: a canny and successful attempt to transform a youthful invention into a durable inheritance. In the *Opticks*, in the exposition of this theory, Newton employed a Euclidean arrangement to create an impression of historical continuity and logical inevitability. In addition, by piling experiment on experiment, and, in each experiment, detail on detail, he created in this work an overwhelming presence for his experimental method. Finally, in the book's last section, he initiated a cascade of rhetorical questions, whose cumulative effect was both to sanction his science and license his speculations.

The Use of Arrangement

"Let me suggest," says David Lindberg, our most perceptive historian on the subject,

> that optics from Alhazen to Newton (and, in some respects, from Aristotle and Ptolemy to Newton) must be seen as a continuously unfolding discipline. Historians of science have become overly conditioned to perceive the sixteenth and seventeenth centuries as a revolutionary period, during which scientists made a clean break with the past. There is, of course, much truth in this view, even in optics; but if pushed too far, it obscures the large measure of continuity in the history of optics from Alhazen onward.[45]

Lindberg's interpretation of optical history has a Newtonian parallel. In February of 1676, in a letter to Hooke, Newton explicitly acknowledges his place in a long optical and scientific tradition: "What Des-Cartes did [the discovery of the sine law] was a good step. . . . If I have seen further it is by standing on ye sholders of Giants."[46] But there is a vast difference in intent. Lindberg presents the essential continuity of optics as a historical interpretation, an end; but, for Newton, in this early letter, historical continuity is a means, the outline of a persuasive strategy. This strategy—so different from that of his early paper—is fully embodied nearly three decades later in his *Opticks*.

In his early paper, the dominant arrangement is narrative; in his *Opticks*, virtually the same material is arranged as a Euclidean deduction. In the early paper, Newton uses narrative to dramatize the clash between past and present; in the *Opticks*, he uses Euclid to display the present as a deductive consequence of the past: "I have now given in Axioms and their Explications the sum of what hath hitherto been treated of in Opticks. For what hath been generally agreed on I content my self to assume under the notion of Principles, in order to what I have farther to write."[47]

In the crucial first book of the *Opticks*, though Euclidean form is particularly strict, the method of all but the last Proposition is analysis, induction rather than deduction; it "consists in making Experiments and Observations, and in drawing general Conclusions from them by Induction, and admitting no Objections against the Conclusions, but such as are taken from Experiments, or other certain Truths."[48] Only in the last Proposition does the method change to synthesis or "Composition": the nature of light and of both radiant and natural color having been proved "by Experiments," they are now demonstrated by deduction.[49] Nevertheless there is throughout this book a firm adherence to Euclidean structure: definitions and axioms are followed by a series of propositions with their proofs. In the *Opticks*, there is a conflict between deductive form and inductive epistemology.

Indeed, to move from the early papers to the *Opticks* is not to move from one science, but from one rhetoric, to another; to pass from a work that continually clarifies the epistemological priority of experiment to one that blurs the distinction between the traditional view of experiment as confirmatory, and Newton's more radical view. That this blurring is deliberate is clear from Newton's suppression of the explicit epistemological introduction to the first edition.[50] For the initiated, the emphasis on experiment underscores its radical epistemological role; but this emphasis can bear a more conservative interpretation: it allows Malebranche to say that Newton's theory "fits (*s'ajuste*) . . . all his experiments."[51] Newton deliberately includes in his scientific audience even those who might differ from him on epistemological grounds.

The Use of Presence

The epistemological ambiguity of the experiments described in the *Opticks* contrasts vividly with their clear and complete presentation. For any who might find experiments, for any reason, persuasive, Newton describes, not two, as in his first paper, but dozens, all with a meticulousness arguably new to science. "I have set down such Circumstances, by which either the Phaenomenon might be render'd more conspicuous, or a Novice might more easily try them."[52] In his first paper, even the *Experimentum Crucis* was meant to have its effect—an effect nothing less than the overturning of traditional optics—without the benefit of either a diagram or sufficient detail for easy replication. In contrast, in the *Opticks*, a typical experiment is accompanied by a detailed diagram, and a description that begins thus: "In the middle of two thin Boards I made round holes a third part of an Inch in diameter, and in the Window-shut a much broader hole being made to let into my darkned Chamber a large Beam of the Sun's Light."[53]

This concern for the persuasive value of meticulous detail extends to measurement. Everywhere in the *Opticks*, measurement, so important in the early papers, increases in importance; in seemingly every case, measurements previously made are remade: in book II, part II, of the *Opticks*, for example, Newton recalculates the values of a table he published in the *Transactions* nearly thirty years before.[54] The differences are sometimes very small indeed. In the earlier work, the thickness of glass at which indigo of the second order is most intense is .0000085; in the *Opticks* the figure is .000008182, a difference of 318 *billionths* of an inch!

Indeed, throughout the *Opticks*, the rhetorical presence of Newton's experimental method is enhanced by the sheer number of experiments described, and by the quantitative meticulousness with which their methods and results are reported. In the *Opticks*, in fact, this presence increases until "the whole field of consciousness [is filled] with [it] so as to isolate it, as it were, from the [reader's] overall mentality." By this means, "presence, at first a psychological phenomenon, becomes an essential element in argumentation."[55]

The Use of Rhetorical Questions

Halfway through the first part of the third book a remarkable event occurs: both the Euclidean structure and the experimental program break off completely, and a long section of Queries is initiated, each in the form of a negative rhetorical question.[56] In the body of the *Opticks*, for example, light ray is defined as "the least Light or part of Light, which may be stopp'd alone without the rest of the Light, or propagated alone, or do or suffer any thing alone, which the rest of the Light doth not or suffers not."[57] The definition is operational: light behaves as if it were composed of rays made up of parts. In the Queries, on the other hand, light rays are unequivocally physical entities: "Are not the Rays of Light very small Bodies emitted from shining Substances?"[58] The question-mark bestows upon Newton a double benefit: it underlines the speculative nature of the Query, but leaves undiminished its rhetorical force: by their very nature, negative rhetorical questions are strong positive assertions.[59]

This rhetorical characteristic of the Queries permits Newton to meet some crucial objections of his early critics. In their early criticisms, Hooke and Pardies had objected to the inference that light was a substance. In the *Opticks*, Newton blunts this objection by a distinction. In the body of this work, he confirms by experiment that light behaves consistently as if it were a substance; in his Queries he makes a plausible, but not a scientific, case that light is, in fact, a substance.

Newton's distinction between science and speculation also blunts Huygens's objection. To Huygens, Newton's explanation was incomplete because it did not include a specific mechanism. But in his *Opticks* Newton shows his willingness to speculate about mechanisms so long as speculations are not confused with science. For example, from the anomalous refractive behavior of light when passed through island crystal, Newton infers a physical property of its rays, one that might account for the anomaly: "every Ray may be consider'd as having four Sides or Quarters, two of which opposite to one another incline the Ray to be refracted after the unusual manner."[60] His Queries allow Newton to be both bold and cautious in inference, to put forward under the same cover explanations that are, at the same time, puritanically narrow and imaginatively deep.

But the Queries serve certainty rhetorically just as do experimentation and Euclidean form: they stake out the limits of certainty. By plainly defining the opposing edges of science and speculation, the Queries affirm the scientific status of the conclusions that precede them, and the methods by which these conclusions were reached.[61] The caution Newton exhibits and the limits he honors in the scientific portion of the *Opticks* are not inadvertent, but principled; they clearly arise because Newton wants to say that he has higher standards of scientific certainty than do his critics. At the same time, the rhetorical forcefulness of the Queries, enhanced by the aging Newton's overwhelming reputation, clearly intend a constraint on the future of optical research. Newton sets his Queries down "in order to a farther search to be made by others."[62] In the body of the *Opticks*, the properties of light are experimentally demonstrated and deductively proved;

in the Queries, the ground-plan for the next century of optical research is successfully laid out.[63]

Solely by means of its rhetoric, by means of its strict Euclidean form, its striking experimental presence, its provocative speculations—Newton's master-piece became the model for optics, and for experimental science, in the next century. Initially, scientists did not have to believe in the epistemological priority of experiment, or in the particle nature of light, to accept Newton's empirical results, and to acknowledge his scientific leadership. But that acceptance and that acknowledgement eventually transformed optics into a modern experimental science along Newtonian lines; eventually, to do optics was to accept particle theory, and to presuppose the epistemological priority of experiment.

Conclusion

"Solely by means of its rhetoric": the triumph of the *Opticks* is wholly rhetorical because science is rhetorically constituted, a network of persuasive structures, patterns that extend upward through style and arrangement to invention itself, to science itself. In his well-known monograph, *The Structure of Scientific Revolutions*, Thomas Kuhn makes the Newtonian moment in optics an example typical of the initial formation of a scientific paradigm, a set of precepts, exemplars, and inference rules I have renamed an argument field. According to Kuhn, optics before Newton was pre-paradigmatic; that is, there was no consensus on methods and theories concerning light and color. After Newton, Newtonian method and theory represent the consensus. In my view, Kuhn's historical reconstruction is unpersuasive. We view optics before Newton as pre-paradigmatic only because of the persuasive effect of his second, evolutionary formulation of his theory; had his first formulation prevailed, we would have seen his optics as revolutionary, as discarding an existing, and well-entrenched, paradigm.

Each of Newton's two reconstructions of his optical past was an attempt to persuade his scientific audience of the truth of his unchanging views. In the early papers, Newton emphasized the discontinuities between his theories and the optical past; he cast his work in a revolutionary mold. In contrast, in his *Opticks*, he emphasized the continuity between his theories and the optical past; he presented his work as an evolutionary development. Was Newton's optics revolutionary, or evolutionary? The question is otiose. Historical continuity and discontinuity are not discovered; they are invented by rhetorical means to suit particular persuasive purposes. The lesson of Newton's optics is that there is no privileged reconstruction of the past; like the rest of us, scientists recreate their past to reflect the importance of the present they favor.

One may admit that the revolutionary or evolutionary character of Newton's optical theory, indeed, of any scientific theory, is a rhetorical *trompe-l'oeil*, but many still insist that the theory itself represents a claim independent of its, or of any, rhetoric. Let us be as strict as possible; let us present scientific claims and arguments in logical form: "If, and only if, *a*, then, necessarily *b*; if, and only if, white light is a compound of all the spectral colors, each consisting of particles in mechanical interaction, then, necessarily, the solar spectrum will be oblong." One may quarrel with this particular schematic representation of Newton's central empirical regularity, and of the theory by which he explains it; but, in fact, any such representation will accomplish the same task; it will discard the style and alter the arrangement of the original only to replace them with a style and arrangement of another, more austere, sort. Style and arrangement in science are not veils that can be removed to reveal theory, the scientific core beneath.

One may concede as much, but point out that, nevertheless, the scientific content of a text is precisely that which survives its legitimate paraphrases; there really is a scientific core, even if it can never actually be freed from all linguistic embodiments. But what is it to say that Newton's *Opticks* is the scientific equivalent of his early papers on light and colors? It is to say that to paraphrase either is to say the same scientific thing.[64] Paraphrase, however, is not linguistic equivalence, the meaning that persists, for instance, through the active-passive transformation; it is, rather, a theory about a text, an equivalence claim whose support is the beliefs of particular audiences concerning science. We can say Newton's optical papers and his *Opticks* share the same paraphrasable content only because we know what experiments and theories are, and can agree in advance on what it means for an experimental result to verify a theory, a theory to be entailed by an experimental result.

We are inclined to say that the scientific content of texts can be paraphrased because they contain a core of meaning. But, in fact, the reverse is the case: the core of meaning is just what we paraphrase consistently. In the particular case of Newton's optics, we would regard as radically defective any paraphrase that omitted either his central empirical regularity or its explanation. But to include these is nothing more than to fulfill an expectation. Paraphrase is theory driven; this theory selects the invariants we call the core of scientific texts. Our ability to paraphrase is only evidence that we have already been persuaded, that rhetoric has done its work.

Paraphrasability is a necessary, but not sufficient, condition of scientific texts; even poems can be paraphrased.[65] To do science, therefore, is to make a claim about paraphrases, to assert that the truth of theories, of the paraphrasable content of scientific texts, is more than an effect of successful persuasion. I take this claim to mean that scientific theories are true if, and only if, they survive the test of predicting certain crucial empirical regularities. The general theory of relativity, for example, triumphed by predicting what classical physics could not: the motion of the perihelion of Mercury, the deflection of light by a gravitational field, and the displacement of spectral lines toward the red.[66] Because such empirical regularities seem generally stable over time,[67] they form a firm evidential base for scientific theories.

But empirical regularities are not science; they become science only through interpretation: their scientific significance depends entirely on their theoretical role. Regularities in cranial topography did not cease to exist when phrenology collapsed as a science; they merely ceased to be indices of scientific truths. The truths embodied in scientific theories, then, are far less stable than the empirical regularities they explain or predict: Ptolemy and Copernicus explain the same stellar motions, but their theories are mirror opposites. In optics, theory followed theory, as century followed century: in the nineteenth, wave succeeded particle; in the twentieth, quantum theory ushered in a whole new physics. These theories provided incompatible explanations of the same phenomenon: the oblong shape of the solar spectrum. But can the triumph of mistaken theories count as success? Can a theory that has been proven false ever have been true?

In the long run, most, if not all, scientific theories prove mistaken. Because there is never an absolute warrant for the move from experiment to theory, being mistaken is less a fault than a necessary property of such theories. In a hypothetical syllogism, if *a*, then *b*, affirming the consequent says nothing necessary about the truth of the antecedent: for example, the empirical regularity of the oblong spectrum says nothing necessary about the truth of Newton's theories. Nor can we move with full confidence from theory to experimental confirmation; the hypostatized real world, the source of empirical regularities, need not conform to *any* of our ways of thinking. Science is no open sesame to this real world; in science, as in all branches of knowledge, we have instead two sorts of sentences: theory sentences that constitute claims and observation sentences that constitute their confirmation.

The relationship between these sets of sentences remains deeply ambiguous: theory does not unerringly point to its experimental instances; experimental instances do not clearly confirm any theory; experimental anomalies do not unequivocally disconfirm any theory. But one of the implications of these truisms of logic and of philosophy of science is routinely, though not invariably, ignored:[68] persuasion must be the mode of scientific knowledge, rhetoric must be its substance. In his optical work, Newton the scientist and Newton the rhetorician cannot be separated. In his work, Newton explained those empirical regularities that seemed most compatible with his theory, and most likely to win assent; he explained away those results that seemed most clearly to challenge his formulations, and weaken his case. In this, he was a typical scientist, undeterred by doubts of future generations, however legitimate. He emphasized, not the tentativeness of his mistaken theoretical conclusions, but their rightness, their inevitability.

Notes

Alan G. Gross read an earlier version of this paper at an Argument Conference at Wake Forest University in 1986. He would like to thank Purdue University for a Scholarly Award and Dr. Harold Fromm for merciless scrutiny.

1. As a basis for its views, this paper assumes without argument numerous philosophical, historical, and scientific claims. Of necessity, given the essentially contested nature of Newton and Descartes scholarship, none of these claims is above contention, though all, I hope, would be defensible on challenge. In general, the reader can discern my biases by following my footnotes. I should like to mention here some works that escaped quotation, but were crucial in forming my views: I. Bernard Cohen, *Franklin and Newton: An Inquiry into Speculative Newtonian Experimental Science and Franklin's Work in Electricity as an Example Thereof* (Cambridge: Harvard University Press, 1966); Alexandre Koyré, *Newtonian Studies* (Chicago: University of Chicago, Press 1968); A.C. Crombie, *Robert Grosseteste and The Origins of*

Experimental Science: 1100-1700 (Oxford: Clarendon Press, 1961); Thomas S. Kuhn, *The Essential Tension: Selected Studies in Scientific Tradition and Change* (Chicago: University of Chicago Press, 1977); David C. Lindberg, *Theories of Vision from Al-Kindi to Kepler* (Chicago: University of Chicago Press, 1976); René Descartes, *Le Monde, ou Traité de la Lumierè*, ed. Michael Sean Mahoney (New York: Abaris, 1979); William A. Wallace, *The Scientific Methodology of Theodoric of Freiberg: A Case Study of the Relationship Between Science and Philosophy* (Fribourg: The University Press, 1959). In addition, I should like to thank Professors Daniel Garber and Howard Stein of the University of Chicago for allowing me to sit in on their classes, and for sharing some conversations. Needless to say, the mention of these works and these scholars is not meant to absolve me from the responsibility for any errors of interpretation.

2. The first principles of an Aristotelian science, though not scientific knowledge, were more certain than such knowledge. In the much disputed last chapter of the *Posterior Analytics*, these principles seem to result from a transcendental abduction. In *Topics* (101^b3), they seem to be the product of dialectic. For a discussion, see Jonathan Barnes, ed., *Aristotle's Posterior Analytics* (Oxford: Clarendon Press, 1975), 248-60.

3. Ch. Perelman and L. Olbrechts-Tyteca, *The New Rhetoric: A Treatise on Argumentation*, trans. John Wilkinson and Purcell Weaver (1958; Notre Dame: University of Notre Dame Press, 1971, 4-5, 54. For an important commentary on the relationship between rhetoric an dialectic, see Edward Meredith Cope, and John Edwin Sandys, *The Rhetoric of Aristotle, With a Commentary* (Cambridge: The University Press, 1877), 1:1-3.

4. Aristotle, *"Art" of Rhetoric*, trans. John Henry Freese (1926; reprint, Cambridge: Harvard University Press, 1975), 13 (1355^b14).

5. Jürgen Habermas, "A Reply to Critics," in *Habermas: Critical Debates*, ed. John B. Thompson and David Held (Cambridge: The MIT Press, 1982), 274.

6. Hans-Georg Gadamer, *Truth and Method*, trans. Garret Barden and John Cumming (1965; New York: Crossroad, 1975), 410.

7. Charles Arthur Willard, *Argumentation and the Social Grounds of Knowledge* (University: University of Alabama Press, 1983), 91.

8. Willard, *Argumentation*, 91.

9. René Descartes, *Optics*, in *Discourse on Method, Optics, Geometry, and Meteorology*, trans. Paul J. Olscamp (Indianapolis: Bobbs-Merrill, 1965), 70. This passage is typical of Descartes's expository technique, and a good example of rhetorical transparency: the analogy is designed to make science clearer.

10. Descartes, *Optics*, 77.

11. Descartes, *Optics*, 80.

12. Throughout, but especially René Descartes, *Principles of Philosophy*, trans. Valentine Rodger Miller and Reese P. Miller (Dordrecht: D. Reidel, 1983/84), 286-88. But when his unaided reason convinced him that Copernicus was right, he was careful to avoid provocation. He suppressed his early physics, *Le Monde*, and presented his later physics in *The Principles* so that it squared with orthodoxy.

13. Descartes, *Principles*, 287-88.

14. Descartes, *Principles*, 69 n.

15. René Descartes, *Philosophical Letters*, trans. and ed. Anthony Kenny (1970; reprint, Minneapolis: University of Minnesota Press, 1981), 38.

16. René Descartes, *Meteorology, in Discourse on Method, Optics, Geometry, and Meteorology*, trans. Paul J. Olscamp (Indianapolis: Bobbs-Merrill, 1965), 337.

17. The persuasiveness of this account depends also on the traditional assumption that light is basic, color derivative, an assumption unchallenged until Newton.

18. In French, *expérience* means both "experience" and "experiment."

19. Descartes, *Optics*, 81.

20. Descartes, *Principles*, 106.

21. "Meditations on First Philosophy"in *The Philosophical Works of Descartes*, trans. Elizabeth S. Haldane and G.R.T. Ross (1931; reprint, Cambridge: Cambridge University, 1983), 1: 164; see also "Rules for the Direction of The Mind" 56 ff., and *Principles*, 76-77.

22. For example, compare Descartes, *Meteorology*, 268 with a letter to Father Mersenne, dated by inference March 1, 1638 in *Oeuvres de Descartes*, ed. Charles Adam and Paul Tannery. (Paris: Léopold Cerf, 1898), Correspondance 2: 29.

23. Compare Stephen Gaukroger, "Descartes' Project for a Mathematical Physics," in *Descartes: Philosophy, Mathematics and Physics*, ed. Stephen Gaukroger (Sussex: Harvester Press, 1980), 134.

24. Descartes, *Meteorology*, 339.

25. Isaac Newton, *The Mathematical Papers*, ed. D.T. Whiteside, M.A. Hoskin and A. Prag (Cambridge: Cambridge University Press, 1969), 3: 543-49.

26. Descartes, *Philosophical Letters*, 46.

27. Descartes, *Meteorology*, 338.

28. Descartes, *Meteorology*, 342.

29. Isaac Newton, "A Letter . . . containing his New Theory about Light and Colours," in *Isaac Newton's Papers and Letters on Natural Philosophy*, ed. I. Bernard Cohen and Robert E. Schofield, 2d ed. (Cambridge: Harvard University Press, 1978), 47-48.

30. Newton, "New Theory," 50.

31. Newton, "New Theory," 51.

32. Newton, "New Theory," 47.

33. Newton, "New Theory," 55.

34. Newton, "New Theory," 53.

35. Newton, "New Theory," 56.

36. Richard S. Westfall, *Never at Rest: A Biography of Isaac Newton* (Cambridge: Cambridge University Press, 1980), 164.

37. Isaac Newton, "A Serie's of Quere's . . . to be determin'd by Experiments, positively and directly concluding his new Theory of Light and Colours . . ." In *Isaac Newton's Papers*, 94.

38. For the relation between rhetoric and Baconian philosophy, see Alan G. Gross, "The Form of the Experimental Paper: A Realization of the Myth of Induction," *Journal of Technical Writing and Communication* 15 (1985): 15-26.

39. Isaac Newton, *The Correspondence*, ed. H.W. Turnbull (Cambridge: Cambridge University Press, 1959), 1: 169 (a paraphrase of the Latin); see also Isaac Newton, "A Serie's of Quere's," 93 and 506 n.

40. Isaac Newton, "Mr. Newton's Answer. . . ." In *Isaac Newton's Papers*, 174.

41. Newton, "New Theory," 57. It is a delicious irony that in his passage Newton uses Aristotelian terminology against itself.

42. Christiaan Huygens, "An Extract of a Letter . . . containing some Considerations upon Mr. Newtons Doctrine of Colors. . . ." In *Isaac Newton's Papers*, 136. The idea that certainty demands a complete causal explanation is as old as Aristotle and Aquinas. Neither Huygens not Newton would have seriously objected to this view. For science, the list of causes was standardly Aristotelian, the final cause being generally excepted. After the Baconian and Cartesian assault, the final cause was always excepted. The full substitution of law-like for causal explanation lay in the future.

43. Newton, "New Theory," 57.

44. Richard S. Westfall, *Never at Rest*, 280.

45. David C. Lindberg, "The Cause of Refraction in Medieval Optics," *The British Journal for the History of Science* 4 (1968): 36.

46. Newton, *The Correspondence*, 1: 416.

47. Isaac Newton, *Opticks, Or A Treatise of the Reflections, Refractions, Inflections, & Colours of Light*. Based on the Fourth Edition, 1730 (1952; reprint, New York: Dover, 1979), 19-20.

48. Newton, *Opticks*, 404.

49. Newton, *Opticks*, 405.

50. Richard S. Westfall, *Never at Rest*, 640-44.

51. Henry Guerlac, *Newton on the Continent* (Ithaca: Cornell University Press, 1981), 110.

52. Newton, *Opticks*, 25.

53. Newton, *Opticks*, 45.

54. Isaac Newton, "Newton's second paper on color and light. . . ." In *Isaac Newton's Papers*, 219; Newton, *Opticks*, 233.

55. Perelman and Olbrechts-Tyteca *The New Rhetoric* 118; 117.

56. The number of Queries rose from a mere sixteen in the first English edition (1704), to twenty-three in the first Latin edition two years later, to the full thirty-one in the second English edition (1718). I. Bernard Cohen, "Preface," in Isaac Newton, *Opticks* xxxi; *Never at Rest*, 641.

57. Newton, *Opticks*, 2.

58. Newton, *Opticks*, 370.

59. Randolph Quirk, Sidney Greenbaum, Geoffrey Leech, Jan Svartvik, *A Grammar of Contemporary English* (London: Longman, 1972), 401. Newton seems to have borrowed this device unacknowledged from Hooke. See Robert Hooke, *Micrographia, of Some Physiological Descriptions of Minute Bodies Made*

By Magnifying Glasses With Observations and Inquiries Thereupon (1665; reprint, New York: Dover, 1938), 233-40.

60. Newton, *Opticks*, 360.

61. Compare "Account of the Commercium Epistolicum" written by Newton and reprinted in A. Rupert Hall, *Philosophers at War: The Quarrel Between Newton and Leibniz* (Cambridge: Cambridge University Press, 1980), 312.

62. Newton, *Opticks*, 339.

63. It may have occurred to the reader that Newton did not publish his *Opticks* until 1704, a year after Hooke, the last living critic of his early paper, died. A neurotic fear of rejection probably fueled this delay and, in large part, drove Newton's rhetorical ingenuity. But my arguments are rhetorical, not biographical. Regardless of Newton's conjectured motives, the *Opticks* remains a rhetorical masterpiece. For a psychological interpretation, see Frank E. Manuel, *A Portrait of Isaac Newton* (Cambridge: Harvard University Press, 1968).

64. We cannot make this claim on the basis of the direct comparison of the early papers with the *Opticks*; in such a comparison, likenesses could be found merely by inspection; by an optical scanner, for instance.

65. In contrast with scientific texts, we tend to say of poems that their paraphrasable content contains none of their essence.

66. Albert Einstein, *Relativity: The Special and General Theory*, trans. Robert W. Lawson (New York: Crown, 1961), 123-32.

67. But Fleck, *passim*, shows the instability of even these; and Singer and Cole, for example, both point to Leonardo's notorious drawing showing non-existent perforations in the inter-ventricular septum (the wall between the ventricles of the heart). See Ludwik Fleck, *Genesis and Development of a Scientific Fact*, ed. Thaddeus J. Trenn and Robert K. Merton; trans, Fred Bradley and Thaddeus J. Trenn (Chicago: University of Chicago Press, 1979); Charles Singer, *A Short History of Anatomy and Physiology from the Greeks to Harvey* (New York: Dover, 1957), 90-92; F.J. Cole, *A History of Comparative Anatomy From Aristotle to the Eighteenth Century* (1949; reprint, New York: Dover, 1975), 53.

68. In *Against Method* (1975; reprint, London: Verso, 1978), Paul Feyerabend adopts the position that science is rhetorical.

The Birth of Molecular Biology:
An Essay in the Rhetorical
Criticism of Scientific Discourse

by S. Michael Halloran

In his introduction to the Norton Critical Edition of *The Double Helix*, Gunther Stent assigns a birthday to the science of molecular biology: April 25, 1953, the publication date of James Watson and Francis Crick's paper sketching the double helical structure they had devised for the DNA molecule.[1] Others have confirmed the view that this paper was pivotal in establishing molecular biology as a science. The editors of the journal *Nature* carried a series of papers under the collective title "Molecular Biology Comes of Age" just twenty-one years and a day after its publication, and they used a facsimile of it as the title page for the retrospective.[2] Horace Freeland Judson's exhaustive history of the field quotes from scores of scientific papers but reproduces *in toto* only this one.[3] More clearly than any single scientific paper in recent years, this one stands near if not precisely *at* the center of what Thomas Kuhn would call a scientific revolution.

I have written elsewhere of the rhetorical implications of Kuhn's view of the nature of science, and in a general way of the rhetorical dimensions of Watson and Crick's work.[4] This essay might be regarded as a sequel to that earlier one, but I don't intend to pursue further the theoretical implications of Kuhn's ideas for developing a rhetorical analysis of science. I want instead to develop a more thorough critical analysis of Watson and Crick's 1953 paper. Rhetoric has traditionally been a strongly empirical field of study in that it places great emphasis on the particular case. The job of the rhetorical critic is to discover what in the particular case were the available means of persuasion, and judge whether the rhetor managed them well or badly. The particular case commands his or her attention as something worth knowing in itself, apart from any general principles that might be abstracted from it. But while a number of scholars have been arguing theoretically that science is rhetorical, very little attention has been paid to particular cases of scientific rhetoric.[5]

This essay comes at the rhetoric of science from a critical perspective; I want to explicate a particular case that is surely worth the effort. Ultimately, I hope to show that the Watson-Crick paper establishes an *ethos*, a characteristic manner of holding and expressing ideas, rooted in a distinctive understanding of the scientific enterprise.

I am here consciously echoing lines from an essay by Edwin Black in which, while he does not use the term *ethos* as I am using it, he develops a very similar concept in connection with nineteenth-century American oratory:

> Groups of people become distinctive as groups sometimes by their habitual patterns of commitment—not by the beliefs they hold, but by the manner in which they hold them and gIve them expression. Such people do not necessarily share ideas; they share rather stylistic proclivities and the qualities of mental life of which those proclivities are tokens.[6]

The most general point I hope to make in this paper is that scientific communities can be bound together in this fashion. While the specific beliefs they hold—the *logos* of the discipline—may be crucial to a scientific community, their identity as a community may rest equally on "stylistic proclivities and the qualities of mental life of which those proclivities are tokens," that is, on what I am calling *ethos*. I will begin by concentrating rather closely on a single scientific paper, then try to place it in a larger context.

"A Structure for Deoxyribose Nucleic Acid" was the first published announcement of the double-helical structure Watson and Crick had devised for DNA, the molecule that had by the early 1950s been identified as the transmitter of genetic information.[7] While the story of how Watson and Crick arrived at their discovery has been told elsewhere,[8] certain facts bear retelling here, by way of outlining the rhetorical situation. First, there was a degree of competition surrounding the work: Linus Pauling in California was known to be working on the problem of DNA's structure; Maurice Wilkins and Rosalind Franklin at King's College in London were also working on it, and there was a vague sense in England that the problem belonged to them; Watson and Crick were supposed to be working on other matters at Cambridge, but they hoped to be first to the solution of the DNA molecule. Second, while the structure of the DNA molecule was regarded as an important research problem, no one knew beforehand just how important its solution would turn out to be. No one knew or even hoped that genetic information would turn out to be transmitted by a straightforward mechanical process, and that knowing the structure of DNA would therefore suggest the possibility of mastering and ultimately manipulating the process.

Because of the competitiveness of the situation and of the unanticipated significance of the discovery, Watson and Crick chose to publish their discovery in *Nature*, a journal that would publish the article promptly and reach a broad scientific audience. *Nature* is published weekly and, like the U.S. Journal *Science*, is not specialized in a particular discipline. An arrangement was made for Wilkins and Franklin to publish simultaneously with Watson and Crick results of their most recent x-ray diffraction studies, which tended to support the proposed model. What appeared in *Nature*, then, was a trilogy of articles under the collective title "Molecular Structure of Nucleic Acids." The first is Watson and Crick's paper; the second is "Molecular Structure of Deoxypentose Nucleic

Acids," signed M. H. F. Wilkins, A. R. Stokes and H. R. Wilson; the third is "Molecular Configuration in Sodium Thymonucleate," signed Rosalind E. Franklin and R. H. Gosling.[9] This apparent attempt to portray the discovery of the double helix as a broad-based team effort failed. The Watson-Crick paper is reprinted in three places that I know of with the title "Molecular Structure of Nucleic Acids," but without the Wilkins and Franklin papers that properly go with it under that title.[10]

The paper consists of fourteen paragraphs totaling just over 900 words. It contains one figure—a "purely diagrammatic" representation of two helices wound around a central axis—and no formulas. The text is organized as follows:

Paragraph 1—introduction: "We wish to suggest a structure for the salt of . . . (DNA). This structure has novel features which are of considerable biological interest."

Paragraphs 2–3—review of selected literature: Models of DNA proposed by Pauling and Corey (paragraph 2) and Fraser (paragraph 3) are considered and rejected.

Paragraphs 4–12—body of the paper:

Paragraphs 4–5 sketch the broad outlines of the model (two helical chains wound around each other).

Paragraphs 6–8 describe the "novel feature" of the structure, the mechanism by which the two chains are bound together. Each chain contains four bases in a sequence that "does not appear to be restricted in any way." The bases on one chain form hydrogen bonds with those on the other according to a fixed pattern (adenine bonds only to thymine, guanine only to cytosine), so that the sequence of bases on one chain automatically determines the sequence on the other.

Paragraph 9 notes that experimentation has shown adenine equal to thymine and guanine equal to cytosine in DNA.

Paragraph 10 speculates that the structure will not be found in RNA.

Paragraph 11 considers the status of the proposed model relative to available x-ray data. The model is "roughly compatible" with the data, "but it must be regarded as unproved. . . . Our structure . . . Rests mainly though not entirely on published experimental data and stereochemical arguments."

Paragraph 12: "It has not escaped our notice that the specific pairing we have postulated immediately suggests a possible copying mechanism for the genetic material."

Paragraphs 13–14—conclusion: Paragraph 13 promises a more detailed picture of the structure to be published elsewhere; paragraph 14 acknowledges the help of a few other scientists, including Wilkins and Franklin.

There are, as I see it, three substantial arguments put forward in support of their model by Watson and Crick. The most important is the great elegance of the model, particularly the base-pairing mechanism (described in paragraphs 6–8) that holds the two helical chains together. This argument is in their paper left entirely implicit, as it probably had to be. They assume that the description in paragraphs 4–8 will appeal strongly to the reader's sense of theoretical elegance. The argument is in effect an enthymeme whose missing premise is a scientific *topos* so basic and powerful that it would be gauche in the extreme to state it openly in a technical paper. But in *The Double Helix* Watson makes the argument from elegance explicit: "a structure this pretty just had to exist."[11] (Future developments in the history of molecular biology would demonstrate that a structure could be extraordinarily "pretty" and still not exist. The story of the comma-free code—"an idea of Crick's that was the most elegant biological theory ever to be proposed and proved wrong"—is too intricate to warrant retelling here, and it is told with appropriate elegance by Judson.[12] I mention it to underscore my point that for the scientist elegance functions not as an absolute principle, but as a rhetorical *topos*, a premise for argument in contingent cases.)

The second argument offered by Watson-Crick is that the proposed model provides a very precise theoretical explanation for what before had been simply a curious fact—the observed ratios of adenine to thymine and guanine to cytosine, referred to in paragraph nine. Given the base pairing mechanism that holds the DNA molecule together according to the model, the ratios become inevitable rather than just curious. What is interesting rhetorically about this argument is that Watson and Crick leave it implicit just as they had the argument from elegance, though in this case they might have made their point explicit without any impropriety. Instead they merely juxtapose a statement of the observed ratios immediately following the descriptonof the base-pairing mechanism, and expect the reader to make the connection. The argument is another enthymeme resting on the *topos* of explanatory power.

The third argument is negative: the proposed model is not inconsistent with any available experimental data. Unlike the first two arguments, this one is laid out explicitly, in paragraph 11. But notice how carefully qualified the statement is.

Argumentatively, then, the paper is understated, and the rhetorical effect is to communicate a sense of supreme confidence. The claim that the proposed model is of "considerable biological interest" is advanced boldly in the first paragraph, and the arguments in support of the model are assumed to be so persuasive that they need no bolstering of emphasis.

Stylistically, the most striking quality of the paper is its genteel tone. Note, for example, the diction of the introductory paragraph: "We *wish* to *suggest* a structure for the salt of deoxyribose nucleic acid (D.N.A.). This structure has *novel features* which are of *considerable biological* interest" (italics added). Note too the delicate fashion in which they reject the model that had been proposed by Linus Pauling and his colleague: "*In our opinion*, this structure is *unsatisfactory* for two reasons: (1) *We believe* that the material which gives the

X-ray diagrams is the salt, not the free acid. Without the acidic hydrogen atoms *it is not clear* what forces would hold the structure together. . . . (2) Some of the van der Waals distances *appear to be* too small" (Paragraph 2; italics again mine). That this is a consciously contrived style becomes apparent in light of *The Double Helix*, from which we know that Watson and Crick regarded the Pauling-Corey model as an incredible blunder, a violation of the most elementary fact of chemistry. They were astonished and jubilant to find the great Pauling guilty of what they regarded as a gross error. If we can believe *The Double Helix*, the genteel style of Watson and Crick's first published paper reflects a rhetorical persona, perhaps fabricated with a bit of intentional, tongue-in-cheek irony; in the flesh they were obstreperous and irreverent.

Note finally the one sentence paragraph that concludes the body of the paper, a sentence in which the genteel style becomes a transparent burlesque: "It has not escaped our notice that the specific pairing we have postulated immediately suggests a possible copying mechanism for the genetic material." One can almost feel the elbow in one's ribs.

The effect of Watson-Crick's self-consciously genteel style is to give the paper a highly personal tone that is somewhat unusual in scientific prose. There are a number of conventional devices by which scientific prose is depersonalized. Simplest and most frequently noted by critical readers is the passive voice construction: "It was observed that . . ." Rather than "I observed . . ." Somewhat more subtle and rhetorically interesting is a device that one finds with increasing frequency in academic writing across the disciplines, a device that amounts to the manufacture of abstract rhetors: "*The data* show that . . ." Or "*This* paper will argue that . . ." The effect of the device is to suppress human agency, to imply that what are essentially rhetorical acts—arguing, showing, demonstrating, suggesting—can be accomplished without human volition. Watson and Crick are noteworthy in that they generally avoid this convention, particularly in putting forward their own case. They claim the argument quite explicitly as their own: "We wish to suggest" "In our opinion" "We believe" "We wish to put forward" "It has not escaped our notice" "We have postulated." By contrast, the Wilkins et al. paper that appears immediately following Watson-Crick in the April 1953 *Nature* is bloodless and impersonal in the manner more typical of scientific prose: "The purpose of this communication is . . ." "It may be shown that . . ." "It must be decided whether . . ." "The . . . significance of a two chain nucleic acid unit has been shown . . ."

Both argumentatively and stylistically, then, Watson and Crick put forward a strong proprietary claim to the double helix. What they offer is not *the* structure of DNA or *a* model of DNA, but Watson and Crick's structure or model. Moreover, in staking their claim they enact a distinctive way of adhering to ideas in public; they dramatize themselves as intellectual beings in a particular style. The paper articulates a recognizable public persona, an *ethos*. The Watson-Crick *ethos* does not necessarily overturn established conventions of scientific rhetoric, though we shall see that it offends at least one authoritative sensibility. What I believe it does is shape a particular image of *the scientist speaking*, within a broader set of more vague and general norms that apply to all scientific discourse. And this *ethos,* I contend, is an important aspect of what Kuhn would call the paradigm offered to the broader scientific community.

This claim is a rather large critical speculation based on my general sense of how rhetorical norms operate in scientific communities. Some evidence in its support can be found by comparing Watson and Crick's rhetoric with that of other biologists, and I will turn now briefly to a paper without which Watson and Crick's own work might not have been possible.

The effort devoted to discovering the structure of DNA—not just by Watson and Crick, but by Wilkins, Franklin, Pauling and a great number of others—was of course based upon a consensus that DNA is indeed the substance that transmits genetic information from one generation of cells to the next. The first published demonstration of this crucial fact was a paper by Oswald Avery and two associates that appeared in 1944 in the *Journal of Experimental Medicine.*[13] Avery and his colleagues had used a well-known experimental procedure in which one strain of Pneumococcus bacteria is transformed into a genetically distinct strain. By a series of tortuously executed procedures, they isolated the "active principle" involved in the transformation and identified it as DNA. Prior to the publication of their work, DNA was thought to be a genetically irrelevant substance, and most biologists assumed that genes consisted of some form of protein. In a sense, then, one might date the "revolution" in molecular biology from the appearance of their paper rather than Watson and Crick's. The simplest reason for not doing so is that Avery's work did not have an immediate revolutionary effect. Although it is now regarded as an air-tight demonstration of DNA's role in heredity, scientists were slow to accept it as such and focus on DNA's structure as a biologically important problem. According to Gunther Stent, Avery's discovery was "premature" because it could not at the time be connected with canonical knowledge in the field.[14]

An obvious feature of the Avery et al. paper is that by comparison with Watson and Crick's on the double helix, it is much longer and more dense with technical detail. Whereas Watson and Crick simply sketch in broad outline the results of their work, Avery and his colleagues rehearse in painstaking detail the experimental technique by which the "active principle" was isolated and then the analytic techniques by which it was identified as DNA. In effect they present their case according to the method of residues, recording a sequence of technical procedures that gradually narrows the explanatory possibilities down to the single conclusion that the substance responsible for the phenomenon in question is DNA. A characteristic point of their argumentative strategy is that the paper does not state its thesis in the introductory section and in fact does not even mention the substance DNA until roughly half-way through its 7500 word length. They make no strong claims about the importance of their discovery, and in fact introduce the paper as simply a "more detailed analysis" of the already well-known transformation phenomenon. They observe all the conventions of depersonalization: events transpire in the passive voice, data suggest conclusions without human assistance, and Avery and his colleagues take on that ultimate *nom de plume*, "the writers."

I am tempted to suggest that the "prematurity" of Avery's work was owing in part to his rhetoric in presenting it to the scientific community.[15] But while I think that a persuasive case could be made for such a claim. I am more concerned here simply to point up the contrast between Avery's *ethos* and that of Watson and Crick. The character that speaks to us from Avery's paper is that of a cautious skeptic who is forced somewhat unwillingly to certain conclusions. That of the Watson-Crick paper is quietly confident, so much so that "he" can indulge in a gentle bit of leg-pulling. I put quotation marks around "he" for the obvious reason that the voice in which the paper speaks is in a sense that of two men speaking in unison, a fact that points up the corporate, conventional, public nature of the phenomenon I am trying to capture. What interests me here is not the unique personality of an Oswald Avery or a Francis Crick, but the public role of *scientist* as dramatized by them. They offer two sharply contrasting versions of that role, two images of a fitting way for scientists to hold ideas.

A larger view of the contrast comes into focus if we consider the notions of form and strategy. At a simple level, the Watson and Crick paper offers something very close to a textbook illustration of Burkean form: the "arrows of desire" are pointed in the opening sentences,[16] which promise that the proposed structure has *novel features* which are of *considerable biological* interest. The promise of novel features is satisfied by paragraphs 6–8, which describe the crucial base-pairing mechanism, introducing it with the phrase, "The novel feature of the structure is ..." The promise of biological interest is partially satisfied by the last substantive paragraph (12), which hints at an explanation of the genetic process. But of course this is simply a further pointing of arrows, and the appetite so aroused is in turn satisfied by a second paper that appeared in *Nature* just five weeks later, this one speculating on how the DNA molecule (as described in "our model") might transmit genetic information by means of an essentially mechanical process.[17] And, since neither of these two brief papers develops a sufficiently detailed picture of the model for other scientists to begin working with it, both together serve to create an appetite for two more technically elaborate papers that followed in more specialized journals.[18] Finally, just eighteen months after the publication of the original brief paper in *Nature*, Crick published an essay in *Scientific American* reviewing the state of the art in molecular biology and placing his work with Watson in this larger context.[19]

The April 1953 paper, then, is really just the initial move in a rhetorical strategy aimed at gaining and holding the attention of an audience. As such, it presumes an understanding of science as a human community in which neither facts nor ideas speak for themselves, and the attention of an audience must be courted. By contrast, Avery and his colleagues present their work in a single technical paper structured in a reportorial pattern which implies that facts *do* speak for themselves. Their strategy seems to presume that the work of the scientist is simply to give oneself up to the facts. Avery speaks from within an essentially positivistic, pre-Kuhnian view of science, Watson and Crick from within what Frederick Suppe calls a *Weltanschauungen* view.[20] They recognize that a discipline includes tacit assumptions about what is and what is not a legitimate question, and that in order to gain a hearing for a new theory, one may have to suggest what use the theory might have, what new questions it might both pose and answer, what new lines of research it might open up.

The success of the Watson-Crick strategy is indicated by the publication of Crick's essay in *Scientific American*, a periodical that addresses the entire scholarly community and carries a very substantial weight of authority. In this piece, Crick is no longer speaking only for himself and Watson, but for the community of specialists in molecular biology as well. When he writes of what "we" know to be the case and what "we" regard as an important research problem, he speaks for an international scientific community, defining their view of the world to an audience that embraces scholars and scientists in all disciplines. The essay does not yet speak of the double-helical model of DNA as one of the established facts of biology; the model is still in effect the private property of Watson and Crick. But as "owner" of that theory, Crick has gained the authority to say what the facts are in biology and to place the theory before the entire scholarly community in the context of those facts. The implication is that the theory is a strong candidate for admission to the canon of established knowledge in biology.

Perhaps I should be explicit here on a point that I hope would go without saying: none of this is meant to deny the importance of Watson and Crick's model in its technical particulars. To say with Perelman that a *fact* is defined by its claim to the adherence of the universal audience is not to deny that a fact must also correspond to an observation of the world.[21] The double-helix and its attendant explanation of genetic information transfer would not have survived as a theory had it failed to work in predicting experimental observations. My point is to bring into focus another aspect of Watson and Crick's work, a rhetorical aspect that falls under the heading of *ethos*. In offering their model of DNA to the scientific world, they simultaneously offered a model of the scientist, of how he ought to hold ideas and present them to his peers. I believe that this ethical aspect of Watson and Crick's work contributed to the speed with which their model of DNA gained prominence as a theory, but I have been more concerned simply to explicate the *ethos*.

One question remaining is whether the confident, personal, rhetorically adept *ethos* of Watson and Crick was effective *as a model*, whether it was adopted by other scientists. A strongly persuasive answer to that question would have to rest upon close rhetorical analysis of the work of later biologists. That analysis remains to be done. In the meanwhile, I can offer two somewhat weaker arguments in support of my belief that Watson and Crick *have* become a rhetorical-ethical model for others.

First, there is today an adventuresome, entrepreneurial, slightly irreverent spirit associated with the field of molecular biology and genetic engineering, a spirit that on its face strikes me as a recognizable offspring of the Watson-Crick *ethos*. The irreverence is apparent in the breezy, somewhat whimsical terminology current in the field: "gene splicing" is done with the assistance of a "gene machine"; segments of "genetic gibberish" on the DNA molecule are thought by some to act as "genetic errand boys."[22] I am inevitably reminded of the touches of delicate irony in the original Watson-Crick paper. The entrepreneurial spirit is evident in the enthusiasm with which researchers have welcomed opportunities for commercial exploitation of knowledge, even to the possible detriment of traditional academic values. According to an article in a recent issue of *Science 81*,

> Currently there is not a single top-ranking molecular biologist at an American university who has not signed up with one of the new genetics companies. The development dismays some biology watchers. One of science's brighter points as a human endeavor has been the traditional willingness of scientists to share time, information, and even specimens and equipment. That scientific knowledge should be considered private property is a concept repugnant to many scientists.[23]

I would contend that this notion of scientific knowledge as private, profit-making property is simply a logical extension of the manner in which Watson and Crick laid proprietary claim to their original discovery.

It is also a reduction to a peculiarly simple form of a tendency present generally in modern science. In an essay on the nature of scientific discovery, Kuhn points out that "to make a discovery is to achieve one of the closest approximations to a property right that the scientific career affords."[24] In her study of the cultural effects of print technology, Elizabeth Eisenstein makes a similar point about the importance of clarifying the proprietary claims of authors as a condition for the development of modern science: the incremental building of knowledge—science—presumes a certain niceness in identifying the individual components in the developing structure.[25] Heretofore the scientist's claim to a discovery was a rather inexact approximation of a property right, in that one exercised this right most fully by having others make free use of the property; a discovery became a "contribution" in the very moment that it became one's own property. There was, perhaps, some residue of the much older rhetorical tradition, within which all knowledge was in a sense *commonplace*—available for use by anyone—and such notions as plagiarism, copyright, and product patent would consequently have made little sense.

My second argument rests on the testimony of one of the original actors in the revolution in molecular history. The chemical ratios in DNA which Watson and Crick offered as evidence for their model (paragraph 9 of the paper) had been established in the laboratory and documented in the literature by Erwin Chargaff. These equalities were in fact known as "Chargaff's ratios," and Watson and

Crick's failure to mention his name (except as a footnote) in connection with this crucial piece of evidence might be regarded as a slight. It is perhaps worth noting that while Watson and Crick generally avoid the passive voice, in paragraph 9 they use it with the result that Chargaff's contribution becomes anonymous: "It has been found experimentally that. . . ." In any event, Chargaff has become a strong critic of the direction that biology has taken since the discovery of DNA's structure, and the tenor of his critique tends to support my view that the Watson-Crick *ethos* has been adopted by others. For example, in the 1974 *Nature* retrospective, "Molecular Biology Comes of Age," he recalls reading the original Watson and Crick papers in this way: "The tone was certainly unusual: somehow oracular and imperious, almost decalogous. Difficulties were brushed aside in *the Mr. Fix-it spirit that was to become so evident in our scientific literature*" (italics added).[26] In Chargaff's mind, Watson and Crick have influenced not just the ideational content of biology, but the manner in which ideas are pursued, the spirit in which science is done. The "Mr. Fix-it spirit" that he deplores is what I have been calling an *ethos*.

Assuming that my analysis of the Watson-Crick *ethos* is at all persuasive, it has one very important implication for rhetorical studies of scientific discourse: except at the most general level, it may be misleading to speak of *the* rhetoric of science. While there is a sense in which what Oswald Avery was doing rhetorically is "the same as" what Watson and Crick were doing, the contrasts are profound, and they suggest the possibility that other particular cases of scientific rhetoric will exhibit their own peculiarities. A detailed understanding of the rhetoric of science will have to include some sense of the permissible range of variation. To achieve this sense, we need a body of critical literature on particular cases of scientific discourse.

I am suggesting the need for a program of what Black calls "emic"criticism, criticism that begins with the particular instance and aims toward the development of theories comprehending more general principles that operate across larger bodies of discourse.[27] To do this with scientific discourse is particularly difficult simply because the discourse is specialized and highly demanding. It is a daunting experience to take up a technical paper in, say, molecular biology, equipped with an education in the hard sciences that ends somewhere around the time of Matthew Arnold and T. H. Huxley. That it can be done responsibly is demonstrated by Horace Freeland Judson, who started as a journalist and wrote what is generally regarded as a definitive history of molecular biology. That it should be done by scholars in rhetoric is suggested by the increasing importance of scientific matters in the arena of public affairs, the traditional realm of rhetoric. Science is itself an increasingly public enterprise, both in the sense that the public supports it financially and in the sense that it offers monumental threats and promises to our well-being. Science also serves as warrant for many of the arguments about traditionally non-specialized, civic questions—war and peace, ways and means for promoting the public welfare. To understand public discourse in the closing decades of this century, we must have some understanding of scientific discourse.

Notes

1. James D. Watson, *The Double Helix: A Personal Account of the Discovery of the Structure of DNA*—Text, Commentary, Reviews, Original Papers, ed. By Gunther S. Stent (New York: W. W. Norton & Company, 1980), xi.

2. "Molecular Biology Comes of Age," *Nature*, 248 (April 26, 1974), 765-88.

3. Horace Freeland Judson, *The Eighth Day of Creation: Makers of the Revolution in Biology* (New York: Simon and Schuster, 1979), 196-98.

4. "Technical Writing and the Rhetoric of Science," *Journal of Technical Writing and Communication*, 8 (Spring 1978), 77-88; reprinted in *Technical Communication* (fourth quarter 1978), 7-10, 13.

5. Here follow some of the more interesting studies of scientific rhetoric: John Angus Campbell, "Charles Darwin and the Crisis of Ecology: A Rhetorical Perspective," *QJS*, 60 (Dec. 1974), 442-49; Paul Newell Campbell, "The *Personae* of Scientific Discourse," *QJS*, 61 (Dec. 1975), 391-405; Joseph Gusfield, "The Literary Rhetoric of Science: Comedy and Pathos in Drinking Driver Research," *American Sociological Review*, 41 (Feb. 1976), 16-34; Carolyn R. Miller, "Technology as a Form of Consciousness: A Study of Contemporary Ethos," *CSSJ*, 29 (Winter 1978), 223-36; Michael A. Overington, "The Scientific Community as Audience: Toward a Rhetorical Analysis of Science," *Philosophy and Rhetoric*, 10 (Summer 1977), 143-64; Herbert W. Simons, "Are Scientists Rhetors in Disguise? An Analysis of Discursive Processes Within Scientific Communities," in Eugene F. White (ed.), *Rhetoric in Transition: Studies in the Nature and Uses of Rhetoric* (Univ. Park: Penn. State U. Press, 1980), 115-30; Philip C. Wander, "The Rhetoric of Science," *Journal of Western Speech Communication*, 40 (Fall 1976), 226-35; Walter B. Weimer, "Science as a Rhetorical Transaction: Toward a Nonjustificational Conception of Rhetoric," *Philosophy and Rhetoric*, 10 (Winter 1977), 1-29. Of these, only Campbell's paper on Darwin, and Gusfield's on drinking driver research are *critical* essays in the sense that they include close analysis of particular rhetorical transactions.

6. Edwin Black, "The Sentimental Style as Escapism, or The Devil with Dan'l Webster," in Karlyn Kohrs Campbell and Kathleen Hall Jamieson (eds.), *Form and Genre: Shaping Rhetorical Action* (Falls Church, Va.: SCA, n.d.), 85.

7. J. D. Watson and F. H. C. Crick, "A Structure for Deoixyribose Nucleic Acid," *Nature*, 171 (April 25, 1953), 737-38.

8. In addition to *The Double Helix* and *The Eighth Day of Creation*, see Robert Olby, *The Path to the Double Helix* (London: Macmillan, 1974).

9. *Nature*, 171 (April 25, 1953), 738-41; both papers are reprinted in the Norton Critical Edition of *The Double Helix*, 247-57.

10. Both the 1974 *Nature* retrospective and Judson's *Eighth Day* reprint the Watson-Crick paper in this somewhat misleading way. The same error is committed by Mary Elizabeth Bowen and Joseph A. Mazzeo (eds.), *Writing About Science* (New York: Oxford University Press, 1979). Stent's Norton edition of *The Double Helix* includes the Wilkins et al. and Franklin-Gosling papers, but it places them following a second Watson-Crick paper that appeared in *Nature* more than a month after the original trilogy of DNA papers.

11. *The Double Helix*, 120.

12. *The Eighth Day of Creation*, 318 ff.

13. Oswald T. Avery, Colin M. MacLeod and Maclyn McCarty, "Studies on the Chemical Nature of the Substance Inducing Transformations of Pneumococcal Types," *Journal of Experimental Medicine*, 79 (1944), 137-58; reprinted in Harry O. Corwin and John B. Jenkins (eds.), *Conceptual Foundations of Genetics: Selected Readings* (Boston: Houghton Mifflin Company, 1979), 13-27.

14. Gunther S. Stent, "Prematurity and Uniqueness in Scientific Discovery," *Scientific American*, 227 (Dec. 1972), 84-93.

15. Judson speaks of the Avery et al. paper as "a model of reasoning from and about experiment" (*The Eighth Day of Creation*, 37). I see no conflict between this view and my own belief that the paper is *rhetorically* weak.

16. I am thinking of Kenneth Burke's notion of form, as developed in the essays "Psychology and Form" and "Lexicon Rhetoricae," both in *Counter-Statement* (Berkeley: University of California Press, 1968), 29-44 and 123-83. My belief that Watson and Crick make use of what Burke calls the psychology of form in presenting their model of DNA would require some qualification of the views Burke expresses about science in these essays.

17. J. D. Watson and F. H. C. Crick, "Genetical Implications of the Structure of Deoxyribonucleic Acid," *Nature*, 171 (May 30, 1953), 964-67; reprinted in the Norton Critical Edition of *The Double Helix*, 241-47, and in Crown and Jenkins, *Conceptual Foundations of Genetics*, 52-55.

18. J. D. Watson and F. H. C. Crick, "The Structure of DNA," *Cold Spring Harbor Symposia on Quantitative Biology*, 18 (1953). 123-31; F. H. C. Crick and J. D. Watson, "The Complementary Structure of Deoxyribonucleic Acid," *Proceedings of the Royal Society*, A, 223 (1954). 80-96. Both papers are reprinted in the Norton Critical Edition of *The Double Helix*, 257-74 and 274-93.

19. F. H. C. Crick, "The Structure of the Hereditary Material," *Scientific American*, 191 (Oct. 1954), 54-61.

20. Frederick Suppe (ed.), *The Structure of Scientific Theories* (Urbana: University of Illinois Press, 1974), 125-220. In addition to Kuhn, Suppe identifies Stephen Toulmin, N. R. Hanson, Paul Feyerabend, Karl Popper, and David Bohm as proponents of *Weltanschauungen* views of science.

21. Ch. Perelman and L. Olbrechts-Tyteca, *The New Rhetoric: A Treatise on Argumentation*, trans. By John Wolkinson and Purcell Weaver (Notre Dame: University of Notre Dame Press, 1969), 67-70.

22. Graham Chedd, "Genetic Gibberish in the Code of Life," *Science 81*, 2 (Nov. 1981), 50-55.

23. Boyce Rensberger, "Tinkering with Life," *Science* 81, 2 (Nov. 1981), 47-48.

24. Thomas S. Kuhn, "Historical Structure of Scientific Discovery," *Science*, 136 (1 June 1962), 760.

25. Elizabeth L. Eisenstein, *The Printing Press as an Agent of Change: Communications and Cultural Transformation in Early-modern Europe* (Cambridge: Cambridge University Press, 1979), 119 ff.

26. Erwin Chargaff, "Building the Tower of Babel," *Nature*, 248 (April 26, 1974), 778. See also Chargaff's "A Quick Climb Up Mount Olympus: A Review of *The Double Helix*," *Science*, 159 (29 March 1968), 1448-49. According to Stent, Chargaff refused permission for this piece to be reprinted together with other reviews in the Norton Critical Edition of *The Double Helix* (p. 168).

27. Edwin Black, "A Note on Theory and Practice in Rhetorical Criticism," *Western Journal of Speech Communication*, 44 (fall 1980), 331-36.

Conflict in Science

Arguing in Different Forums:
The Bering Crossover Controversy

by Jeanne Fahnestock

"The initial entry of humans into the New World remains one of the major unsolved problems of American archaeology" (Fladmark 1983, 13). In this opening sentence of a review article, Knut R. Fladmark, an archaeologist of Simon Fraser University, at once reflects and perpetuates a polarizing controversy among American archaeologists about when and how the first humans migrated across the Bering Straits and dispersed through the Americas. Some favor a date no earlier than about 12,000 BP (before present) and others would push the date back much earlier. Proponents on both sides of the controversy have published scholarly pieces in specialist journals, review articles that address broader audiences in the discipline, and popular pieces including those that appeared recently in the magazine *Natural History*. Thus a number of arguments exist written by the same authors on the same subjects, but intended for the different audiences reached by different publications. As such, articles on this controversy constitute something of a test case for the rhetorician interested in how written arguments vary according to the audience addressed.

An analysis of texts from a rhetorical perspective asks what tactics and topics of argumentation are used, and how the arguments are arranged sequentially as a series of effects, and how they are actually expressed, their precise wording, their qualifications (or lack thereof), their indirection, their use of figures (tropes and schemes). The rhetorician is primarily interested in explaining textual features as an arguer's creative response to the constraints of a particular situation. A thorough analysis of even one text would be considerably longer than that text itself. In this article I can only mark some of the devices of argumentation used in the review articles and the popular pieces. The preliminary observations offered here suggest that, when arguing in popular forums, scientists use appeals that are rarely made explicit in scholarly arguments addressed to colleagues.

The Rhetoric of the Reviews

The debate over early or more recent migration into the New World has persisted for at least thirty years if not longer. The perpetuation of the controversy is perhaps in part due to the special nature of archaeology as a discipline. Evidence in archaeology is not like evidence in experimental sciences, for luck plays an enormous part in whether there is any evidence at all. Researchers cannot easily design experiments and thereby will data into existence the way particle physicists or molecular biologists can. Though archaeologists can investigate likely sites, they cannot control what turns up. The Meadowcroft Rockshelter, frequently cited as the best source of evidence of a pre-12,000 BP human presence (Shutler 1983, 11), was initially investigated simply to give archaeology students at the nearby University of Pittsburgh a promising place to practice their craft (Adovasio and Carlisle 1986, 20). In addition to the fortuitous nature of the data, deep disagreement exists about what actually constitutes evidence of human occupation. In a review of the case for an early human presence in the far west, Roy L. Carson (1983, 73) summarized "those problems that plague archaeological interpretation everywhere." "Disputes arise in three areas again and again: artifact recognition; validity of particular dates and dating techniques; and degree or type of association of dated material with manmade materials, or of the latter with particular geomorphic units." The persistence of disputes about what constitutes evidence and what can be inferred from it has also forced many archaeologists to become aware of modes of arguing in their field. "Lacking a symbolic language such as mathematics," archaeologists "employ a mixture of natural-scientific and historical patterns of thought that seems to be unique to the discipline" (Dincauze 1984, 293, 290). Thus most arguments in the field cannot demonstrate certainty but only establish some degree of probability, a standard with which many in the field seem, perhaps unreasonably, uncomfortable, because it always leaves room for disagreement.

Given all these potential sources of disagreement and the impossibility of reaching certainty in specific cases, arguers on both sides of the controversy use appeals based on the status and alignments of advocates and opponents, the ethos and pathos identified by Aristotle 2500 years ago and elaborated by many rhetoricians over the centuries. Such appeals appear frequently in review articles whose purpose is to summarize and weigh the evidence for a claim or set of claims. For example, one tactic that an arguer can use to dissociate sides in a controversy is to create or suggest the existence of a new field or specialty that one side is insufficiently aware of. The early daters have claimed special knowledge of "taphonomy, the science of embedding," "the study of all factors affecting individual animal remains and assemblages from the time of death until the moment of discovery and collection" (Morlan and Cinq-Mars 1982, 356). Presumably this special knowledge warrants inferences that those without it cannot make.

Another tactic of opponent labeling used in the dating controversy seems at first somewhat surprising. Each side has cast the other as the majority opinion, the establishment, the prevailing view, evidently trying to gain for themselves the status of being in the minority. According to one late dater, "Despite the unformalized state of the data that form the basis for claims of human occupancy in the Americas prior to about 12,000 years ago, and despite the frequency of criticism of particular cases, there seems to be a covert consensus that such occupancy is an established fact" (Dincauze 1984, 276), and "the strength of numbers . . . is with the proponents [of early dating]" (282, 293). Another writes, "The vast majority of archaeologists, if they do not believe the point yet proven, accord a high probability to the presence of humans in the Americas during Wisconsinan times and perhaps earlier" (Owens 1984, 521). Meanwhile the early daters claim the opposite: "Then [1967], as today, archaeologists were more comfortable with artifacts dated no more than about 12,000 years old" (Irving 1987, 10); "most of the field of American archaeology has in this period clung to a 12,000 year first entry date" (Carter 1980, ix). These characterizations are not based on surveys of opinion in the field. They are impressions that serve a rhetorical purpose. Illustrating this point further is one collection of articles that does include papers by both early and late daters. In the introduction to a set of articles in this collection, the two sides are characterized as "at least equally numerous" (Hopkins 1982, 327).

Along with this characterization of their own position as in the minority comes the collateral move to depict their view as one on a growth curve gaining desirable adherents while their opposition declines in numbers. One early dater appeals to "a younger generation, better trained in lithic technology, [who] can recognize the early lithic industries" (Carter 1980, x). Another writes that "a well-established 'school' of archaeological thought exists whose dwindling number of proponents firmly and sometimes dogmatically adhere to the notion that human beings did not enter the New World from Asia prior to the end of the Wisconsinan glaciation" (Adovasio et al. 1983, 171). Meanwhile a late dater claims that "more recently, other Americanist archaeologists have come to question the validity of a pre-Clovis human presence" and claims that his side is attracting a "growing number" (Owen 1984, 523, 553). In roughly the same time period, each side claims to attract more adherents, especially among the young and sophisticated, the most prized group of advocates.

One would think that it would be more desirable to depict a position as favored by the majority in a field given that the goal in any science is the agreement of all one's peers to the unrefutable self-evidence of the arguments presented for their assent. But in this controversy, invoking the ethos of the minority seems more advantageous. One explanation of this tactic may be the high value placed on apparent newness and progress in scientific thought. Although it seems like the height of foolishness to claim that most people in your field disagree with you, it is obviously a way to represent your opponents as stodgy and your own view as progressive.

Another appeal found in these arguments reveals that contemporary archaeologists have, perhaps unfortunately, read *The Structure of Scientific Revolutions* (Kuhn 1970). Both sides cite the notion of a "paradigm shift" to explain their endeavors. One early dater says that fields run in ruts, and that change comes like "a train jumping the tracks," praising Kuhn for catching the essence of this pattern and, of course, implying that all who represent the opposing point are in the rut (Carter 1980, xii, 317). And a late dater has written a review article in which he divides the field of early American archaeology into four paradigms, because, as he explains, "the controversy as to who and when were the first Americans has reached what Kuhn has called the 'paradigm debate' state in the development of a science" (Owen 1984, 549). He then lists the assumptions required from the believer in each camp because he wishes to demonstrate that only the favored 12,000 BP date does not demand ludicrous leaps of faith from its adherents. (He specifically invokes the value of "scientific parsimony and probability.") It is possible that the desire to invoke the Kuhnian model is behind the attempts of both sides, mentioned above, to characterize themselves in the minority. Nevertheless, when arguers on both sides of a scientific controversy claim to be the movers and shakers in a paradigm shift, and use that explicit appeal as persuasive in and of itself in review articles, their discipline has reached a high degree of self-consciousness.

When the evidence is uncompelling on an issue, the extent to which an individual researcher commits him- or herself to an incomplete case and thus makes an implicit appeal to his or her status can be striking. Instances of such commitment come from the collection *Early Man in the New World*, edited by a pre-Clovis proponent and filled primarily with papers supporting that view, though not with equal stridency. The papers were originally presented at a 1981 symposium held during the meeting of the Society for American Archaeology. R. S. MacNeish was given the task of summarizing progress over the last ten years in establishing and characterizing the presence of early humans in Mesoamerica. Although MacNeish carefully hedges claims and acknowledges the lack or incompleteness of evidence or corroboration, he nevertheless personally testifies to the ultimate outcome: "Once the lower levels of this cave are dug again with the same techniques I used in my initial testings . . . I'm sure we will have a more adequately documented example of stage II" (1983, 127). "While none of these finds have yielded large inventories, and though the stage is ill defined, there are a number of sites which, upon excavation, will give us better samples of artifacts and ecofacts so that the whole matter may be settled shortly to everyone's satisfaction" (127). Elsewhere MacNeish uses popular maxims, "Who knows what tomorrow will bring?" (128) and "thus, the best is yet to come from people digging in the northwest of Mesoamerica" (135). Of course, the eternal possibility of undiscovered evidence in archaeology makes such appeals possible, but their persuasiveness also depends on the reader's understanding of the status of the speaker, a status nowhere explicitly claimed in the article.

The author who was given the privilege of writing the final summary of *Early Man in the New World*, H. Marie Wormington, opens with a distinctly dour assessment of the pre-Clovis position:

Had I been asked a decade ago what I thought the principal change in Early Man studies would be by 1980, I would have said: "By then we will have conclusive evidence of the presence of human beings for considerably more than 12,000 years." I would have been wrong. We still lack universally acceptable proof of occupation before Clovis times (Wormington 1983, 191).

In the absence of such universally acceptable evidence, workers in the field, and this writer in particular, invoke broad alignments shaped not by what they find more credible but by what they find less incredible. In other words, they justify their allegiance to an inadequately supported view on the grounds of avoiding a worse alternative. This position in a controversy creates another set of potential "lesser of two evils" arguments. Wormington cites the apparently early dates of sites in Central and South America (all challenged by late daters) and the improbability of humans migrating the length of two continents within 500 to 1,000 years. "I cannot believe," Wormington (1983, 192) writes, invoking sarcasm with a *reductio ad absurdum*, "that the Paleoindians rushed off to the south singing 'Patagonia, here we come.'" Throughout this assessment, Wormington uses statements of personal doubt and belief: "My feeling is," "I cannot believe," "I have great faith." The review ends by invoking maxims of reassurance, "the unknown far exceeds the known," and, like MacNeish, "the best is yet to come." These are followed by a statement little short of the words of a creed: "This I believe implicitly" (195). The arguer who makes strong statements of personal belief relies greatly on ethos to carry the day. One might even identify an ethical *a fortiori* argument here: if I can believe then so should you. Because Hannah Wormington has had a long and distinguished career in American archaeology, evidenced in the introduction and in the dedication of the volume to her, it seems likely that her ethical appeal is warranted for the particular audience addressed by the collection.

Not all the authors participating in the controversy have been as frank as Wormington in delineating which claims they hold by evidence and which by conviction despite inadequate support. One of the most forceful of the pre-Clovisites has been Richard E. Morlan, an archaeologist with the National Museum of Man in Canada. His career has been devoted to northern Yukon finds, a staple of the pre-Clovisites and a frequent target of the late daters. The most controversial aspect of these finds has been the status of certain fractured and fragmented bones of mammoth extinct for 11,000 years. The early daters claim that these are part of a human tool kit; the late daters, that these are artifactual. In fact late daters have coined a word for these putative pieces of evidence; they are "geofacts" not "artifacts."

The debate over these finds has led to one of the rare attempts to perform a contemporary experiment that will contribute to settling an issue in archaeology. Early daters have argued that the fractures on the mammoth bones could only have occurred when the bones were "green" or fresh and that further they could only have been produced by such concentrated and unusually applied pressure that they must have been made intentionally by humans. In the famous "Ginsberg" experiment (which did make the newspapers), two archaeologists butchered a zoo elephant, killed accidentally, and attempted to reproduce the critical fractures on its bones by primitive means. Their success was not complete enough to change anyone's mind, though it was sufficient to strengthen the already predisposed. In the field as a whole, however, the potential cause of a certain variety of mammoth bone fracture and thus the status of these bone fragments as artifacts or geofacts remain highly debatable.

But that debatable status is not evident in Morlan's (1983) review article on "Pre-Clovis Occupation North of the Ice Sheets." Latour and Woolgar (1979) and Myers (1985) have pointed out that scientists negotiate the status of their claims in the review process. In general they press along a scale toward fact status, trying to avoid mitigation and hedging in the endeavor to increase the certainty of their claims. In writing for progressively larger mixed audiences, however, they are proportionately less accountable for the status of their claims. So it is not surprising that things seem far more certain when they are summarized in review articles.

Morlan makes the human agency in the creation of the mammoth bone fragments a matter of fact by casually referring to it in subordinate clauses: "Among the vertebrate fossils are altered bones, antlers, and tusk fragments, some of which were artificially worked when in fresh condition" (1983, 52); and "since most of the artificially altered vertebrate fossils have been recovered from the modern banks and bases of the Old Crow River Valley"(54). An attribution that can be left for the aside of a relative or subordinate clause is well on its way to being one of the unquestioned assumptions of a field. Morlan also exploits the inherent ambiguity of the word *flake*, in the phrases "cores and flakes of mammoth bone," and "mammoth bone flake" (54). The term *flake* can refer to a slice of stone or bone removed from another piece either intentionally or unintentionally, but in the context of scholarly archaeology, it is usually applied to a deliberately used tool.

Curiously, at the end of his article assessing the evidence for early human occupation in the northern Yukon, Morlan suddenly acknowledges that "all Old Crow bone artifacts in the following discussion [sic] should be read with their hypothetical status in mind. The interpretation of artificially fractured and flaked mammoth bones from redeposited context will always be hypothetical" (Morlan 1983, 61). It seems quite likely that this caveat was a response to a reviewer's criticism of the unmitigated and indeed offhand manner in which Morlan transformed the debatable into the factual. Needless to say, this disclaimer is not worded so as to undo what has been done; the reference should point to the previous presentation of evidence, not to the following discussion. Furthermore, Morlan has made what we might call a Freudian grammatical slip when he speaks of "the interpretation of *artificially* fractured and flaked mammoth bones" instead of "the interpretation of fractured and flaked mammoth bones *as artificial*."

In general the crossover controversy has proceeded by the early daters finding evidence and the late daters debunking it. The appeal to quantity inherent in this process is definitely on the side of the pre-Clovisites. The more sites and bits of fossils the early daters claim to find, the more convincing their *prima facie* case.

And the more the late daters have to deny evidentiary status to the detritus of human occupation in very old layers or the odd bones from a redeposition, the more curmudgeonly they appear in their dismissals, the more hollow their repeated appeals to "site contamination," "insufficient stratigraphic data," or "awaiting further tests." In other words, there is a built-in rhetorical disadvantage to one-by-one refutation.

To obviate this one-by-one refutation, the late daters have tried two strategies. First they have proposed four criteria that a site must fulfill simultaneously in order to be accorded status as evidence (Griffin 1979, 44). No pre-Clovis site fits all four, so not surprisingly these standards have come under attack by early daters as arbitrary. Second the late daters have carried out their own fresh experiment to once and for all debunk their opponents. In the so-called Nebraska experiments, a group of paleontologists excavated several sites known to be several million years old where no one claimed there could be human habitation. They found broken bones resembling the supposed bone artifacts recovered elsewhere (Myers et al. 1980). Late daters can now potentially refute any pre-Clovis evidence by claiming that there is a certain "background noise" (their metaphor) in the earth that their adversaries continually misread as signs of human agency (Owen 1984, 538).

In 1985 the early daters were dealt a telling blow and, were it not for the recency of the event, a swing of the uncommitted to the late daters might be obvious. The early daters' prized piece of evidence, obtained from the Old Crow site in the Yukon, was a caribou tibia unmistakably carved by human hand into a tool for scraping hides (see also below). The first carbon-14 dating on this implement suggested that it was 27,000 years old, a stunning anomaly when first reported (Irving and Harington 1973). This date, however, was obtained using the inorganic or apatite fraction of the bone, a technique now believed unreliable. New radiocarbon dating using organic carbon from the collagen fraction of the bone revealed the flesher to be only about 1300 years old; the original date was 26,000 years off. Thus the early daters lost the only piece of evidence that was both indisputably more than 12,000 years old and indisputably a human artifact. An article reporting the redating appeared in *Science* in May 1986 but the manuscript was submitted in October 1985 and its results were probably widely known for months before publication.

The Rhetoric of Popularization

It is somewhat surprising then that, in the face of this challenge to their views, the early daters apparently launched an extensive attempt to argue their case in public in a series of articles that began in November 1986 in the magazine *Natural History*, a publication of the American Museum of Natural History in New York. The Editor's Note opening the series speaks of "tantalizing data from scattered sites [which] suggest the New World was peopled earlier [than 12,000 years ago]," and of the anticipation of "a turning point in archaeological opinion." Most of the articles give support to the early daters' position by discussing evidence of earlier openings in the glaciers, experiments in bone percussion, and excavations of sites presumably demonstrating very early human occupation in Pennsylvania, Venezuela, and Brazil. The articles are written in variable styles, but they generally conform to what one would expect of occasional pieces in a glossy magazine. Still they are not incidental. Written by professional archaeologists, not popularizers, they argue cases. But in the forum of a widely circulated magazine intended for nonspecialist readers, these articles use supporting appeals not usually found in research reports or review articles.

To cite two examples for illustration: first, personal references do appear in scholarly reports and reviews, but in a popular forum, the personal authority and accomplishments of an individual researcher, and hence that researcher's ethos, can be persuasively heightened by personal attribution. In a review article, anthropologist C. G. Turner, who specializes in teeth, could support the evidential basis of a generalization by citing in a nonrestictive participial phrase buried in the middle of a sentence "4,000 personally studied individuals" (Turner 1983, 148). In his *Natural History* article, Turner can claim boldly in the opening of a paragraph, "Over the past twenty-five years, I have studied more than 200,000 prehistoric teeth from the New World, representing the remains of some 9,000 individuals" (Turner 1987, 6). Second, scholarly formats generally dictate the passive voice, diminishing a grammatical role for human agents in sentences. In contrast, the journalistic style of a magazine encourages the active voice, and thus literally demands the creation of agents as grammatical subjects in order to achieve a lively style. In the *Natural History* account of his research on a South American site called Monte Verde, Tom Dillehay concludes with reservations about whether the site gives evidence of a human presence. But earlier in the piece, the demands of active voice bring into existence the "Monte Verdeans" who "foraged in thirteen different ecological zones," and who thus become real by virtue of their role as grammatical subjects.

Each of the articles in the *Natural History* series repays close scrutiny, but the two articles written by important figures on each side of the debate deserve the most attention for the rhetorical skill of their authors and the exploitation of appeals available in popularizations.

The most outspoken of the articles in favor of the pre-Clovis thesis was written early in the series by William N. Irving, a professor of anthropology at the University of Toronto. Irving was present at the original discovery of the mammoth bones and flesher in the Yukon in the late 1960s and he devoted much of his career to the study of these finds and the northern Yukon sites. How does Irving handle the refutation of the dating of the flesher published in *Science* in 1986? Because he was perfectly aware when he wrote the *Natural History* article that in a sense his life's work had been challenged, he has serious repair work to do, and it is perhaps strange that he chooses to do it in the format of a nonspecialist's publication. But publication in specialist journals is essentially closed to him unless he has new data to present, and the occasions for review articles are infrequent. Irving has the quickest access to the largest audience in a popular forum, though precisely what he can accomplish in such a forum is unclear.

We can appreciate Irving's skill as a rhetorician if we consider his dilemma fully. The majority of his *Natural History* audience knows nothing of the dating controversies in which he has played so prominent a part; furthermore, their purpose for reading is entertainment or a kind of scientific voyeurism. Yet Irving can also predict that old partners in contention will read his column, readers who are aware of the current state of the evidence. Irving's solution to the dilemma of disparate audiences is to adopt a narrative structure and to give a strict chronological accounting of research events and the process of investigation. Such a historical narrative cannot be faulted as inaccurate by critics; at the same time it is inherently appealing to lay readers and is in fact a frequent choice in popularizations. Furthermore, the narrative approach places Irving's argument in a well-known rhetorical genre: the apologia. His article is more than just support for the earlier crossover date; it is also in some ways a defense of his entire career. In this defense Irving can construct a persuasive ethos for himself as an archaeologist and an arguer. The appeal from the status of the speaker, implicit in the review articles, is in the popularization made explicit.

Irving's narrative account emphasizes two things: the slowness with which he came to believe in an early date and the careful and rational process of investigation that he and his colleagues conducted. "As Harington unwrapped his specimens," he writes, "we were very skeptical," "we puzzled over the flesher and the fragments of broken mammal bone found with it." "Of course, there were other possible explanations to consider." "We realized how much more work would be needed . . . if we were ever to answer the most important questions" (1987, 8, 10). The appeal behind such phrases could be paraphrased as something like the following: "If I, who know so much more, had my significant skepticism overcome by the evidence, so should you."

The narration of the careful research process gives the argument chronological structure as well as persuasiveness. The reader learns of the thought processes of Harington and Irving back in Ottawa after the field season of 1966, their sending samples to the British Museums for uranium and fluorine testing that proved "inconclusive"; an account of the following summer's fieldwork when they learned that the deposit from which they took the flesher represented a jumble of redeposited material, their "next recourse" of sending bones and flesher samples for radiocarbon dating—with, of course, the surprising result of a date "about 27,000 years old." The article goes on to tell the reader what happened in 1972, 1975, and 1977. The impression created is one of years of patient work and accumulating evidence. In the world of ordinary rhetorical arguments, the amount of time and effort that reputable people place in an inquiry is persuasive in and of itself.

The public conceives of scientists as workers who painstakingly accumulate data and pile up inference, but it also, and perhaps more prevalently, imagines scientists as thinkers who have immediate, stunning insights or "Eureka events." Irving (1987,

10) includes one. According to his account, "In 1972, I made one of many trips to visit colleagues and solicit their opinions on the nature and purpose of the Old Crow bone-breaking activity." (The last phrase reifies hypothesis into fact; the whole sentence characterizes the activity of scientists in a curious way, as though they were Laputans who go around with evidence in sacks slung over their backs asking colleagues what to make of it, instead of as arguers who have to submit claims about their evidence, no matter how tentative and hedged. Irving's version, though ingenious, seems somewhat disingenuous; if he didn't know what to make of the evidence, why did he consider it important at all?) While at the University of Alberta, probably giving an invited seminar, Irving encountered a graduate student who had been studying Paleoindian stone tools and the "uses of animal bone by modern Indian hunters." This graduate student "took about a minute to appraise the situation and deliver an opinion: the pieces of mammoth bone that I had presented for his mystification were evidence of the use of percussion to produce cutting, scraping, and other tools from fragments of large mammal bones" (Irving 1987, 10). This immediate interpretation is achieved when the prepared mind encounters the evidence, and its very suddenness is convincing. Irving has achieved a wonderful dissociation in his article between two styles of discovery and conviction, which both nevertheless yield the same conclusion.

But what about the redating? Irving (1987, 13) in a sense dismisses the flesher; he even makes it disappear: "most of the flesher has been sacrificed over the years to accommodate all the tests." Instead, Irving puts all his chips on the chips, as it were, repeating the point made so casually by Morlan. The stakes are very high for this interpretation, for these curiously fractured mammoth bones have survived the redating and emerged as genuinely earlier than 12,000 BP. Basically Irving uses two lines of argument: that nothing else could have produced these bone artifacts and that similar bone artifacts are to be found elsewhere, an argument that presupposes that such bone fragments are indeed artifacts. He also suggests a new field of expertise in "paralithic technology" for the special study of these implements (1987, 13).

Essentially the pre-Clovis argument as Irving expresses it has successfully maneuvered itself into the position of being unprovable but unrefutable. It has and presumably will continue to maintain that the oddly fractured mammoth bones are more likely to have been created by human intervention than by any force in nature, whether predators, river ice, or erosion. No one can prove that the bones were altered by humans, but no one can disprove it either. Thus believing in the pre-Clovis occupation of the New World begins to look like a matter of personal allegiance or a leap of faith.

In the October 1987 *Natural History*, rebuttal came in the twelfth article in the series from Paul S. Martin, a professor of geosciences at the University of Arizona, long a vocal opponent of the early daters (Owen 1984, 552). It is very likely that Martin volunteered his remarks to *Natural History*, convincing the editors that they had to present the other side in the controversy. Writing in refutation of someone else's work, Martin cannot use Irving's narrative technique nor Irving's defense based on his ethos as a careful scholar. But Martin is also a master of the genre of the magazine science article, a skillful rhetorician who understands in particular how to manipulate the persuasive technique of *copia* to which Erasmus devoted an entire treatise. The term *copia* refers generally to fullness or amplification, the opposite of brevity, but it can also be applied to the technique of enumeration or listing, creating a series that suggests a large number of things, too many for the writer or speaker to specify. This stylistic device is especially useful when the arguer's purpose is to appeal to quantity, the presumption that more of something has greater value and significance than less of it (Perelman and Olbrecht-Tyteca 1969, 85-89). Martin wants to suggest that the preponderance of evidence, the sheer quantity of it, favors a more recent peopling of the New World, and wants to demonstrate a catastrophic loss of evidence on the other side to prevent their making the same appeal.

> While a number of sites described as older than 12,000 years have been presented in this series, I and many others [another appeal to quantity] view the latest finds as less than conclusive. The reason is our sense of *deja vu*. In the last thirty years, various discoveries have been reported to be 15,000, 30,000 and even 200,000 years old. One after another of these once-popular claims, however, have crumbled when subjected to modern techniques of geological and archaeological analyses. For example, the Koch mastodon site in the Missouri; Sandia points from Sandia Cave in New Mexico; the Holly Oak pendant from Delaware; Tule Springs, Nevada; Smith Creek Cave, Nevada; Lewisville, Texas; and Gypsum Cave, Nevada—all once widely touted—have been reinvestigated by a new generation of scientists. The results have not yielded any confirmations (Martin 1987, 13).

There are many persuasive techniques at work in this paragraph, but the list of examples exemplifies the technique of copia perfectly. The first three items in the series are fuller, specifying the artifact as well as the location. From then on only sites are named. The impression created is one on increasing quickness as though the arguer were speedily recalling instances from a vast store. The speed creates the impression that the author could name many more sites if either he or the audience had the time. That is precisely the impression copia is supposed to create.

Copia is the opposite of the related figure *climax*. Employing climax, the writer also creates a series, but the purpose of this series is to build to a crescendo so that enormous emphasis falls on the last item. In copia, on the other hand, the series is to diminish hastily; the last item receives little attention, presumably because it is just one of many that the writer could have recalled in a miscellaneous catalogue; it is only accidentally the last mentioned.

Martin is so fond of copia that he uses it frequently in his article, not as a mere stylistic excrescence, the *sprezzatura* of eloquence, but as an integral technique of persuasion in an argument that uses quantity to refute anomaly. Martin's first paragraph consists of three sentences that are virtually just lists of living large animals, extinct mammals, and then predators. Three of the four sentences in the second paragraph also end in lists like the following: "If, as I believe, the first Americans were foragers who mainly hunted game, they would have been drawn to the habitats of these animals—to grasslands or woodlands adjoining flood plains, to mineral springs, lake shores and coastal marshes. Large herbivores frequented these places because they provided edible foliage, fruits, seeds, roots and tubers" (Martin 1987, 10). (The diminished parallelism of prepositional phrases in the first sentence, its covering two related items and then three related items, is the tell-tale sign of copia in this passage.) The third paragraph provides the reader with lists of "sites famous for their late Pleistocene bones" grouped in categories of "such fresh-water and mineral springs," "coastal deposits such as," "numerous alluvial deposits such as" "various damp caves," and "an assortment of dry caves and sink holes" (10). Each of these categories is followed by one or two examples giving the impression of a series within a series, and again the overwhelming effect is that of copia, of an extensive set of evidence or details from which a number have been randomly selected for mention. All of these lists precede the short last sentence of the third paragraph (a place reserved for maximum emphasis, and also the place where expectation tells most readers they will find a thesis): "Yet at none of these sites do human remains or artifacts turn up until after 12,000 years" (10). Strictly speaking this dramatic sentence follows only the lists of sites just given in the third paragraph. But the reader has been conditioned by three paragraphs of copia into thinking that there are vast arrays of all kinds of things and that somehow all this multiplicity of stuff mitigates claims for a migration earlier than 12,000 years ago.

Martin also uses copia to set the amount of evidence of the Stone Age peopling of the Old World against the claims for the New World. As he puts it with no fear of overstatement: "The difference between the Old World and the New in the quantity of such artifacts deposited before 12,000 years ago is awesome" (12). In a master stroke of accommodation to his *Natural History* audience, Martin catalogues some of the exhibits of European Stone Age artifacts displayed in an American Museum of Natural History exhibition the previous year. "If," he asks, "Paleolithic hunters entered America long before 12,000 years ago, why didn't they leave us similar trophies of equivalent age?" (12). In the words of the rhetorical figure *chiasmus*, the absence of evidence is evidence of absence.

After Martin's rebuttal, no articles in the "First Americans" series appeared for two months. Then the January 1988 issue carried an article on Siberian sites by a Russian archaeologist. The following month *Natural History* published "Geofacts and Fancy," a homily against finding what you want to find, by C. Vance Haynes, Jr., long an outspoken critic of early daters' claims. An editorial note embedded in the article proclaimed, "This is the concluding essay in the series exploring archaeological sites and other lines of evidence bearing on the peopling of the New World. *Natural History* will continue to explore this controversial subject on an occasional basis" (Haynes 1988, 4). Perhaps the prestigious magazine repented of its decision to air the views of the early daters and, out of a total of fourteen articles, allowed the late daters to have the last word—twice. It is difficult to know what effect, if any, the series has had on the magazine's lay readers who have no expertise or need to take sides in the disagreement. Some may believe vaguely that humans have been in North and South America rather a long time; they may be more receptive to any further evidence of an early human presence.

It is doubtful that any specialists involved in the controversy have had their minds changed by the *Natural History* articles. If anything, the two sides may have become more polarized as a result of this public airing of their differences. The persistence of the controversy has much to do, as noted above, with the open-ended nature of research in archaeology, the ever-present possibility of new discoveries that postpones closure on an issue. W. N. Irving, for example, has gone to Siberia to find datable evidence of the people who migrated from there.

Two other explanations can be offered for the stalemate. First, though the public's "romance" with archaeology of all types is perennial, there is little social pressure on or public need for archaeologists to resolve their differences in order to go on to a next step or to bring a line of speculation to fruition. In rhetorical terms, there is little "exigence" or push, either from within or without the profession, to settle the matter (Bitzer 1971, 385-87). Thus Dincauze (1984, 290) complains in a review article that the controversy "roils at the level of evidence and argument, with few of the contested cases settled one way or another. It is worth noticing that while some few of the cases have been definitively retired from serious contention (see below), none has been wholly vindicated at the level of scientifically acceptable demonstration. Most simply remain in the literature, cited by some, ignored by others."

Dincauze's lament suggests a second explanation. The true exigence may favor not agreement but continuation of the controversy. Though no archaeologists would acknowledge or even feel this motive, it is nevertheless true that part of the real work of any discipline is its own perpetuation. For archaeology that perpetuation, which may involve mounting expeditions to remote sites, can be very costly. An unresolved question hanging over a discipline justifies further research, further funding, and hence continuation of the field. The conclusion of Dennis Stanford's (1983, 72) review of evidence for "Pre-Clovis Occupation South of the Ice Sheets" makes the case perfectly.

> We can find fault with the interpretation of all of these sites, and we cannot as yet decisively push back the time of human occupation beyond the 12,000 year B.P. time marker in North America south of the ice sheets. Collectively, there appears to be enough evidence to encourage continu-

ation of the search for a pre-Clovis occupation in this area, although perhaps the materials in question are only "background noise" created naturally during and after the final millennia of the Pleistocene. Only continued research will resolve the problem.

Works Cited

Adovasio, J. M., and R. C. Carlisle. 1986. "Pennsylvania Pioneers." *Natural History* 95 (December): 20-27.

Adovasio, J. M., J. Donahue, K. Cushman, R. C. Carlisle, R. Stuckenrath, J. D. Gunn, and W. C. Johnson. 1983. "Evidence from Meadowcroft Rockshelter." In *Early Man in the New World*, ed. R. Shutler, Jr., 163-189. Beverly Hills, CA: Sage.

Aristotle. 1954. *The Rhetoric and Poetics of Aristotle*. With an introduction by E.P.J. Corbett. New York: Modern Library.

Bitzer, L. 1971. "The Rhetorical Situation." In *Contemporary Theories of Rhetoric: Selected Readings,* ed. R. L. Johannesen, 381-93. New York: Harper & Row.

Carlson, R. L. 1983. "The Far West." In *Early Man in the New World*, ed. R. Shutler, Jr., 73-96. Beverly Hills, CA: Sage.

Carter, G. 1980. *Earlier than You Think: A Personal View of Man in America*. College Station: Texas A & M Univ. Press.

Dillehay, T. D. 1987. "By the Banks of the Chinchihuapi." *Natural History* 96 (April): 8-12.

Dincauze, D. F. 1984. "An Arcaeo-logical Evaluation of the Case for Pre-Clovis Occupations." Vol. 3, *Advances in world archaeology,* 275-323. New York: Academic Press.

Erasmus. 1963. *On Copia of Words and Ideas*. Trans. D. B. King and H. D. Rix. Milwaukee, WI: Marquette Univ. Press.

Fladmark, K. R. 1983. "Times and Places: Environmental Correlates of Mid-to-Late Wisconsinan Human Population Expansion in North American." In *Early Man in the New World*, ed. R. Shutler, Jr., 13-41. Beverly Hills, CA: Sage.

Griffin, J. B. 1979. "The Origin and Dispersion of American Indians in North America." In *The First Americans: Origins, affinities, and adaptations,* ed. W. S. Laughlin and A. B. Harper, 43-55. New York: G. Fisher.

Haynes, C. V., Jr. 1988. "Geofacts and Fancy." *Natural History* 97 (February): 4-12.

Hopkins, David M. 1982. "Man in Ancient Beringia." In *Paleoecology of Beringia*, ed. D. M. Hopkins, J. V. Matthews, Jr., C. E. Schweger, and S. B. Young, 327-28. New York: Academic Press.

Irving, W. N. 1987. "New Dates from Old Bones." *Natural History* 96 (February): 8-13.

___, and C. R. Harington. 1973. "Upper Pleistocene Radiocarbon-dated Artifacts from the Northern Yukon." *Science* 179:335-40.

Kuhn, T. S. 1970. *The Structure of Scientific Revolutions*. 2nd ed. Chicago: University of Chicago Press.

Latour, B., and S. Woolgar. 1979. *Laboratory Life: The Social Construction of Scientific Facts*. Beverly Hills, CA: Sage.

MacNeish, R. S. 1983. "Mesoamerica". In *Early Man in the New World*, ed. R. Shutler, Jr., 123-35. Beverly Hills, CA: Sage.

Martin, P. S. 1987. "Clovisia the Beautiful!" *Natural History* 96 (October): 10-13.

Morlan, R. E. 1983. "Pre-Clovis Occupation North of the Ice-sheets." In *Early Man in the New World*, ed. R. Shutler, Jr., 47-63. Beverly Hills, CA: Sage.

___, and J. Cinq-Mars. 1982. "Ancient Beringians: Human Occupation in the late Pleistocene of Alaska and the Yukon Territory." In *Paleoecology of Beringia*, ed. D. M. Hopkins, J. V. Matthews, Jr., C. E. Schweger, and S. B. Young, 353-81. New York: Academic Press.

Myers, G. 1985. "The Social Construction of Two Biologists' Proposals." *Written Communication* 3:219-45.

Myers, T. P., M. R. Voorhies, and R. G. Corner. 1980. "Spiral Fractures and Bone Pseudotools at Paleontological Sites." *American Antiquity* 45:483-90.

Nelson, D. E., R. E. Morlan, J. S. Vogel, J. R. Southon, and C. R. Harington. 1986. "New Dates on Northern Yukon Artifacts: Holocene not Upper Pleistocene." *Science* 232:749-51.

Owens, R. C. 1984. "The Americas: The Case Against an Ice-Age Human Population." In *The Origins of Modern Humans: A World Survey of the Fossil Evidence*, 517-63. New York: Alan R. Liss.

Perelman, C., and L. Olbrechts-Tyteca. 1969. *The New Rhetoric: A Treatise on Argument.* Notre Dame, IN: Univ. of Notre Dame Press.

Shutler, R., Jr., ed. 1983. *Early Man in the New World.* Beverly Hills, CA: Sage.

Stanford, D. 1983. "Pre-Clovis Occupation South of the Ice Sheets." In *Early Man in the New World*, ed. R. Shutler, Jr., 65-72. Beverly Hills, CA: Sage.

Turner, C. G., II. 1983. "Dental Evidence for the Peopling of the Americas." In *Early Man in the New World*, ed. R. Shutler, Jr., 147-57. Beverly Hills, CA: Sage.

___. 1987. "Telltale teeth." *Natural History* 96 (January):6-10.

Wormington, H. M. 1983. "Early Man in the New World: 1970-1980." In *Early Man in the New World*, ed. R. Shutler, Jr., 191-95. Beverly Hills, CA: Sage.

"Punctuated Equilibria": Rhetorical Dynamics of a Scientific Controversy

by John Lyne and Henry F. Howe

Reviewing some recent work on rhetoric and science, Trevor Melia drew the following conclusion:

> In their disposition of philosophical issues, Kline, Munévar, and Weimer establish the *possibility*, in the most fundamental sense, for a rhetoric of science. Along with Kuhn, Feyerabend, Hanson, Polanyi, Bohm, et. al., they breach the once impenetrable wall of hard science. Inside these walls lies *terra incognita* for rhetoric. And no amount of debate about the work of the philosophers of science, whatever its merit, will secure scientific territory for rhetoric.[1]

It is something of an injustice to early explorers of the rhetoric of science to call the hard sciences *terra incognita* for rhetoric, but there is little question that the territory remains inadequately mapped.[2] Melia is also correct in maintaining that the mapping cannot be done upon philosophical principles alone, although a philosophy, like a compass, may prove useful in providing an orientation.

The "rhetoric of science" is both old and new. It is old in the sense of antedating the positivistic model of science against which the above mentioned theorists are reacting.[3] In the context of that model's decline, however, it is newborn and still in a stage of asserting its own possibility, typically through philosophical relativism.[4] A mature rhetoric of science should tell us something more concrete about how scientists use language and authority to engage audiences and lure them into sharing their view of things. It should tell us something about the strategies and implications of that engagement, including how understanding is distorted as well as how it is enhanced. In short, it should highlight the ways that scientific discourses are relative to and compelling for audiences. One way to begin constructing a rhetoric of science, therefore, is to follow what happens when specific arguments go before different audiences. We will trace the dynamics of one such rhetorical process.

Most scientific debates rage within the pages of technical journals, never emerging into the light of public scrutiny. The "experts" and their "audiences" in such cases can, especially to an outsider, appear to be simply the same population.

69

The diversity of expert-audience relationships within "the scientific community" can thus be underestimated.[5] But when a scientific theory becomes news to the public, an obviously expanded and diverse readership forces to disclosure what had been the case all along—scientific discourses confront different audiences in different ways. In the case of a theory originating in paleontology called "punctuated equilibria," we have found a series of different audiences engaged, some within the scientific community, some without. And this, we believe, makes our case study a "representative anecdote" of the way science and rhetoric condition one another.

Readers of popular magazine articles with such titles as "Darwin's Mistake," "Darwin on the Run Again," and "Science Contra Darwin" might suspect that a major upheaval is wracking the scientific community.[6] The spectacle of evolution under attack is not new, of course. Darwin-baiting has long been a favored pastime of fundamentalist preachers, textbook critics, and other guardians of public morals. But such exercises, at least since Darwin's triumph in the last century, were generally perceived as rear guard actions from those who did not know or did not accept the authority of science. What is different in the current debate over "punctuated equilibria," or "jumps" in the fossil record, is that the public may perceive devastating attacks on Darwinism from within the scientific community. "The alert reader of scientific magazines," writes zoologist Richard Dawkins, "can hardly fail to be aware of a widespread lay rumour of something rotten in the state of Darwinism."[7]

This controversy, ultimately one over evolutionary theory, reveals the problematic relationship between public discourse and the discourses of science, a growing concern of rhetorical critics and theorists, and surely an issue for the rhetoric of science.[8] Certain features of this controversy complicate any picture of science versus non-science as two discursive worlds in conflict.

First, distinct fields of discourse complicate communication *within* the sciences. Because researchers within different specialities hold different assumptions, a scientist's venture beyond the strict confines of a research specialty will sometimes lead to misunderstanding.[9] In the case of punctuated equilibria, the experts in paleontology alternately play the role of authorities and naifs, as their respective audiences know either more or less than they do about aspects of the subject matters addressed. There is, therefore, no simple way for a curious public to demarcate expert from non-expert discourse.

Second, expertise functions as both a scientific and a rhetorical construct. Its impact is determined partly by user-audiences, not just by a scientific writer's position within a field of knowledge. Readers with different interpretive frameworks manipulate the same expertise differently in a process over which authors may lose all control. The authors of the theory of punctuated equilibria, for instance, ironically find their work, and more importantly, their authority, appropriated to the creationist cause.

Our account of this debate is a narrative, which begins with a single essay within the sovereignty of paleontology. The debate begun there would swell to a many-sided controversy in which the various participants—paleontologists, geneticists, creationists, and journalists—employ very different rhetorics in a competition for adherents.

The First Audience Engaged

Our story begins with a 1972 essay entitled "Punctuated Equilibria: An Alternative to Phyletic Gradualism."[10] Paleontologists Niles Eldredge, of the American Museum of Natural History, and Stephen Jay Gould, of Harvard, attempted to explain what paleontologists since before Darwin had considered a problem: the gaps in the fossil record. Darwin himself felt forced to explain why we should not expect a complete record of transition among life forms. A spotty fossil record, continually eroded and buried by natural processes, allows nature to cover its own tracks. In Darwin's metaphor:

> . . . I look at the natural geological record, as a history of the world imperfectly kept, and written in a changing dialect; of this history we possess the last volume alone, relating only to two or three countries. Of this volume, only here and there a short chapter has been preserved; and of each page, only here and there a few lines. Each word of the slowly-changing language, in which the history is supposed to be written, being more of less different in the interrupted succession of chapters, may represent the apparently abruptly changed forms of life, entombed in our consecutive, but widely separated, formations.[11]

In speaking of the "apparently abruptly changed forms of life," Darwin thought the abruptness only apparent. It was an essential tenet of his theory that nature does not make jumps (*"Natura non facit saltum"*). Indeed, so central to his theory was this premise, he thought, that any documented jump in organic evolution would, in modern Popperian terms, falsify his theory: "If it could be demonstrated that any complex organ existed, which could not possibly have been formed by numerous successive, slight modifications, my theory would absolutely break down."[12] To Darwin, the very notion of natural selection required gradualism. "Why should not Nature have taken a leap from structure to structure?" he asks the readers of *On the Origin of Species*. "On the theory of natural selection," he answers, "we can clearly understand why she should not; for natural selection can act only by taking advantage of slight successive variations; she can never take a leap, but must advance by the shortest and slowest steps."[13] All the while, Darwin noted the enormous difficulty in documenting these crucial intermediate steps in the fossil record. This was one difficulty, he wrote, that "pressed so hard" on his theory.[14]

Darwin knew neither the mechanisms of inheritance not the time scale implied by the fossil record.[15] "Slight successive variations" are now the domain of geneticists who count generations rather than eons. Ten thousand or one hundred thousand years, while "instants" to geologists accustomed to musing about millions or hundreds of millions of years, are very long periods of genetic time. Virtually any process of interest to a geneticist, including "slight successive changes" or the formation of new species, could fit nicely within a period of no recorded activity (or "gaps") of one or two hundred thousand years. To traditional paleontologists, these gaps might seem either long or short, depending on how much species change could be observed on either side. But gaps, instants or not, remain for paleontologists to fill, either with fossil specimens or scenarios accounting for their absence.

The first audience engaged by the theory of punctuated equilibria was the paleontologists, in a phase of the controversy that pitted Eldredge and Gould and a few allies against tradition. The traditionalists, like Darwin, assumed similar rates of evolution in both recorded and unrecorded segments of fossil lineages. The rebels claimed that evolution was more rapid in the gaps than in the documented record. The issue was the *regularity* of evolutionary tempo. But Eldredge and Gould chose to frame the issue in terms of a general outlook, which they called "gradualism."

The Gradualist Picture Framed

The original piece by Eldredge and Gould was as much a metatheoretical as a theoretical statement. Their general point was that one's theoretical models shape the process of data collection and may even determine what counts as data. Eldredge and Gould took this premise as well established in the philosophy of science, but insufficiently appreciated among practicing scientists. "Inductivist notions continue to control the methodology and ethic of practicing scientists raised in the tradition of British empiricism," they observed, and these notions produced a kind of naivete about the way information is accumulated and interpreted.[16] Most paleontologists, they found, were held captive to a "picture" of evolution as a steady, gradual process. Thus, the fossil record would ideally consist of a long sequence of continuous, finely graded intermediate forms linking ancestors to descendants. Any break in that sequence would be taken as an imperfection in the record—it could not be seen as a jump in the evolutionary sequence itself. The *expectation* of a faulty record generated a kind of institution-alized response to discontinuities in the fossils. Eldredge and Gould found "a nearly-ritualized invocation of the inadequacy of the fossil record" whenever a graded continuum could not be documented (p. 97). They wished to awaken their fellow paleontologists to the influence of unexamined gradualistic imagery.

The bias toward gradualism had discernible historical roots. In Darwin's time, anti-evolutionists held the "catastrophist" doctrine, which was that geologi-cal change had always been abrupt and that the creation of species also had been abrupt. Although organic evolution could be conceived as either a continuous or a discontinuous process, the opponents of evolution were united in their belief that species were created in a divinely inspired instant. This unified opposition is what led Darwin to see his theory as inseparably bound to the gradualist position.[17] Gradualism, according to Eldredge and Gould, was further institution-alized in the 1930's and 1940's, when the hitherto feuding sciences of genetics, natural history, paleontology, morphology, and systematics gelled in the "modern synthesis," which explained all variation in living things in light of rules of inheritance derived from Mendelian inheritance and Darwinian natural selection.[18]

Eldredge and Gould would prove successful in highlighting the danger of treating gaps in the fossil record as "no data." They showed that fossil strata often showed real plateaus, interrupted by far shorter intervals than any gradualist would expect. The gaps were more likely, therefore, to represent a few thousand hard-to-sample years than immense periods of eroded rock. Re-analysis of existing fossil data has shown, to the increasing satisfaction of the paleontological community, that Eldredge and Gould were correct in identifying periods of evolutionary stasis which are interrupted by much shorter periods of evolutionary change. Gould and Eldredge made a point in the paleontological community: "stasis is data." But was the new non-gradualist picture to apply only to the rate of evolutionary change ("tempo")? Or was it likewise to affect the way one saw the very mechanisms of change ("mode")? Eldredge and Gould reached beyond their field for an answer.

Sharpening the Alternative

While gaps in the fossil record have prompted resignation in most paleontolo-gists, the periods of unrecorded history offered Gould and Eldredge a theoretical challenge. The phenomenon to fill the gaps, the phenomenon which could produce species in "a flash" and remain undetected by geologists, was the theory of allopatric speciation formulated by Gould's colleague, biologist Ernst Mayr, in 1963.[19] Promoted by Gould and Eldredge as *the* theory of speciation among biologists, the significance of Mayr's theory of allopatric speciation seemed lost on most paleontologists. So, Gould and Eldredge became the interpreters of a theory which seemed to hold significant implications not only for understanding the origins of species, but for where and how one should expect to find fossil evidence of speciation. A section from the 1972 essay provides a useful summary of these implications:

> The central concept of allopatric speciation is that new species can arise only when a small local population becomes isolated at the margin of the geographic range of its parent species. Such local populations are termed *peripheral isolates*. A peripheral isolate develops into a new species if *isolating mechanisms* evolve that will prevent the re-initiation of gene flow if the new form re-encounters its ancestors at some future time. As a consequence of the allopatric theory, new fossil species do not originate in the place where their ancestors lived. It is extremely improbable that we shall be able to trace the gradual splitting of a lineage merely by following a certain species up through a local rock column.

Moreover,

> ... Most morphological divergence of a descendant species occurs very early in its differentiation, when the population is small and still adjusting more precisely to local conditions. After it is fully established, a descendant species is as unlikely to show gradual, progressive change as is the parental species. Thus, in the fossil record, we should not expect to find gradual divergence between two species in an ancestral-descendant relationship. Most evolutionary changes in morphology occur in a short period of time relative to the total duration of species.[20]

Here, it seemed, was the "burst" of evolutionary change that catapults a species from one morphological plateau to another. The gaps were short periods of species proliferation. And not only were the ephemeral transitional forms *not* found in the fossil record, they *should not* be found. Their tenure was too short, and besides, they formed elsewhere, on the periphery of the parental species range. Only after reinvasion of old habitat, now represented in rock columns, would the novel species be obvious. Thus the venerable explanation for gaps in the record, poor sampling due to erosion and other processes, was replaced by a new and compelling story.

The notion of the biological species evolved in peripheral isolates proved persuasive to paleontologists.[21] Little thought had been given in their discipline to the biological meaning of species, because species names were more or less arbitrarily assigned to fossils that were, to the taxonomist, "different enough." No theory of species existed there. Mayr's polished scenario, complete with what to a paleontologist must have appeared a highly sophisticated genetic rationale, explained the gaps. If specimens could not be found to fill the holes in the fossil record, a coherent idea, resplendent in its newness and apparent power, could. A rhetoric of punctuational inquiry had been forged.[22]

Dialectical Dichotomies

The use of biology to solve a problem in paleontology was a bold but eventually controversial move. From the perspective of biological sciences accustomed to debates about species change, it had the effect of oversimplifying. Eldredge and Gould would be criticized for contrasting only the extreme alternatives in what might be a continuum of possible positions. The terms "phyletic gradualism" and "punctuated equilibria," after all, had been created by the authors to pose a contrast. Writing in *Science*, Charles Harper congratulated Eldredge and Gould for introducing the theory of rapid species formation in small peripheral isolates to the paleontological community, but he found that the rebels had posed the alternatives too starkly: few paleontologists would rule out species formation in peripheral isolates, and few biologists would reject the idea that occasional phyletic evolution occurs within established lineages.[23] Harper also took to task an ally of Eldredge and Gould, paleontologist Steven Stanley, for reducing an empirical continuum to a rhetorical choice between two extremes: "His argument is analogous to a defense of laissez-faire capitalism by saying 'but surely you don't prefer Maoist communism.'" Ruling out the "extreme" versions of the two viewpoints, Harper concluded that the real issue would be the relative frequency of species originating by rapid, as contrasted with slow, changes. This, he admonished, would be a task for population geneticists.

Writing again in 1977, and probably heeding such criticisms, Gould and Eldredge (their order of authorship now reversed) backed away from the notion that punctuated equilibria were really revolutionary. They cited George Gaylord Simpson, Gould's former doctoral advisor and the greatest paleontologist of the twentieth century; they wrote:

> For all the hubbub it engendered, the model of punctuated equilibria is scarcely a revolutionary proposal. As Simpson (1976, p. 5), with his unfailing insight, recognized in three lines (where others have misunderstood in entire papers), our model tries to "clarify and emphasize ideas nascent in previous studies." We merely urged our colleagues to consider seriously the implications for the fossil record of a theory of speciation upheld by nearly all of us. . . .[24]

The "nascent ideas" clarified were those of Simpson himself, who had in 1944 explored evolutionary rates in one of his most important works, *Tempo and Mode in Evolution.*[25]

Stephen Toulmin has observed that theoretical controversies in science often move from the positing of dichotomies to the progressive softening of apparent extremes.[26] In the great nineteenth century controversy between geological "catastrophists" and "gradualists," for instance, repeated analyses revealed less and less catastrophic catastrophies and less and less gradual graduations. Ultimately the two camps were divided more by their rhetorics than by their data. The example is suggestive here, as the debate between catastrophists and gradualists set the stage for Darwin's thinking. Its exaggerated oppositions echo through the current controversy about punctuated equilibria. Consider the 1982 testimony of George Gaylord Simpson, he of "unfailing insight":

> The authors [of punctuated equilibria] . . . have created a dichotomy which, like most dichotomies, is merely the dialectic separation of the two extremes of a continuum. Slow evolutionary change or no change at all in a span of geological time has certainly occurred and so has exceptionally rapid change in an unusually shorter span, but so have all the intermediates between those extremes and all the combinations of the two.[27]

The revolution seemed to be losing both its rationale and its rhetorical bite.

The controversy might have run its course toward some reconciliation of extreme positions had Gould not published what in 1980 appeared to be a manifesto for a new general theory of evolution. As co-authors, Eldredge and Gould convinced their peers that *tempos* of evolution varied, and through their promotion of Mayr's speciation theory made some strong inferences about *modes* of species formation. On his own, Gould rejected what he perceived to be in the "modern synthesis" of evolutionary biology, and pushed more ambitiously into theorizing about general modes of evolution.

In an article entitled "Is a new and general theory of evolution emerging?," Gould boldly announced that the modern neo-Darwinist synthesis had broken down on its fundamental tenets: extrapolationism (deriving macroevolutionary explanation by extrapolating from microevoutionary changes), and exclusive reliance on natural selection as the engine of evolutionary change.[28] Evolution, he postulated, was hierarchically organized, with different modes of operating at each level, thus making it impossible to extrapolate from one level to another. Natural selection might play a role at some levels, chance at others. Furthermore, macroevolution was governed by higher-order selection, operating on groups of species, not subject to individual adaptation. "A new and general evolutionary theory," he wrote,

> will embody this notion of hierarchy and stress a variety of themes either ignored or explicitly rejected by the modern synthesis: punctuational change at all levels, control of evolution not only by selection, but equally by constraints of history, development and architecture—thus restoring to evolutionary theory a concept of organism.[29]

Gould's 1980 essay shifted both his persona and the interpretive frame within which the anti-gradualistic stance would be received. This shift follows the rhetoric of a manifesto, as Gould positions himself against the "orthodoxy," "dogma," and unwarranted "faith" in the Old Order, whose death he pronounces. Adopting the role of a storyteller, Gould couches his account in a historical narrative, reconstructing the inherently corrupt foundations on which the old synthesis rested, and revealing the strategies by which its proprietors maintained their power: "A synthesist could always deny a charge of rigidity by invoking . . . official exceptions, even though their circumspection, both in frequency and effect, actually guaranteed the hegemony of the two cardinal principles." The exclusive dependence on adaptation had "infected our language" and become "virtually impossible to dislodge because the failure of one story leads to invention of another rather than abandonment of the enterprise." He gives personal testimony as to how the synthetic theory had "beguiled" him as a graduate student. Then he had watched as its intellectual justification slowly unraveled, leaving it to be carried on by an orthodoxy. Gould reveals that the orthodoxy is saturated with Lamarckian error and Western themes of "ranking by intrinsic merit." He tells of the excitements of the new punctuational approach, and offers appealing new imagery of organic, non-deterministic processes. "Organisms are not billiard balls, struck in deterministic fashion by the cue of natural selection, and rolling to optimal positions on life's table," he concludes. Rather, they are more like polyhedrons, whose many facets keep them from "rolling" unidirectionally. Organisms "influence their own destiny in interesting, complex, and comprehensible ways. We must put this concept of organism back into evolutionary biology." Even for the layman, this is stirring prose. But who were the "we" being addressed here?

The Second Audience Provoked

Gould was no longer speaking just to paleontologists and to lay readers of his popular essays in *Natural History* magazine. If his revolutionary banner had bobbed above the horizon with his earlier embrace of Mayr's speciation theory, it had not drawn much attention. If anything, such a partisan view of speciation eroded Gould's credibility among biologists, many of whom never had thought much of allopatric speciation.[30] But now a parade of Gouldian banners waved across the turf of evolutionary geneticists and population biologists.

Of the receptions that followed, perhaps the most damaging was that of two leading geneticists, G. L. Stebbins and F. J. Ayala, in a 1981 article in *Science*.[31] According to these authors, not only had Gould erected a straw man in his version of the fundamental claims of the synthesis, he had misunderstood seventy years of evolutionary genetics. Here is where Gould adherents would pay for crossing disciplinary bounds unprepared.

Stebbins and Ayala gently informed Gould that non-selective processes—mutation, genetic drift, migration—had been integral to the modern synthesis since its inception. In fact, a living architect of the original genetic theory that underpinned the synthesis, Sewall Wright, had made a career of exploring random and developmental influences on the direction and rate of evolution.[32] If anyone was responsible for the notion of "organism" in evolutionary genetics, it was Wright and his many followers. It was simply incorrect, Stebbins and Ayala charged, to claim that the synthetic theory concerned only adaptive genetic change under the influence of natural selection. This century has been marked by major controversies over the relative importance of random and selective processes in evolution, and over the relative force of selection on genes, entire organisms, and populations. Gould, it would seem, had been a peripheral isolate himself.

Key misunderstandings, Stebbins and Ayala found, arose from grafting a paleontologist's rhetoric onto biological subject matter. The first of these can be traced to different criteria for what a species is. A paleontologist distinguishes two fossils as different species if they look different. A geneticist distinguishes living populations as species if they fail to interbreed, regardless of appearance. Many species of fruit flies, for instance, look alike but cannot interbreed. Two significant implications follow. First, Stebbins and Ayala noted, paleontologists document only substantial morphological change, *"because only when such change has occurred is the paleontologist able to recognize the presence of a new species*" (p. 968). In a sense this guarantees "gaps" in the fossil record. By whatever processes substantial morphological change occurs, the products will be *defined as* fossil species on either side of a gap. The other side of the coin is that active and extensive speciation may occur without being noticed by paleontologists. For these geneticists, "bursts" of species formation seems a curious place-holder in the paleontologists' vocabulary, a tendentious term to cover what the method is unequipped to see.

The second source of misunderstanding was the two different vocabularies of time. How gradual is gradual? How instantaneous is instantaneous? To paleontologists, sixty thousand years is a geological "instant." Yet reproductive isolation can evolve in hundreds of years. According to Stebbins and Ayala, Gould had been enticed by his notion of instantaneous change into thinking that something more than tempo had to be at issue—that there *must* be something radical about the *mode* of change to explain its abruptness. Hence Gould was compelled to abandon his view of the synthesis (actually Mayr's version of it) and proclaim a general revolution in evolution, even though any time span that one might detect in a rock column was ample for speciation as biologists knew it. Could it be that Gould's manifesto had been prompted by an unexamined vocabulary of time—that he was himself held captive to a "picture"?

Not all was bad news for Gould in this landmark response. If Gould's scenarios for refiguring the gaps with rapid speciation were not the only possible ones, at least the geneticists had many other possible scenarios which could accomplish the same result. If Gould was not offering a revolutionary new emphasis on mutation, genetic drift, and developmental constraints, at least large audiences already existed within the army of "synthesists" which were interested in listening to his ideas and could probably improve upon them. In short, the biologists found nothing fundamental in Gould's views of evolution which could not be easily accommodated by the contemporary incarnation of "the modern synthesis."

If the story ended here, the tale of punctuated equilibria would be one of a revolution within one subdiscipline of science, in this case paleontology, which was unable to transform a broader discipline, in this case, evolutionary biology. Gould has continued to clarify his hierarchical theory, but it has lost much of its power as an unsettling challenge to the scientific community.[33] Ironically, however, the public presentation of his work as a general challenge to Darwinism has become increasingly visible. Having weathered the battles within the scientific literature, Gould now finds his expertise used in new and different ways.

The Third Audience Aroused

When scientific talk catches the ear of the laity, the experts may completely lose control of the use of their expertise, as they did when the advocates of a position called "scientific creationism" appropriated the expertise of Gould to further their own ends. In a widely distributed book by creation science's leading spokesman,

Duane Gish, *Evolution? The Fossils Say No!*, Gould is made to play the creation-ist ally by criticizing the received theory of evolution.[34] By a logic of "if they're wrong, we must be right," the creationists are able to put the critique of phyletic gradualism to good use.

Consider the following examples of Gish's quotations from Gould, followed by his interpretation:

> Gould: All paleontologists know that the fossil record contains precious little in the way of intermediate forms; transitions between major groups are characteristically abrupt.

> Gish: Gould is thus arguing that the fossil record . . . does not produce evidence of the gradual change of one plant or animal form into another and that, again, . . . each kind appeared abruptly (p. 172).

Or again:

> Gould: Even though we have no direct evidence for smooth transitions, can we invent a reasonable sequence of intermediate forms, that is, viable, functioning organisms, between ancestors and descendants? Of what possible use are the imperfect incipient stages of useful structures? What good is half a jaw or half a wing?

> Gish: The argument here, that gradual evolutionary change of one form into another is impossible because the transitional forms, being incom-plete, could not function, is a argument that has long been suggested by creationists (pp. 172-173).

and again:

> Gould: There has been no steady progress in the higher development of organic design. We have had, instead, vast stretches of little or no change and one evolutionary burst that created the whole system.

> Gish: Eliminate the words "evolutionary burst" and substitute the words "burst of creation" and one would think he was reading an article by a creationist (p. 177).

Indeed, played in this context, Gould's words sound as if they might have come from the anti-evolutionist sermons of Herbert W. Armstrong, head of the Worldwide Church of God. The fact that he has written a series of essays disputing the basis of creationist claims seems no impediment whatsoever to the creation-ists' purposes. Gould's credibility for their targeted audience derives less from his overall research program than from his saying what serves the rhetorical purpose at hand—to weaken the authority of the scientific consensus in the public mind by exposing dissent within. There is a paradox here, of course. Gould's credibility as an evolutionary scientist is invoked to weaken the credibility of evolutionary science. But this is really nothing more remarkable than the time-honored rhetorical device of turning the tables.[35] As evidence of the very arguability of prevailing scientific views, punctuated equilibria proved invaluable to the rhetoric of "scientific" creationism. Gould's weapons had fallen into enemy hands.

Few scientists communicate with anyone outside of the small international fraternity that investigates the same vanishingly narrow specialty. Perhaps one reason the debate over punctuated equilibria reached the public ear is that Stephen Jay Gould is an exceptional writer as well as scientist; he regularly contributes to *Natural History*, a monthly magazine that is popular with the educated, though not necessarily scientific, public. Hence when Gould hints that he is, in his professional life, promoting a revolutionary perspective that may profoundly revise evolutionary thought, the educated public, including journalists, listens.

Creationists, on the other hand, write and speak to Everyman. Moreover, the creationist perspective is represented on television, daily radio shows, and now and then by heads of state. While the total audience of the scientific debate over punctuated equilibria probably never exceeded a few thousand observers and a few dozen engaged participants, the public audience of the creationist version of the debate numbers in the tens of millions. When presidential candidate Ronald Reagan comments that evolutionists no longer seem to agree on the inerrancy of Darwin's theory, many millions of people hear the comment and wonder.[36] The claim that evolutionary science is in disarray is at least far more comprehensible, and is far more accessible to the public, than the subtleties of allopatric speciation. The public version of this debate has resoundingly amplified the apparent uncertainty within evolutionary science.[37]

A Rhetorical Bind

Positioned by the popular press as a would-be falsifier of Darwin, Gould has responded by portraying himself as "essentially" a good Darwinian, despite the differences that he wishes to make so much of.[38] He finds himself in something of a rhetorical bind insofar as his celebrity as a theoretical innovator rests on his dissent from the "Darwinian synthesis," and yet his scientific self-respect (as contra the creationists) derives from his being a part of that very tradition. One rallies around "Darwin" in declaring against anti-scientific forces. It is a curiously fine line that his different purposes have forced him to walk. For one purpose, he must reject what he claims are the basic tenets of the Darwinian synthesis. For another purpose, he must remain a theorist working in the spirit of The Founder. However much distance Gould may need to put between himself and Darwin in order to clear a space for himself in history, he cannot afford to lose access to the considerable cultural capital invested in Darwin. Theories may come and go, but Darwin is the sign under which business in the evolutionary sciences is conducted—the eponym for an age.

So how does Gould maintain his tenuous position? He does so partly by way of a "loose constructionist" strategy, whereby the strict tenets of Darwinism (the letter, as he sees it) are transcended in order to remain true to the basic program (the spirit). Oddly enough, the spirit is upheld despite Gould's challenge to what he now calls "the essence of Darwinism": the claim that natural selection on individual organisms is the primary creative force in evolutionary change. The matter of tempo in evolution, the focus of earlier debates is quite in the background. Gould calls for a theory about the *mode* of evolution that departs from the received synthesis, but one, he says, which "would embody, in abstract form, the essence of Darwin's argument expanded to work at each level." This hierarchically restructured theory of evolution might thus be seen as "higher Darwinism."[39]

In taking this strategy, Gould is obliged to show that the gradualism upon which Darwin placed such emphasis is ultimately unimportant to Darwinism. Accordingly, he quotes T. H. Huxley, who wrote to Darwin on the day before the publication of *Origin of Species*: "You load yourself with an unnecessary difficulty in adopting *Natura non facit saltum* so unreservedly."[40] The hierarchical theory, Gould hopes, would take that onus off evolutionary science, as well as promote the notion that selection can work on groups, species, or even lineages of organisms, rather than merely on individual organisms, as Darwin and his "synthesist" followers maintain.[41] Darwin would maintain a prominent role in evolutionary biology, largely as a source of inspiration for challengers of smug orthodoxy.

What will be the ultimate influence of punctuated equilibria? What of Gould's continuing promise to reorganize evolutionary biology? Can Gould be expected to break through the ponderous defenses of competing scientific specialties, touching off a scientific revolution on the scale he so obviously desires? The philosophy of science is filled with talk of scientific revolutions. But evolutionary biology is not now living with the incongruities of nineteenth century English science, and Gould may not—for all his public eloquence—be a Darwin. Even though the concept of punctuated equilibria has made a fundamental difference in the way paleontologists regard their science, evolutionary biology seems undisturbed.[42]

Conclusions

The story of punctuated equilibria shows an interpenetration of science and rhetoric such that a "two worlds" notion of technical and social domains at loggerheads is an insufficient conceptual apparatus for understanding what actually happened.[43] This controversy took the dimensions that it did partly because the scientists made aggressive rhetorical moves; and their audience grew and changed accordingly. As Gould spoke beyond his area of technical expertise, his status and authority would be perceived differently by the different segments of his audience. This highlights the difficulty of sorting out when an arguer is speaking within his or her range of expertise and when not. Are Gould's arguments about the hierarchical nature of evolution those of an expert? An educated lay reader might say yes, while a geneticist might say no. Each would be right in a sense.

Under Eldredge's and Gould's, then Gould's sponsorship, a punctuationist rhetoric evolved to include new ways of graphing and otherwise "picturing" the fossil record; of grafting biological notions of species formation onto paleontological evidence; of grafting paleontological time onto biological evolution; of characterizing the opposition as rigid and blindly reductive; of addressing new and broader audiences. This punctuational rhetoric invited a rhetorical response in the scientific community that devalued the novelty of the approach; stressed the diversity of the modern "evolutionary synthesis;" and asserted disciplinary prerogatives. This response within the scientific community triggered a response in the press that dramatized the apparent conflict and created new opportunities for non-scientific appropriation by creationists. All of these manifestations of the controversy involve some admixture of science and rhetoric.

As this case shows, one does not need to jump the chasm between scientific knowledge and social purposes to experience misunderstanding, provocation, and distortion. Any move to a new audience holds the peril of rhetorical conflict. The kind of inter-audience in this debate make clear that expert knowledge is not just a possession of speaker or writer, but is appropriated in different ways by different audiences.[44] Gould was an expert on biological issues to paleontologists, but not to biologists. To laymen, he is an imminent expert on "scientific" matters. To creationists, he is a convenient tool.

We began by arguing that a "rhetoric of science" cannot rest its case simply on a philosophical stance, at least not if it is to tell us something about how scientists use language and authority to engage audiences. Case histories of scientific controversy need to be analyzed in order to begin identifying the ways that scientific discourses and audiences condition one another. In analyzing the process by which the theory of punctuated equilibria was propounded and tested against various discourse communities, we have found several dimensions along which rhetorical dynamics seemed to have played a significant role, including the following:

1. *The persona of the writer.* The persona of the scientific writer as a construction of the writer has received some scholarly attention.[45] Considered here as a construction not wholly under an author's control, persona turns out to be an important factor in shaping the reception of the scientific discourse. Eldredge and Gould acted first as importers of outside knowledge into the field of paleontology. Then in successive formulations of the theory, Gould became a revolutionary with a bolder mission. No longer acting as a narrow specialist, Gould was able to broaden his audience, which simultaneously brought attacks from specialists in other disciplines and caught the imagination of others. He ended up in the paradoxical position of carrying the Darwinian banner against its critics while taking on the persona of challenger to Darwin.

2. *The changing audience for the theory.* Scientific claims generated in one disciplinary environment can sometimes influence response beyond that original environment, often productively, yet often with a distortive effect. The interpretive frames of a given discipline can filter out critical scrutiny from the outside, but only to a point. As theoretical claims grow more ambitious, scientists court both the promise and the risk of capturing the attention of other specialties. As the scope of the non-gradualistic, and finally hierarchical, view of evolution became greater, it strongly invited response from geneticists and population ecologists. At the same time, the implied sweep of this theory made it fit for trans-disciplinary and even popular appeal.

3. *The picturing strategies of the discourse.* Eldredge and Gould were apparently quite right in calling attention to the ways that "picturing" can influence theory and method. One aspect of such picturing they did not notice lay in the different vocabularies of the various sciences. Different conceptions of time and ways of marking its passage seemed in this case to influence how "gaps" in the record were conceived, for instance, as did different methods of marking biological change. Obviously, each science will have its special tools and methods; but these are not the only unities within a scientific discourse. Discursive inventions such as metaphors and imagery will also help shape the interpretation of findings. Gould was acting as an astute rhetorician in "unmasking" the forms of expression and "invented stories" he found among synthesists, as he was also when he invented new images for picturing evolutionary process. Organisms, for instance, would be envisioned as "polyhedrons" rather than the supposed "billiard balls," while alleged ideological biases would be redirected. Such efforts as creating or revising a "picture" of a phenomenon under description are part of what gives a discourse, in science as elsewhere, its connection to "common sense" understanding. They also provide a nexus between science and the web of ideology, and hence a means of harnessing affective energies to rhetorical ends.

4. *Strategies of contrast.* New scientific theories gain attention partly because of the contrast they pose to existing theories. The theory of punctuated equilibria drew criticism for posing that contrast too sharply, and yet this was also a source of its rhetorical potency. In a sense, it made itself attention-worthy by positioning itself in dialectical opposition to accepted views. When Gould moved on to his "new synthesis" manifesto, his rhetoric could be construed as an assault on the scientific establishment itself, and hence its appearance in the popular press and creationist tracts should have been predictable. In the rhetoric of antigradualism, as well as in the counter-rhetoric it provoked, one finds implied definitions of the opposition which help determine how the criticized position would be dealt with—whether it will be corrected or rejected outright, absorbed as unremarkable, or otherwise diffused. In general, it would seem useful for analysts of scientific rhetoric to examine not only how a discourse presents itself, but also how it implicitly or explicitly presents its opposition.

The rhetoric by which the theory of punctuated equilibria was presented carried it through several different discursive formations, involving different audiences and different issues. There are, therefore, a variety of ways that its authors could be called to account. In this case, and in general, it is a worthy challenge for the new rhetoric of science to clarify the shifting frameworks to which scientific discourses of public interest should be held publicly accountable. This clarification would pay attention not only to the different criteria for truthfulness in the various scientific disciplines, but to the conditions under which writers, audiences, and disciplines gain and lose proprietorship of scientific discourses.

Notes

1. Trevor Melia, "And Lo the Footprint . . . Selected Literature in Rhetoric and Science," *Quarterly Journal of Speech*, 70 (1984), 311.

2. Melia does credit some of the pioneering work in his essay, including John A. Campbell, "The Polemical Mr. Darwin," *Quarterly Journal of Speech*, 61 (1975), 375-90. M. Finochario, *Galileo and the Art of Reasoning: Rhetorical Foundations of Logic and Scientific Method* (Holland: Reidel, 1980); Donald N. McCloskey, "The Rhetoric of Economics," *Journal of Economic Literature*, 21 (1983), 482-517; and Joseph R. Gusfield, "The Literary Rhetoric of Science: Comedy and Pathos in Drinking Driver Research," *American Sociological Review*, 41 (1976), 16-34. More recent trespasses onto the ground of science include: S. Michael Halloran and Annette Norris Bradford, "Figures of Speech in Rhetoric of Science and Technology," in Robert Connors et. al., eds., *Essays On Classical Rhetoric And Modern Discourse* (Carbondale: Southern Illinois Press, 1984, pp. 179-92; and several studies appearing in *Argument in Transition: Proceedings of the Third Summer Conference on Argumentation*, ed. Jack Rhodes, et. al. (Annandale, VA: Speech Communication Association, 1983); notably Anne Holmquest, "Rhetorical Argument in Science: The Function of Presumption, pp. 257-71; John Lyne, "Ways of Going Public: The Projection of Expertise in the Sociobiology Debate," pp. 400-15; and Donald N. McCloskey, "Notes on the Character of Argument in Modern Economics," pp. 170-87. In *Argument and Social Practice: Proceedings of the Fourth SCA/AFA Conference on Argumentation*, (Annandale, VA: Speech Communication Association, 1983), ed. J. Robert Cox et. al., see Anne Holmquest, "Rhetoric and Semiotic in Scientific Argument: The Function of Presumption in Charles Darwin's *Origin of Species Essays*," pp. 376-402; John Lyne, "Punctuated Equilibria: A Case Study in Scientific and Para-Scientific Argument," pp. 403-19; and Gonzalo Munévar, "Rhetorical Grounds for Determining What is Fundamental Science: The Case of Space Exploration," pp. 420-35.

3. Rhetoric in science has been approached in many different ways over the centuries, from stylistic, dialectical, to literary perspectives. The study of style among Renaissance scientists, for instance, might be said to constitute a rhetorical interest—Francis Bacon constructed an entire theory of communication for scholars and researchers. See James Stephens, "Rhetorical Problems in Renaissance Science," *Philosophy and Rhetoric*, 8 (1975), 213-29, and *Francis Bacon and the Style of Science* (Chicago: University of Chicago Press, 1975). In the Nineteenth Century, Charles Peirce wrote prolificly on rhetoric and communication within the various sciences, and even proposed a set of rhetorical principles to govern both the generation and the dissemination of scientific knowledge. See John R. Lyne, "Rhetoric and Semiotic in C.S. Peirce," *Quarterly Journal of Speech*, 66 (1980), 155-68; and "C.S. Peirce's Philosophy of Rhetoric," *Rhetoric Re-Valued*, ed. Brian Vickers (Binghamton, NY: Center for Medieval and Early Renaissance Studies, 1982), pp. 267-76.

4. This focus on relativism in its various guises (including loose interpretations of paradigms, perspectivism, Weltanschauungen, etc.) was clearly evidenced in many of the papers presented at the 1984 Iowa Symposium on Rhetoric and the Human Sciences. See Herbert W. Simons, "Chronicle and Critique of a Conference," *Quarterly Journal of Speech*, 71 (1985), 52-64; John Lyne, "Rhetorics of Inquiry," *Quarterly Journal of Speech*, 71 (1985), 65-73.

5. See Lyne, "Ways of Going Public."

6. Tom Bethell, "Darwin's Mistake," *Harper's*, Feb. 1976, pp. 70-5; "Darwin is on the Run Again," *People*, 8 Dec. 1980, pp. 151 ff.; Sharon Begley, "Science Contra Darwin," *Newsweek*, 8 April 1985, pp. 80-1.

7. Richard Dawkins, "What was all the fuss about?," *Nature*, 316, 22 August (1985), p. 683.

8. This issue reflects a general concern that deeply fissured bases of knowledge and authority hamper reasonable judgment—public, private, and scientific. Thomas B. Farrell and G. Thomas Goodnight, for instance, explore the chasm between what they call "social knowledge" and "technical knowledge" in "Accidental Rhetoric: The Root Metaphors of Three Mile Island," *Communication Monographs*, 48 (1981), 271-300.

Goodnight poses a contrast between the "public sphere" and the "technical sphere" of argument in "The Personal, Technical, and Public Spheres of Argument: A Speculative Inquiry into the Act of Public Deliberation," *Journal of the American Forensic Association*, 18 (1982), 214-27. Alan G. Gross identifies a "technical ideology" in collision with a "social ideology" in "Public Debates As Failed Social Dramas: The Recombinant DNA Controversy," *Quarterly Journal of Speech*, 70 (1984), 397-409. Walter R. Fisher distinguishes two sets of discourse rules governing technical reasoning and public moral argument—a rational paradigm and a narrative paradigm, in "Narration as a Human Communication Paradigm: The Case of Public Moral Argument," *Communication Monographs*, 50 (1984), 1-22.

9. See Charles Arthur Willard's treatment of the problems posed by the relative closure of various argumentative fields, in *Argumentation and the Social Grounds of Knowledge* (Tuscaloosa: University of Alabama Press, 1983), esp. pp. 270-79.

10. Niles Eldredge and Stephen Jay Gould, "Punctuated Equilibria: An Alternative to Phyletic Gradualism," in *Models in Paleobiology*, ed. T. J. M. Schopf (San Francisco: Freeman, Cooper and Co., 1972), pp. 83-115.

11. Charles Darwin, *On the Origin of Species: A Facsimile of the First Edition*, (Cambridge, MA: Harvard Univ. Press, 1964), pp. 310-11.

12. Darwin, p. 189.

13. Darwin, p. 194.

14. Darwin, p. 302.

15. Michael Ruse, *The Darwinian Revolution: Science Red in Tooth and Claw* (Chicago: University of Chicago Press, 1979).

16. Eldredge and Gould, p. 85.

17. Ruse.

18. For a historical account of this synthesis, see Ernst Mayr, "Prologue: Some Thoughts on the History of the Evolutionary Synthesis," in *The Evolutionary Synthesis*, ed. E. Mayr and W.B. Provine (Cambridge, MA: Harvard University Press, 1980), pp. 1-48.

19. Ernst Mayr, *Animal Species and Evolution* (Cambridge, MA: Belknap Press of Harvard Univ. Press, 1963).

20. Eldredge and Gould, pp. 94-5.

21. For instance, geographical speciation is by far the dominant mode discussed in S. M. Stanley's influential text, *Macroevolution* (San Francisco, CA: Freeman and Company, 1979).

22. On the notion of "rhetorics of inquiry," see Lyne, "Rhetorics of Inquiry."

23. Charles W. Harper, Jr., "Origin of Species in Geologic Time: Alternatives to the Eldredge-Gould Model," *Science*, 3 Oct. 1975, pp. 47-8. Responding to Harper, Steven Stanley pointed out that he agreed that intermediate positions were possible, and reiterated his view that phyletic gradualism should be identified when phyletic change is "the clearly dominant mode of evolution." The "punctuated equilibrium" model, by contrast, should represent the view that "much more than 50 percent of evolution occurs through sudden events in which polymorphs and species are proliferated." What had begun as contrasting "pictures" was on its way to more prosaic comparisons of percentages. See Steven M. Stanley, "Stability of Species in Geologic Time," *Science*, 16 April 1976, p. 268.

24. Gould and Eldredge, p. 117.

25. George Gaylord Simpson, *Tempo and Mode in Evolution* (New York: Columbia University Press, 1944).

26. Stephen Toulmin, *Human Understanding: The Collective Use and Evolution of Concepts* (Princeton: Princeton University Press, 1972), pp. 119–20.

27. *The Book of Darwin*, ed. with introduction and commentary by George Gaylord Simpson (New York: Washington Square Press, 1982), p. 207.

28. *Paleobiology*, 6, No. 1 (1980), 119-30.

29. Gould, 1980, abstract.

30. Mayr's theory of allopatric speciation claimed that reproductive isolation required a total "genetic revolution" (p. 621). The genetical rhetoric is unsupported by evidence, and not believed by geneticists. However, Gould and his allies evidently first accepted Mayr's claim at face value, mistook it as the consensus of the "modern synthesis," and then rejected it (and with it what they perceived as the modern synthesis itself) in favor of the idea that differences in a few genes could cause reproductive isolation. Geneticists have long been aware that minor genetic differences can result in reproductive isolation.

31. G. Ledyard Stebbins and Francisco J. Ayala, "Is a New Evolutionary Synthesis Necessary?" *Science*, 28 Aug. 1981, pp. 967-71.

32. Sewall Wright has been a prolific contributor to evolutionary genetics since his first paper in 1917. Still active, Wright recently summarized some highlights of his extraordinary career, while bluntly criticizing Ernst Mayr's view of evolutionary genetics. [Professor Wright regrettably passed away in 1988.–RAH]

This is relevant to the "punctuated equilibria" controversy because Mayr, an ornithologist, appears to be the source of much of Gould's understanding of genetics. See Sewall Wright, "Genic and Organismic Selection," *Evolution*, 34, No. 5 (1980), pp. 825-43.

33. Stephen Jay Gould, "The Meaning of Punctuated Equilibrium and its Role in Validating a Hierarchical Approach to Macro-evolution," in *Perspectives on Evolution*, ed. R. Milkman (Sunderland, MA: Sinauer Associates Inc., 1982), pp. 83-104.

34. Duane T. Gish, *Evolution: The Fossils Say No!* (San Diego: Creation-Life Publishers, 1979).

35. Creationists turn the tables on evolutionists in other ways. A favored tactic is to set up strict criteria for what would constitute a legitimate theory, e.g. that a scientific theory must be falsifiable. See Gish, p. 13. The claim that evolutionary theory is not falsifiable is interesting, as one cannot readily imagine science rejecting the concept of evolution, although evolutionary scientists accept substantial revisions in Darwin's original thinking. To creationists, a pervading belief in evolution smacks of dogma. But this is because creationists fail to understand that the weight of rational and empirical support for evolution is distributed across many scientific specialities, and that *in principle* data can be imagined that could falsify the theory, even if such data never actually existed. Darwin frequently set up such possible scenarios for the destruction of his theory, including one criterion that, with a century of hindsight, appears to be false, the notion that evolution by natural selection requires phyletic gradualism. In 1859, when *On the Origin of Species* was published, "punctuated equilibria" might have been too much for the theory to bear. But "punctuated equilibria" are fully compatible with the much-revised "modern evolutionary synthesis." See Stebbins and Ayala.

36. At a press conference during the 1980 election campaign, Ronald Reagan was asked if he thought that the theory of evolution should be taught in public schools. He responded, "Well, it is a theory, it is a scientific theory only, and it has in recent years been challenged in the world of science and is not yet believed in the scientific community to be as infallible as it once was believed." Asked if he believed in the theory of evolution, the presidential candidate replied: "I have a great many questions about it. I think that recent discoveries down through the years have pointed up great flaws in it." See "Republican Candidate Picks Fight with Darwin," *Science*, 12 Sept. 1980, p. 1214.

37. For example, see James Gleick, "Stephen Gould: Breaking Tradition with Darwin," *New York Times Magazine*, 20 Nov. 1983, pp. 48 ff.; "Darwin is on the Run Again," *People*, 8 Dec. 1980, pp. 151 ff.; See, then, Stephen Jay Gould, "Darwinism and the Expansion of Evolutionary Theory," *Science*, 216, 23 April 1982, pp. 380-87.

38. Gould, 1982 in *Science*, p. 381.

39. Gould, 1982 in *Science*, p. 386.

40. Gould, 1982 in *Science*, p. 386, n. 29.

41. Vigorous debates about the role of chance and the force of selection on individuals, groups, populations, and species predate Gould's notion of hierarchies by at least 40 years. Sewall Wright summarizes circumstances under which chance might be important (see note 32). Evolutionary biologist George C. Williams put much of the controversy to rest with an influential book, *Adaptation and Natural Selection: A Critique of Some Current Evolutionary Thought* (Princeton University Press, 1966), in which he used intuitive arguments to demonstrate that the differential success of groups could be traced to the differential success of their members. Geneticist Richard Lewontin came to the same conclusion in 1970 with a more tightly reasoned argument in the lead article, "Units of Selection," of a prestigious new periodical, *Annual Review of Ecology and Systematics*, 1 (1970), 1-18. It is unlikely that biologists are skeptical of Gould's claim to a revolutionary insight about "hierarchical evolution" because the insight is new or revolutionary. More likely, most biologists probably see Gould as trying to sell a dead horse as a champion stud.

42. See Roger Lawin, "Punctuated Equilibrium Is Now Old Hat," *Science*, 231 (1986). 672-73.

43. We do not suggest that social and technical discourses are indistinguishable. We concur with Farrell and Goodnight (299-300) and others that the distinction can be made. Our point is that there are different ways to credit expertise. This is especially obvious when experts misread each other's technical dialects and fail to share key assumptions. What is expert in one scientific discipline may well be naive in another, even if to the laity it all carries the authority of being "scientific."

44. See: Lyne, "Ways of Going Public."

45. Paul N. Campbell, "The *Personae* of Scientific Discourse," *Quarterly Journal of Speech*, 61 (1975), 391-405.

The Rhetorical Construction
of Scientific Ethos

Lawrence J. Prelli

Aristotle stressed that a central means of persuasion is a rhetor's perceived character or *ethos*.[1] To inspire confidence in claims advanced discursively, a rhetor must display the qualities of intelligence, moral character, and good will that are held in esteem by an intended audience.[2] Research on the sociology of science makes clear that scientific rhetors are also subject to the constraints of *ethos*.

The groundbreaking work on scientific *ethos* was undertaken by Robert K. Merton.[3] Merton sought to identify the binding institutional norms that constrain the behavior of scientists, and facilitate establishing and extending certified, objective knowledge of the physical world.[4] The norms of science, Merton believes, encourage behaviors that minimize distortion during systematic observation and maximize efficient dissemination of certified knowledge:

> The ethos of science is that affectively toned complex of values and norms which is held to be binding on the man of science. The norms are expressed in the form of prescriptions, proscriptions, preferences, and permissions. They are legitimized in terms of institutional values. These imperatives, transmitted by precept and example and reenforced by sanctions are in varying degrees internalized by the scientist, thus fashioning his scientific conscience or, if one prefers the latter-day phrase, his superego. Although the ethos of science has not been codified, it can be inferred from the moral consensus of scientists as expressed in use and wont, in countless writings on the scientific spirit and in moral indignation directed toward contraventions of the ethos.[5]

The scientific *ethos* binds both technically and morally. It is technically binding because it prescribes efficient procedures for securing the extension of certified knowledge. It is morally binding because it is believed to assert what is right and good.[6] The norms identified in Merton's original formulation are as follows: *Universalism* requires that knowledge claims be subjected to pre-established, impersonal criteria that render them consonant with observation and previously established knowledge. *Community* or *communism* prescribes that research is not a personal possession but must be made available to the community of scientists. *Disinterestedness* requires that scientists strive to achieve their self-interests only through satisfaction in work done and prestige accrued through serving the interests of a scientific community.[7] *Organized skepticism* mandates that scientists temporarily suspend judgment in order to scrutinize beliefs critically against empirical and logical criteria of judgment.[8]

Merton added to his original list the norms of *originality* and *humility* when he studied priority claims in scientific discovery.[9] *Originality* is a source of esteem for one's work because through this quality scientific knowledge advances. *Humility* ensures that scientists will not misbehave at the rate that they would if importance were assigned only to originality and to establishing priority in scientific discovery.

Much research done in the wake of Merton's studies has supported his functional norms, but there has also emerged evidence that there exists in science a second set of "counter-norms" wholly incompatible with those identified initially by Merton.[10] For example, in a well-documented study of NASA "moon scientists," Mitroff identified some of these counter-norms.[11] The list includes *particularism*, which counters universalism. Scientists often regard as legitimate judgments about research reports and proposals that are based on personal criteria, such as the ability and experience of authors, rather than on the strictly technical merits of the research claims themselves. Similarly, *solitariness* opposes communality. Scientists often do exercise property rights regarding their work. From this counter-normal vantage point, secrecy is appropriate conduct and, indeed, helps scientists avoid disruptive priority disputes and ensures that they do not waste colleagues' time by rushing immature work into print. *Interestedness* opposes disinterestedness. This counter-norm promotes conduct that serves scientists' special communities of interest (for example, their invisible colleges). Finally, *organized dogmatism* counters organized skepticism. This counter-norm prescribes that scientists do not incessantly doubt their own and others' findings, but instead assent fervently to their *own* findings while doubting the findings of others.[12]

Evidence of conflicting commitments among scientists has led some sociologists to question whether the constituents of Merton's scientific *ethos* should be considered "normative" at all. This question remains unsettled, but I do not intend to address it here.[13] My contention is that when scientists resort to these common themes in discussing, justifying, or evaluating actions, the alleged "norms" and "counter-norms" of science serve a *rhetorical* function, regardless of whatever other functions they might be said to serve. Specifically, the constituents of scientific *ethos* function like rhetorical *topoi* for inducing favorable or unfavorable perceptions of scientific *ethos*. Scientific *ethos* is not given; it is constructed rhetorically. Rhetors respond to, or seek to avoid creating, ambiguities and conflicts about their scientific credibility. They do this by choosing from among a range of strategic options those that are best suited to situational contingencies. What sociologists of science have been calling the "norms" and "counter-norms" of science are effectively conceived as rhetorical *topoi* that index the available range of discursive strategies for establishing negative or positive audience perceptions of scientists' *ethos*.

First, we need to recognize that attention to the constituents of scientific *ethos* becomes salient only when the discourse of one scientist is made and evaluated by others in scientific situations that are rhetorical; that is, in problematic or ambiguous situations that involve inducing adherence to ideas presented as "scientific."[14] Barnes and Dolby make the point that scientists stress the significance of such intellectual qualities as "rationality" or "skepticism" in situations that involve celebration, justification, or conflict.[15] Within such situations successfully resolving problems or clarifying ambiguities hinges partly on the perceived qualifications of the proposing thinker or researcher. What scientific qualities will be most valued can vary from rhetorical situation to rhetorical situation.

Considerations such as secrecy, communality, objectivity, and emotional commitment may or may not be situationally relevant to evaluating a scientist's professional work. The situation in which scientific discourse is being evaluated determines the relevance and salience of each such quality. This is to say that when a scientist makes knowledge claims to colleagues and implies that these claims are "scientific" it *may* be relevant, though it will not always be imperative, to weigh the claimant's objectivity or emotional involvement, openness or secrecy, skepticism or enthusiasm, and the like. In various situations *either* of such opposed qualities *may* be relevant and be judged a *scientific* virtue or vice.

The professional *ethos* of a rhetor as a scientist becomes specially relevant when there is reason to believe that his or her primary aims are tied to such "nonscientific" pursuits as securing personal celebrity with lay audiences, achieving political or religious aspirations, or perpetuating beliefs that have occult or supernatural implications.[16] Even those seeking explicitly to popularize science risk jeopardizing their *ethos* with expert audiences. They must find "common ground" with technically unskilled audiences, leaving themselves open to charges that they are pursuing objectives other than those that are properly "scientific" or "educational."[17]

In science as elsewhere, when the status of claims is problematic the *ethos* of the claimants can become pertinent. As Aristotle said, "We believe good men more fully and readily than others: this is true generally whatever the question is, and absolutely true where exact certainty is impossible and opinions are divided."[18] Then, a claimant's "rationality" or "skepticism" can become important considerations for audiences charged with adjudicating technical claims.[19] Similarly, one might dwell on a scientist's reputation, experience, and technical skill and on such grounds give problematical technical claims the benefit of the doubt.[20]

Whatever is said or done to influence perceptions of a scientist's *ethos* will arise from a finite set of values implied by the notion of "doing good science"; for example, actions or words can indicate appropriate enthusiasm, objectivity, or skepticism. Scientific values of this sort constitute themes that can be treated verbally or implied by actions; they supply scientists with persuasive means for inducing adherence to their aims and claims as scientifically "reasonable" or "legitimate."

As soon as we recognize that we are dealing here with rhetorical *topoi*, the seeming conflict between alleged "norms" and "counter-norms" dissolves. The themes having to do with scientific *ethos* are not law-like or even "rules"; they are lines of thought that bear on a scientist's credibility in this or that rhetorical situation. Mulkay's description of how the constituents of scientific *ethos* actually come into play reflects their rhetorical topicality:

> In science...we have a complex moral language which appears to focus on certain recurrent themes or issues; for instance, on procedures of communication, the place of rationality, the importance of impartiality, and of commitment, and so on. But...no particular solutions to the problems raised by these issues for participants are firmly institutional-ized. Instead, the standardized verbal formulations to be found in the scientific community provide a repertoire or vocabulary which scientists can use flexibly to categorize professional actions differently in various social contexts.[21]

Topoi for building or diminishing a scientist's *ethos* are imprecise in the sense that any of the "recurrent themes" can be applied and developed in many ways. It is the abstract quality of *topoi* that makes them useful generative devices for yielding multiple specific applications. Most scientists will have some degree of commitment to such *topoi* as impartiality, objectivity, commitment, novelty, humility, and communality, and their opposites. Each scientist identifies with the *topoi* in the abstract; and then by moving in thought from the abstract to concrete applications, each can discover an array of specific arguments supportive of his or her overall position. Conflict often turns on whether one is for or against a specific *application* of a *topos*, say, humility, or whether one is for or against a *comparative ranking* of the *topoi* according to their alleged significance in a situation. For instance, there can be disagreement about whether novelty or skepticism is most important. Although the terminology of topical theory is not used, Mulkay's explanation of how "standardized verbal formulations" operate is in fact a discussion of how rhetorical *topoi* relating to scientific *ethos* are chosen:

> A major influence upon scientists' choice of one verbal formulation [what I am now calling a "topic"] rather than another . . . is likely to be their interests or objectives. It can be assumed that, for a given scientist or group of scientists, these interests will vary from one social context to another. Thus . . . when researchers were frustrated by the apparent reluctance of others to make significant findings available to them, they tended to select principles favoring communality which justified their condemnation of the others' behavior and added weight to their exhor-

tations. In contrast, those scientists who had made the discovery were able to find principles in favor of personal ownership of results [secrecy]. In different circumstances, a person's or a group's choice of rules can be entirely reversed. Not only is it possible to vary one's choice of formulations as one attempts to identify the evaluative characteristics of different acts, but it is also possible to apply different formulations to the same act as one's social context changes.[22]

One kind of rhetorical situation in which concerns about professional, scientific *ethos* become particularly salient is marked by questions about demarcation criteria. If it were possible to draw a sharp line of demarcation between science and nonscience, there would be little ambiguity involved when classifying discursive aims and claims as "scientific" or other; hence, there would also not be any need for rhetoric to clarify the scientific standing of those aims and claims. However, whenever we seek to differentiate "science" from "nonscience" there will always be working ambiguities. In these rhetorical situations, scientists will likely choose rhetorical strategies that help them construct "boundaries" that are favorable to their own professional goals and interests and unfavorable to their competitors. For instance, researchers might make rhetorical appeals that construct "narrow" or "rigid" boundaries, inducing adjudicating audiences to distinguish their "scientific" work from the research of those they allege are unorthodox or unscientific "outsiders." In Gieryn's view, scientists engage in "boundary-work" not for the lofty epistemological reasons philosophers often cite (for example, preservation of scientific truth), but as a rhetorical means of solving practical problems that can block achievement of professional goals.[23] Scientists draw sharp contrasts between themselves and "nonscientists" to enhance their intellectual status and authority vis-a-vis the "out groups," to secure professional resources and career opportunities, to deny these resources and opportunities to "pseudo-scientists," and to insulate scientific research from political interference. We should notice, too, that researchers with "unorthodox" aims and claims will also compose rhetoric about scientific *ethos*; but they will typically seek to "broaden" or "soften" the "boundaries" of science as they are defined by defenders of scientific orthodoxy, attempting to show that they, too, are scientists and that their claims should also be taken seriously as reasonable scientific contributions.

Audiences called upon to adjudicate "boundary disputes" are often comprised of scientifically illiterate laypersons, who nevertheless have great respect for the authority of science and its practitioners.[24] When "experts" disagree about matters of public policy, the laity usually cannot decide issues on technical grounds. Rhetors will therefore seek to settle demarcation exigencies by constructing "boundaries" in view of some ideal, public image of science and its practitioners. Both orthodox and unorthodox scientists will choose from among the special *topoi* of scientific *ethos* to construct public images favorable to their respective interests and objectives.

The Topoi of Scientific Ethos: A Case Study

What follows is a case study of *topoi* used to attack and defend scientific *ethos*. The case concerns Francine Patterson's and Eugene Linden's *The Education of Koko*, and Thomas A Sebeok's review of that book.[25] In the book, Patterson explains how she taught Koko, a gorilla, to use American Sign Language. Most scientists would agree that Koko makes some kind of gestures but many also conclude that Patterson has gone far beyond her data when she claims that the animal has learned its lessons so well that it can ask questions, lie, insult, joke, apologize, and even express grief.[26] Sebeok, Professor of Anthropology, Linguistics, and Semantics at Indiana University, is one of the most outspoken critics of language acquisition studies of apes in general and of Patterson's research in particular. In his review, "The not so sedulous ape," Sebeok attacks Patterson's scientific *ethos* on the grounds that she lacks technical and communal qualities that should be displayed by "credible" scientists. Specifically, Sebeok implies that Patterson and her co-author are not "real" scientists; they are scientific "outsiders" because they display qualities of thought and behavior that are, at best, in conflict with the Mertonian "virtues" of scientific *ethos: universality, communality, skepticism,* and *disinterestedness.*[27]

Sebeok suggested that Patterson is not a "real" scientist because her technical claim-making is, in a word, "bizarre"; he repeatedly illustrated how Patterson's claims for Koko's linguistic cleverness conflict with an implied universal consensus about what is already known about the language behavior of apes. He referred approvingly to Herbert S. Terrace's conclusion, in his book *Nim*, "that there is no evidence at all that apes can either generate or interpret sentences." Sebeok buttressed his claim by arguing that Terrace's "hardly surprising" conclusion, in direct contrast with Patterson's claims, is consistent with what has been revealed by "informed" linguists, and "responsible" ethologists: "Terrace's results are . . . in perfect conformity with the long held judgment of informed linguists from Max Muller (1889) to Noam Chomsky. They accord equally well with the view of responsible ethologists, such as Konrad Lorenz, who declared, in 1978, 'that syntactic language is based on a phylogenetic program evolved exclusively by humans,' and that anthropoid apes '. . . give no indication of possessing syntactic language.' "[28] The conclusion that Sebeok would have us draw from this passage is clear. Patterson's research claims are not grounded in and consistent with established research conclusions in respectable fields of scientific inquiry. Accordingly, as a scientist, she can only be held in contempt as "uninformed" and "irresponsible."

Sebeok also criticized Patterson's research claims on the ground that she is an incompetent experimenter. Specifically, he charged that Patterson had not accounted for possibilities that the animal was behaving in response to clues given by experimenters, an experimental problem popularly known as the "Clever Hans" phenomenon. According to Sebeok, this fallacy is one "by which Koko's entire ten-year curriculum has been errantly nag-ridden. . . ." To support this contention, Sebeok appealed to authority:

The eminent Bristol neuropsychologist, Richard Gregory, also concluded, in 1981, that apes do not exhibit either "human language or intellectual ability", and wisely admonished: "There are so many experimental difficulties and possibilities of the animals picking up clues from the experimenters, given unwittingly, that extreme caution is essential."

Sebeok repeatedly assailed Patterson's research with arguments that questioned the reliability of both her data and her interpretations of data. For instance, Patterson's competence became an issue when Sebeok challenged the interpretive claim that Koko was able to tell lies:

Much is made of her aptitude for lying, which, according to the authors, "of course, is one of those behaviors that shows the power of language." Here, however, lurks a terminological confusion, one that, furthermore, begs the question. Many kinds of animals—the most remarkable case on record is that of the Arctic fox, *Alopex lagopus*—give, or give off, deceptive messages, in a word, prevaricate. But a lie must, by definition, be "stated," which Koko simply cannot do.[29]

From this and other instances, Sebeok would have us conclude not only that Patterson's data are suspect but also that her ability to render "scientific" judgments is highly questionable. The claims she advances cannot be accepted unless one replaces "pre-established impersonal criteria" of science with Patterson's idiosyncratic standards of judgment. In a word, the qualities of Patterson's thought and conduct lack affinity with the scientific virtue of "universality."[30]

Sebeok continued his attack on Patterson's *ethos* by drawing arguments from the powerful *topos* of *organized skepticism*. For example, Sebeok amplified both Patterson's and Linden's proclivity to minimize the importance of emotional detachment and systematic doubt in their research by associating them with proponents of a field whose scientific status has been thought questionable. They are said to be "addicted to the use of ploys familiar from parapsychology, such as that the presence of a skeptic tends to ruin experimental results." Among these "ploys" is their emphasis on emotional rapport between experimenters and animal subjects as a precondition for making successful experimental results. Sebeok quotes directly the authors' belief "that one cannot really understand the mental workings of other animals or bring them to the limits of their abilities unless one first has true rapport with them." He then asserted that the "obverse of this claim is that the intimacy between Patterson and her beloved Koko had hopelessly overclouded her scientific objectivity and judgment. . . ." In this way, Sebeok would lead us to conclude that Patterson's research claims must be technically unacceptable because her conduct resonates with an obstinate, dogmatic will to believe that Koko makes and uses language.

A scientist's professional *ethos* depends not only on how audiences perceive his or her competence, but also on how that scientist's "place" within a legitimating scientific community is viewed. To the extent that a rhetor's connection with his or her scientific community becomes confusing, so too does the legitimacy of his or her aims and claims. I have so far shown that Sebeok sought to discredit Patterson's and Linden's *ethos* by arguing that they lack the scientific "virtues" needed to make credible, technical claims. He also repeatedly challenged the authors' credibility by making their relationship with a professional scientific community seem ambiguous to the reader. Sebeok did this by using three lines of argument based on the *communality topos*. First, Patterson and her co-author do not have memberships in any legitimate research institution, so neither is a "real" scientist. Patterson had written that she was not able to analyze Koko's spoken language in detail despite an "enormous" amount of data collected. Sebeok offered this "translation": "In plain text, this citation means that since Miss Patterson's connections with Stanford University has been severed, she no longer enjoys free access to its computers." In other words, Patterson has no scientific standing because she has lost the resources of a legitimizing institution. Sebeok attacks the scientific *ethos* of Patterson's co-author, Eugene Linden, with even less subtlety. Linden is depicted as "a wrestler-turned-journalist, perhaps best known to the public for his *Apes, Men, and Language* (1974, 1981), surely the most gullible, as well as defensively emotion-laden, popular account of attempts at linguistic communication with any of our collateral ancestral species so far published...." The implication here is that Linden lacks connections with legitimizing research institutions; he is, at best, a popularizer of science—and not a very credible one at that.

In addition to arguing that Patterson and her co-author do not "belong" to the scientific community by virtue of position, Sebeok "reads them out of" the community of scientists on grounds that Patterson has been unable to secure ample public funds to support her research. Sebeok was explicit: "While millions of dollars in federal funds were being squandered on the futile search for language in chimpanzees and orangutans, Patterson continued her work, without a proper institutional base, with the support of private sources, including a large, so-called 'non-profit' commercial enterprise,[31] supplementing her income by minor grants from small Foundations." He further asserts that the lack of public subsidy is "one respect" in which Patterson's Project Koko differs "sharply" from other studies of ape language capacities. The reader is presumably to understand that any minimally competent student of language in primates would have had access to and used the all-too-available resources for doing the job "right," scientifically.

A third line of argument involves questioning Patterson's legitimacy as a scientist because she has failed to submit her claims for authorization by competent scientists.[32] She had success reaching popular audiences through such channels as *National Geographic* magazine, *Reader's Digest* and a documentary film,[33] but her publishing in more technical sources has been minimal, giving Sebeok warrant for asserting, "If Penny Patterson tried to publish in a scientific atmosphere, then she would be laughed out of court."[34] She was attacked as having a "warped perspective" and a "lack of receptivity to well-intentioned criticism."[35] Accordingly, the reader is left to conclude that Patterson is an "outsider" who lacks credible standing within the professional scientific community.[36]

Sebeok also challenges Patterson's and Linden's scientific *ethos* with remarks that imply they are not sufficiently *disinterested* to be pursuing "legitimate" scientific objectives with the publication of their book. "Real" scientists identify their self-interests with winning community recognition for "making contributions to the development of the conceptual schemes which are of the essence in science."[37] In contrast, Sebeok makes innuendo about Linden's "real" interests: "Her co-author's stake in this enterprise—as well as, of course, his bond of personal relationship with the gorilla—is clearly of a different order." The implication one can draw from this assertion is that Linden's primary motivations are not scientific, but are tied to making money. Sebeok asserts that Patterson's motives for conducting research amount to nothing more than a "desperate reaching out for media recognition (of which this unfortunate book represents but one example)."

My initial focus in this study has been on examining topical choices that Sebeok made when evaluating Patterson's and Linden's scientific *ethos*. At this point, I want to show how the two authors rhetorically constructed a "revolutionary" public image for science and scientists that sharply contrasts with the public image prescribed by Mertonian "normal" science. Patterson argued that objections to claims that nonhuman primates have language abilities are symptomatic of a deeper, intellectual revolution within the behavioral sciences concerned with language. After R. Allen Gardner and Beatrice T. Gardner published findings that the chimpanzee Washoe was able to use language, *Science* published rebuttals written, according to Patterson, by "the most distinguished names in the behavioral sciences." In the face of vigorous resistance by eminent scientists, Patterson claimed that the Gardners' "success" with Washoe presented "one of the most basic tenets of modern life" with an "anomaly;" there was now evidence that humans are not unique in their possession of language. She contended that the hostile reaction to these anomalous claims indicated that something like Kuhn's notion of scientific revolution was taking place in studies of language acquisition in primates.[38] Once readers embrace this revolutionary scenario, they are easily led to conclude that "credible" scientists can and do display such qualities as *individuality, particularity, dogmatism,* and *interestedness.*

Once readers agree that the Gardners have hurled fields like linguistics into "crisis" there is no warrant for esteeming *communality* as a special, scientific virtue; there is no legitimizing community consensus during periods of "crisis." Judgments about "legitimate" research institutions, "appropriate" sources of research funding, and "qualified" adjudicating audiences become partisan points of contention during "revolutionary" science. This image of non-human primate research allows Patterson to extol qualities of *individuality* as characterizing credible scientific *ethos.*[39] We are to conclude that Patterson, by following the Gardners, displays intellectual courage by making bold judgments to abandon outmoded but comfortable notions about language, and by pursuing a pioneering line of thinking even when confronted with eminent hostile opinion. Thus, the conflict of her position with conventional knowledge is raised to the heights of scientific virtue.

Throughout the book, *universality* is minimized as an important quality of scientific thought and conduct. In revolutionary contexts, scientists cannot turn to "pre-established impersonal criteria" to settle controversy. In the case of linguistic science, they cannot rely on previously certified knowledge because, Patterson says, "there is very little that can be said about language today that is not open to question or controversy."[40] Nor can scientists settle disputes by making carefully constrained observations and using empirically rigorous experimental designs. Patterson contends: "There is much about language that does not lend itself to reduction to statistics and hard data, and some linguists have recently reacted against the rigid, formalized treatment of language. . . . This is not to justify vagueness but to illustrate that it is very difficult to speak with any confidence of 'facts' about language."[41] What scientists must do, says Patterson, is supplement the "rigid" experimental work with "interpretive" case studies.

Case studies require special interpretive sensibilities that allow researchers to achieve what I shall call, for lack of a more fitting expression, *experimental rapport* with subject beasts. The line of thought, based upon the *topos* of *particularity*, is that adequate scientific investigations of the behaviors of sentient beings require that researchers possess special abilities for developing emotional attachments with animal subjects. Patterson uses this line of thought to amplify her scientific credibility and minimize the scientific credibility of her critics. Only those scientists able to establish true rapport with the animal whose behavior is being studied will meet a necessary condition for gleaning positive evidence of language acquisition in non-human primates; those who retain rigidly objective experimental stances toward the animal will confound possibilities for discovering significant language use by the animal. Accordingly, the "best" research with primates encouraged development of rapport between scientist and animal:

> In none of these cases did the experimenter [Terrace and Premack] allow himself to develop a true, close rapport with his chimp. This was justified in the laudable name of objectivity, but given the sensitivity of the animals involved—Koko's signing is affected by even slight disruptions in her routine—it is hard not to wonder whether the different conclusions about ape language abilities reached by these scientists ultimately trace back to the different relationships between experimenters and subjects and to the persistence that has marked the efforts of those of us who have established close rapport with our subjects. If this is the case, I am reaffirmed in my belief that one cannot really understand the mental workings of other animals or bring them to the limits of their abilities unless one first has true rapport with them. Even the critics admit this possibility. What they fail to see is that the problem really is a misunder-

standing of the purpose of language. Once that misunderstanding is straightened out and we accept language as a communicative behavior, the evidence of Koko's abilities is compelling for those who want to see it.[42]

Implicit in Patterson's discussion is a view of researchers as fundamentally *dogmatic*. Scientists cannot skeptically avoid dogmatic adherence to assumptions, so the best that they can do is choose and adhere to assumptions that allow them to comprehend the full range of non-human primate linguistic ability. The choices are clear for Patterson: researchers either assume that Koko is a "dolt" and dismiss all apparently innovative and intentional linguistic acts as "mistakes," or they recognize that the animal is "a bright, playful, creative creature capable of quite sophisticated innovation."[43] "Rigid" experimentalists take the former stance, while those who supplement formal testing procedures with anecdotal, case-study material choose the latter, Patterson asserts. For Patterson, only the second alternative can yield persuasive evidence for Koko's linguistic creativity.

In Patterson's view of scientific *ethos*, credible scientists must be intensely *interested* in the consequences that their research claims have beyond the knowledge-oriented interests of scientific communities. The interested scientist addresses research claims about nonhuman primates' language abilities beyond scientific communities to expose and to falsify religious and philosophical systems that make humans the uniquely "languaging" animal. One consequence of exposing this "false" belief is an ethical challenge to humans' "rights to experiment with or harvest natural resources. . . ."[44] In sum, "virtuous" scientists are interested in the religious, philosophical and ethical implications of technical claims for society at large. They are not focused narrowly and solely on the knowledge-oriented concerns of their specialized communities.[45]

When Patterson's discourse is contrasted with Sebeok's we can see how topical selection and development can influence how situated, lay audiences will judge an individual's or a group's scientific ethos. Sebeok constructed a Mertonian image for science, and on the basis of that image implied that Patterson and Linden were not "real" scientists because they lacked the scientific virtues of communality, universality, skepticism and disinterestedness. Patterson and Linden constructed a revolutionary image for science, and worked from that image to show that Patterson's thought and conduct displayed the esteemed qualities of individuality, particularity, dogmatism, and interestedness and, moreover, that her critics, Sebeok included, lacked those "virtues." Both Seboek and Patterson and Linden strategically selected topoi in order to construct a scientific ethos that was favorable to securing and justifying their respective professional interests and objectives, given the constraining influences of their particular rhetorical situation.

Topoi and Rhetorical Boundaries

The case study illustrates how scientific "insiders" and "outsiders" can rhetorically construct conflicting perspectives on scientific *ethos* when responding discursively to demarcation problems. Sebeok composed a review that invoked sharp "boundaries" between science and non-science, implying that Patterson and Linden were not "real" scientists because their thought and conduct did not adhere to the "virtues" of universality, communality, skepticism, and disinterestedness. Readers are urged to dismiss Patterson and Linden as incompetent or pseudo-scientific outsiders on the basis of these "scientific" qualities and, consequently, to conclude that *The Education of Koko* must be a scientific ruse. Patterson and Linden's book created a perspective on scientific *ethos* that blurred the boundaries between science and nonscience, suggesting that credible scientists adhere in thought and conduct to virtues of particularity, individuality, enthusiasm or dogmatism, and interestedness—qualities of character that do not sharply differentiate scientists from nonscientists. I surmise that this perspective on scientific *ethos* is often created when those who view themselves as revolutionary outsiders want to challenge and to overturn the claims of orthodoxy, but more evidence is needed to substantiate this conjecture. Nevertheless, in this instance both the rhetoric of insiders and outsiders was addressed to the laity which was left with the task of deciding what qualities credible scientists do and do not have.

Sebeok and Patterson and Linden put forward contrasting public images of credible science and scientists. Sebeok depicts science as a unique intellectual activity characterized by a high degree of consensus and agreement about what "counts" as science. "Normal" scientists—those possessing a credible scientific *ethos*—will adhere to Mertonian standards of thought and conduct. Only pseudo-scientists or eccentrics working at the lunatic fringe will neglect those standards. Patterson and Linden portray science as characterized by dissensus and disagreement. "Revolutionary" rather than "normal" scientists possess qualities that are virtuous given that image. In revolutionary circumstances, Mertonian vices are shaded into virtues. Both public portrayals are *idealizations* of science and scientists. Specifically, linguistic science is neither as "normal" as Sebeok implies through his review, nor is it as "revolutionary" as Patterson and Linden would have their readers believe. These idealized visions of science and scientists can be explained only partly by the rhetors' need to address a scientific laity that is incapable of following the relevant technical arguments. I say "partly" because even scholarly commentators cannot agree about whether the actual scientific endeavor is best characterized by consensus or dissensus.[46]

The best explanation is that neither idealized vision can alone prescribe situationally transcendent and uniformly applicable standards constituting *the* scientific *ethos*. Science is both consensual and divisive, cooperative and competitive; and scientists can possess both "normal" and "radical" qualities. The alternative public "faces" of science underscore the point that scientific *ethos* is a *rhetorical* construction. Contrasting images of scientific *ethos* show that either of the opposed qualities of universality or particularity, communality or individuality, skepticism or enthusiasm, and disinterestedness or interestedness can be shaded into a scientific virtue or vice. Rhetors often include among compositional decisions those concerning which strategies are most useful for constructing a scientific *ethos* that can best advance their professional interests, given the proble... they must face and the constraints of their audience and situation.

It may be that the demarcation exigence and the standard kinds of rhetorical strategies that "insiders" and "outsiders" use in response are generalizable. For instance, demarcation exigences emerge when mainstream scientists argue that parapsychology or "scientific" creationism are not "real" sciences and that their practitioners are not "real" scientists. Analysis of these disputes shows that critics of parapsychology and creationism turn to the kinds of *topoi* Sebeok used to construct perspectives on scientific *ethos* in their rhetorical situations.[47] This suggests that the *topoi* identified in this study are among the standard and finite sets of themes that scientists choose from when composing discourse in response to demarcation exigences.

When scientists address ambiguities about what it means to think and to act like scientists, they will turn to those kinds of arguments catalogued by the *topoi* related to scientific *ethos*. However, scientists also respond to technical kinds of exigences, and topical analysis of discourse can reveal the kinds of arguments scientists use when trying to resolve those exigences as well. For instance, scientists encounter ambiguities about the scientific significance of evidence or of the constructs that they use to articulate and apply theory. Ambiguities about the intrinsic or comparative scientific value of evidence and theory constitute a special *kind* of technical exigence which demands that scientists choose the right kind of *topoi* to make an effective rhetorical response. What some commentators call "good reasons" for making scientific judgments might actually function like rhetorical *topoi* that are specially useful for making arguments in response to problems of *scientific* "value" or "significance." A preliminary list would include the scientific "values" of accuracy, consistency, scope, simplicity, and fruitfulness.[48] Future case studies can with profit focus on revealing the kinds of *topoi* that scientists use when responding to this and other kinds of situational exigences.[49] This might lead to the compilation of topical inventories, organized according to the kinds of situational exigences that scientists confront. As these lists are amended and refined, rhetorical critics can use them heuristically to distinguish the kinds of topical choices rhetors *could* have made from those that they *did* make, and to assess the comparative persuasive efficacy of the rhetors' choices given the *kind* of situational exigence they were seeking to resolve.

The "rhetoric of inquiry" is itself a rhetorical effort at heralding a demarcation crisis in the human sciences; it seeks to indict the present "boundaries" among substantive fields of inquiry as largely artificial and points to the need to redraw or traverse those boundaries in fruitful ways. Scientists must confront concerns about professional *ethos* when addressing demarcation exigences; so, too, must scholars conducting research on the "rhetoric of inquiry." However, before we go too far in our rhetorical "boundary-work" we must first clarify what the present boundaries are. Does each field of inquiry have its own technical expectations and substantive concerns which prescribe for its practitioners special "virtues" of professional thought and conduct? What does it mean to think and to act like, say, a poet, sociologist, physicist? Do the "habits" that artists and scientists display show that the "two cultures" are converging or diverging? Topical analysis provides the field-independent methodological approach that allows us to cut across the currently fragmented areas of substantive inquiry and provide answers to these and other questions. By answering such questions, we can become more self-reflective about what it means to be a member of a particular academic community and discern better whether the "boundaries" we invoke to demarcate ourselves from members of other communities are desirable for intellectual or practical purposes, or are merely artificial.

Notes

1. Aristotle, *Rhetoric*, tr. W. Rhys Roberts (Random House, New York, 1954), 1356a5.
2. *Ibid.*, 1378a5-19.
3. Robert K. Merton first published his views on the scientific *ethos* in his essay "Science and the Social Order," *Philosophy of Science*, 5 (1938), 321-37. A more systematic rendering of his ideas appeared in the essay, "Science and Technology in a Democratic Order," *Journal of Legal and Political Sociology*, I (1942), pp. 115-26. I am using the reprint of this essay, "The Normative Structure of Science," in Robert K. Merton, *The Sociology of Science: Theoretical and Empirical Investigations*, ed. Norman W. Storer (Chicago: University of Chicago Press, 1973), pp. 267-78. Unless otherwise indicated, all references to Merton's articles will be to this useful compilation.
4. Merton, "Normative Structure of Science," p. 270. In Merton's view, extending certified knowledge is the institutional goal of science.
5. *Ibid.*, pp. 268-69. Merton's scientific *ethos* is the sociological complement to the epistemological position known as the "standard view." For a critique of this philosophical position which grounds Merton's sociology of science see Michael Mulkay, *Science and the Sociology of Knowledge* (London: Allen, 1979).
6. Merton, "Normative structure of science," p. 270.
7. I am relying on Barber's interpretation of disinterestedness. Storer amplifies this idea by explaining that it encourages pursuit of "science for science's sake." See Bernard Barber, *Science and the Social Order* (1952; reprinted Greenwood, Westport, 1978), p. 92; and Norman W. Storer, *The Social System of Science* (New York: Holt, 1966), p. 79.
8. For Merton's discussion of the norms see "Normative Structure of Science," p. 270-78. Disinterestedness, universalism, communality and skepticism imply that *emotional neutrality* toward ideas and actions is a scientific virtue. Although Merton did not raise objectivity to normative status, other commentators have discussed emotional neutrality as a scientific value. See Barber, *Science and the Social Order*, pp. 88-9; and Storer, *Social System of Science*, pp. 79-80.
9. Merton, "Priorities in scientific discovery," in *Sociology of Science*, especially pp. 239-305.
10. For examples of research strongly supportive of Merton's normative structure of science see John R. Cole and Stephen Cole, *Social Stratification in Science* (Chicago: University of Chicago Press, 1973); Jonathan R. Cole, *Fair Science: Women in the Scientific Community* (New York: Free Press, 1979); Jerry Gaston, *The Reward System in British and American Science* (New York: Wiley, 1978); Barber, *Science and the Social Order*; and Storer, *Social System of Science*.
11. See Ian I. Mitroff, *The Subjective Side of Science: A Philosophical Inquiry into the Psychology of the Apollo Moon Scientists* (Amsterdam: Elsevier, 1974).
12. Particularism, solitariness, interestedness, and dogmatism each imply that *emotional commitment* to one's ideas and actions is a necessary ingredient of science. Some commentators have gone so far as to call emotional commitment the counter-norm to emotional neutrality. See Mitroff, *Subjective Side of Science*, p. 276. For lists containing these and other norms and counter-norms see ibid., p. 79; and Ian I. Mitroff and Richard O. Mason, *Creating a Dialectical Social Science: Concepts, Methods, and Models* (Dordrecht: Reidel, 1981), pp. 147-8.
13. Stehr provides a useful overview of the central issues involved. Barnes and Dolby and Mulkay are among Merton's most vociferous critics. Gaston and Zuckerman issue strong defenses. See Nico Stehr, "The Ethos of Science Revisited: Social and Cognitive Norms," *Sociological Inquiry*, 48 (1978), pp. 172-96; S.B. Barnes and R.G.A. Dolby, "The Scientific Ethos: A Deviant Viewpoint," *Archives Européenes de*

Sociologie, 11 (1970), pp. 3-25; Mulkay, *Science and the Sociology of Knowledge*, especially pp. 63-73; Gaston, *The Reward System*, pp. 158-84; and Harriet Zuckerman, "Deviant Behavior and Social Control in Science," in Edward Sagarin (ed.), *Deviance and Social Change* (Beverly Hills: Sage, 1977), pp. 123-8.

14. I am following Lloyd F. Bitzer's idea that rhetorical situations are characterized by exigences or ambiguities that can be solved, clarified, or modified through discourse. See "The Rhetorical Situation," *Philosophy and Rhetoric*, 1 (1968), pp. 6-7.

15. Barnes and Dolby, "The scientific ethos," p. 13.

16. As examples, consider some standard attacks on the scientific *ethos* of creationists and parapsychologists. See Thomas I. Gieryn, George M. Bevins, and Stephen C. Zehr, "Professionalization of American Scientists: Public Science in the Creation/Evolution Trials," *American Sociological Review*, 50 (1985), especially pp. 399-405; and H.M. Collins and T.J. Pinch, "The Construction of the Paranormal: Nothing Unscientific is Happening," in Roy Wallis (ed.), *On the Margins of Science: The Social Construction of Rejected Knowledge*, Sociological Review Monographs 27 (University of Keele, Staffs, UK, 1979), especially pp. 246-7.

17. On this point, see Thomas M. Lessl, "Science and the Sacred Cosmos: The Ideological Rhetoric of Carl Sagan," *Quarterly Journal of Speech*, 71 (1985), p.176.

18. Aristotle, *Rhetoric*, 1356a6-8.

19. When there is ambiguity or cause for doubt about technical claims, qualities exhibited through presentations and prior reputation are factors of *ethos* that will be weighed in formation of final judgments. Consider some examples. A leading advocate of the molecular memory transfer hypothesis exhibited qualities through presentations that some thought were less than "scientific." The scientist conducted scientific work without being sufficiently "earnest." Another scientist, involved in the discovery of the solar neutrino anomaly. drew upon his established reputation for being careful, modest, and open with results as an important persuasive resource for establishing the anomalous claim. See David Travis, "On the Construction of Creativity: The 'Memory Transfer' Phenomenon and the Importance of Being Earnest," in Karin D. Knorr, Roger Krohn, and Richard Whitley (eds), *The Social Process of Scientific Investigation*, Sociology of the Sciences Yearbook 4 (Dordrecht: Reidel, 1981), pp. 177-8; and T.J. Pinch, "Theoreticians and the Production of Experimental Anomaly: The Case of Solar Neutrinos," in ibid., pp. 94-5.

20. According to Zuckerman, she and Merton found evidence for this point when they examined referees' reasons for recommending or rejecting submissions to the physics journal, *Physical Review*. Specifically, referees were more likely to endorse unorthodox ideas when authored by established scientists than young or rank-and-file scientists. Zuckerman interprets this as evidence for the influence of "performance-based authority" on reviewers' judgments. Unfortunately, this idea was not developed in Zuckerman's and Merton's article on the subject. See Harriet Zuckerman, "Theory Choice and Problem Shoice in Science," *Sociological Inquiry,* 48 (1978), p. 70; and Robert K. Merton, "Institutionalized Patterns of Evaluation in Science," in Merton, *Sociology of Science*, especially pp. 476-91.

21. Mulkay, *Science and the Sociology of Knowledge*, p. 71.

22. *Ibid.*, pp. 71-2.

23. Thomas E. Gieryn, "Boundary-work and the Demarcation of Science from Nonscience: Strains and Interests in Professional Ideologies of Scientists," *American Sociological Review*, 48 (1983), pp. 781-95.

24. Gieryn affirms Mulkay's claim that what I am calling the special *topoi* of scientific *ethos* become especially useful when scientists address professional rhetoric to lay audiences. See ibid., p. 783; Gieryn et al., "Professionalization of American scientists," especially pp. 403-4; and Michael Mulkay, "Norms and Ideology in Science," *Social Science Information,* 15 (1976), p. 646.

25. Francine Patterson and Eugene Linden, *The Education of Koko* (New York: Holt, 1981);Thomas A. Sebeok, "The Not So Sedulous Ape: Review of *The Education of Koko* by Francine Patterson and Eugene Linden," *Times Literary Supplement,* 10 September 1982, p. 976. Unless otherwise indicated, all references to Sebeok's rhetoric are to this review.

26. For instance, see Laura A. Petitto and Mark S. Seidenberg, "On the Evidence for Linguistic Abilities in Signing Apes," *Brain and Language*, 8 (1979), pp. 162-83. Also see the critical review of the techniques and conclusions of language experiments with nonhuman primates, including Patterson's, in Mark S. Seidenberg and Laura A. Petitto, "Signing Behavior in Apes: a Critical Review," *Cognition*, 7 (1979), pp. 177-215.

27. Most of Sebeok's review is directed toward discrediting Patterson's *ethos*, with less attention devoted to Linden's. This is appropriate given that the book is based on Patterson's research, which is presented to the reader in Patterson's voice.

28. Sebeok's appeal to Terrace's study has added rhetorical force because Terrace had initially believed that his chimpanzee, "Nim Chimpsky," was capable of creating a sentence. Terrace admitted that he had "fantasies" about what he could accomplish with his communicating chimp, including using the animal as a translator of ape communication in the wilds; but after further analysis of his data he concluded that Nim said little on his own was merely imitating behavior in an effort to get rewards—Nim was not able to construct sentences. When those holding "heretical" opinions publicly reform their ideas their very example provides conservative defenders of orthodoxy with powerful means of persuasion; the act of reformation casts doubt on the legitimacy of heretical beliefs and renders ambiguous the authenticity of those who hold them. Sebeok used the reformed Terrace's "true confessions" as persuasive means for questioning the legitimacy of Patterson's claims and the scientific authenticity of her motives. Terrace explained his changed views in *Signs of the Apes, Songs of the Whales* (film, produced by Linda Harrar, Nova, WGBH-Boston, PBS, 1983).

29. Sebeok sought to discredit Patterson's scientific *ethos* by asserting throughout his review that her interpretations of Koko's behavior are riddled with "anthropomorphic" (read: "unscientific") tendencies. After all, gorillas do not willfully make false statements; only humans can lie. However, Sebeok was inducing readers to make a sharp dichotomy between "anthropomorphic" and "scientific" interpretation that is not always so easily drawn. Gould suggests that there will always be anthropomorphic tendencies when humans investigate animal behavior because "we cannot write, study, or even conceive of other creatures except in overt or implied comparison with ourselves." Human entanglements with animals is a fact for Gould; what is at issue is the nature and degree of those entanglements. See Stephen Jay Gould, "Animals and Us: Review of *The Chimpanzees of Gombe: Patterns of Behaviour* by Jane Goodall", *New York Review of Books*, 25 June 1987, p. 20.

30. Judgments about scientists' *ethos* often overlap with considerations pertinent to appraising the technical reasonableness of research claims, or what classical rhetoricians would have called the *logos* of scientific discourse. Although closely interrelated, I believe that the arguments just reviewed are best read as an attack primarily on Patterson's *ethos* rather than as a simple critique of technical *logos*. Sebeok is trying to induce readers to doubt Patterson's *technical virtuosity* as a scientist. The *topos* of *universality* recommends the fitting line of thought: that "real" scientists willingly and capably test their claims against pre-established, impersonal standards. Patterson fails to display her technical virtuosity by neglecting two such impersonal standards: (1) the intellectual consensus among scientists regarding what is accepted and rejected knowledge; and (2) empirical observations (due to her alleged experimental incompetence).

31. I surmise that Sebeok is here referring to Patterson's Gorilla Foundation. In her book Patterson makes a tacit but obvious appeal to the laity for financial support of the Gorilla Foundation. See Patterson and Linden, *Education of Koko*, p. 213.

32. Of the three arguments that Sebeok draws from the *topos* of communality, only this one approximates Merton's idea of intellectual communism. From Merton's perspective, Patterson's scientific conduct is questionable because she does not willingly "share" her research by publishing in legitimate journals. Merton's idea of communism makes "secrecy" a vice because it is assumed that what is not being shared with the community is *worth* knowing. However, during demarcation crises *ability* to share can become as much an issue as willingness to share. I found it necessary to reinterpret the *topos* of communality as recommending this more encompassing line of thought: that "real" scientists actively participate in the intellectual life of their community. This is the persuasive basis of Sebeok's three arguments. Patterson fails to display community participation because she lacks a credible position, fails to secure "legitimate" research support, and will not or cannot share technical claims with scientific audiences.

33. Francine Patterson, "Conversation with a gorilla," *National Geographic*, 154, Oct. (1978), pp. 438-65 (condensed in *Reader's Digest*, 114, Mar. (1979), pp. 81-6; *Koko, A Talking Gorilla* (a ninety-minute 16mm film by Barbet Schroeder, available from New Yorker Films).

34. Quoted in Cynthia Gorney, "Gorilla Koko hasn't convinced everybody that she can talk," *Houston Chronicle*, 4 February 1985, section 5, p. 5.

35. Another way of putting this is that she lacked "good will" toward members of relevant knowledge communities. Sebeok recounts how Patterson and Linden responded to Terrace's conclusion in *Nim* that there is no evidence showing that apes can generate or interpret sentences by hurling "such epithets as 'muddle-headed' (Patterson), 'apostate' (Linden), and worse."

36. All of these statements imply that Patterson violated a major unwritten taboo in scientific life: she appealed "to the populace at large" rather than to the "well-defined community of the scientist's professional compeers," as Kuhn had put it. Scientists are especially vigorous in their public condemnations of those who seek to gain lay acceptance for unorthodox and unauthorized claims. The ferocity of Sebeok's review

provides a case in point. See Thomas S. Kuhn, *The Structure of Scientific Revolutions,* 2nd edn (Chicago: University of Chicago Press, 1970), p. 168.

37. Barber, *Science and the Social Order,* p. 92.

38. Patterson and Linden, *Education of Koko,* pp. 24-6.

39. Individuality illuminates Patterson's and Linden's arguments better than solitariness, the purported counternorm of Merton's intellectual communism. Rather than using "secrecy" to bolster their scientific *ethos,* they instead displayed their unwillingness to sacrifice truth to the authority of tradition and its prominent spokespersons. When anti-authoritarianism is treated as a virtue it often is called *individualism,* as some Mertonian commentators have noted. This *topos* conflicts with the *topos* of communality in the sense that I reformulated that *topos* earlier. See Barber, *Science and the Social Order,* pp. 89-90.

40. Patterson and Linden, *Education of Koko,* p. 194.

41. *Ibid.,* p. 79.

42. *Ibid.,* pp. 210-11. Patterson is not without distinguished company when making arguments of this kind. Barbara McClintock, winner of the 1983 Nobel Prize in Medicine or Physiology, offered a philosophical vision of science based not on objective experimentation but on what she called developing "a feeling for the organism." For her, intimacy with and sympathetic understanding of the objects of knowledge is required to have genuine scientific knowledge. Lest this application of "rapport" be dismissed merely as McClintock's philosophical speculation, let us also turn to Jane Goodall's studies of a chimpanzee colony at Gombe in Tanzania. According to Gould, Goodall's work exemplifies a scientist's internal struggles to strike the proper balance in research between emotional involvement with the chimpanzees and the need to secure "maximal distance" so that the chimpanzees' behavior does not become distorted through the scientist's intrusions. These internal struggles illustrate that rapport can have important influences on scientific research. Gould says of Goodall's work: "You can't just march off into dense foliage and find chimps; you must first make contact and build trust in order to win acceptance and establish the possibility of following in the wild" ("Animals and Us," p. 24). Goodall's experience also underscores the fact that topical choices among skepticism and enthusiasm, or objectivity and rapport, are not merely arbitrary. There are circumstances in which scientists can clarify ambiguities about their professional conduct through appeals to "rapport" as a virtue comparatively superior to "objectivity." See Gould, "Animals and us," pp. 23-5; and Evelyn Fox Keller, *A Feeling for the Organisms: The Life and Work of Barbara McClintock* (New York: Freeman, 1983), pp. 197-207.

43. Patterson and Linden, *Education of Koko,* p. 207.

44. *Ibid.,* p. 25.

45. This argument is a revolutionary variation on the theme that virtuous scientists can serve special communities of interest. Patterson and Linden are virtuous not because their interests coincide with those possessed by members of an informal college, but because their interests are linked with issues that concern general society.

46. For an overview of the key points-at-issue see Larry Laudan, "Two Puzzles about Science: Reflections about Some Crises in the Philosophy and Sociology of Science," *Minerva,* 20 (1982), pp. 253-68.

47. When orthodox scientists attack parapsychology and creationism as "pseudo-science" they include among their means of persuasion efforts to discredit the scientific *ethos* of practitioners. They do this by claiming that parapsychologists and creationists do not think and act like "real" scientists. Typically the standard Mertonian arguments are adduced in support. Parapsychologists and creationists are said to be: (1) openly defiant of both the "universal" consensus on accepted and rejected knowledge and the need for empirical confirmation of technical claims (*universality*); (2) pursuing extra-scientific motives including advancement of beliefs in the supernatural (*disinterestedness*); (3) dogmatically attached to their allegedly "scientific" claims (*skepticism*); and (4) incapable of participating in the "real" scientific community, as indicated by their inability to secure visible positions, "legitimate" research funds, and publications in orthodox journals (*communality*). Parapsychologists have been far more successful than the creationists at legitimizing their scientific *ethos,* and they did this largely by emulating Mertonian virtues. Nevertheless, like the creationists and Patterson and Linden, they also attempt to diffuse orthodox criticism by creating a revolutionary scenario for advancing their "radical" claims. However, more critical work is needed before we can safely generalize that these and other "intellectual revolutionaries," like Patterson and Linden, draw arguments from the *topoi* of *particularity, interestedness, dogmatism* and *individuality.* On the use of Mertonian *topoi* against the creationists, see Gieryn, et al., "Professionalization of American Scientists," pp. 401-3. My claims about the parapsychologists are based on Paul D. Allison, "Experimental Parapsychology as a Rejected Science," in Wallis, *On the Margins of Science,* especially pp. 277-88; Collins and Pinch, "Construction of the paranormal," *passim;* and Jim Palmer, "Why is Science Spooked by 'psi'?" *Washington Post,* 8 Mar. 1987, sec. B, p. 3.

48. Kuhn describes functional features of these "good reasons" that reflect their rhetorical topicality: "Individu-
    ally the criteria are imprecise: individuals may legitimately differ about their application to concrete cases.
    In addition, when deployed together, they repeatedly prove to conflict with one another," (Thomas S.
    Kuhn, "Objectivity, Value Judgment, and Theory Choice," in *The Essential Tension: Selected Studies in
    Scientific Tradition and Change* (University of Chicago Press, Chicago, 1977), p. 322). Like the *topoi*
    related to scientific *ethos*, these technical *topoi* have evocative, inventional powers due to their working
    ambiguities, thus allowing generation of varied and sometimes conflicting arguments both within and
    across rhetorical situations. See *ibid.*, pp. 321-5.
49. For an example of how topical analysis can be applied to technical articles see S. Michael Halloran, "The
    Birth of Molecular Biology: An Essay in the Rhetorical Criticism of Scientific Discourse," *Rhetoric
    Review*, 3 (1984), especially pp. 73-4. In this volume, see Halloran. pp. 42-3.

Public Science

Dialectic and Rhetoric
at Dayton, Tennessee
by Richard M. Weaver

Dialectic and rhetoric are distinguishable stages of argumentation, although often they are not distinguished by the professional mind, to say nothing of the popular mind. Dialectic is that stage which defines the subject satisfactorily with regard to the *logos,* or the set of propositions making up some coherent universe of discourse; and we can therefore say that a dialectical position is established when its relation to an opposite has been made clear and it is thus rationally rather than empirically sustained. Despite the inconclusiveness of Plato on this subject, we shall say that facts are never dialectically determined—although they may be elaborated in a dialectical system—and that the urgency of facts is never a dialectical concern. For similar reasons Professor Adler, in his searching study of dialectic, maintains the position that "Facts, that is non-discursive elements, are never determinative of dialectic in logical or intellectual sense. . . ."[1]

What a successful dialectic secures for any position therefore, as we noted in the opening chapter, is not actuality but possibility; and what rhetoric thereafter accomplishes is to take any dialectically secured position (since positive positions, like the "position" that water freezes at 32°F., are not matters for rhetorical appeal) and show its relationship to the world of prudential conduct. This is tantamount to saying that what the specifically rhetorical plea asks of us is belief, which is a preliminary to action.

It may be helpful to state this relationship through an example less complex than that of the Platonic dialogue. The speaker who in a dialectical contest has taken the position that "magnanimity is a virtue" has by his process of opposition and exclusion won our intellectual assent, inasmuch as we see the abstract possibility of this position in the world of discourse. He has not, however, produced in us a resolve to practice magnanimity. To accomplish this he must pass from the realm of possibility to that of actuality; it is not the logical invincibility of "magnanimity" enclosed in the class "virtue" which wins our assent; rather it is the contemplation of magnanimity sub specie actuality. Accordingly when we say that rhetoric instills belief and action, we are saying that it intersects possibility with the plane of actuality and hence of the imperative.[2]

A failure to appreciate this distinction is responsible for many lame performances in our public controversies. The effects are, in outline, that the dialectician cannot understand why his demonstration does not win converts; and the rhetorician cannot understand why his appeal is rejected as specious. The answer, as we have begun to indicate, is that the dialectic has not made reference to reality, which men confronted with problems of conduct require; and the rhetorician has not searched the grounds of the position on which he has perhaps spent much eloquence. True, the dialectician and the rhetorician are often one man, and the two processes may not lie apart in his work; but no student of the art of argumentation can doubt that some extraordinary confusions would be prevented by a knowledge of the theory of this distinction. Beyond this, representative government would receive a tonic effect from any improvement of the ability of an electorate to distinguish logical positions from the detail of rhetorical amplification. The British, through their custom of putting questions to public speakers and to officers of government in Parliament, probably come nearest to getting some dialectical clarification from their public figures. In the United States, where there is no such custom, it is up to each disputant to force the other to reveal his grounds; and this, in the ardor of shoring up his own position rhetorically, he often fails to do with any thoroughness. It should therefore be profitable to try the kind of analysis we have explained upon some celebrated public controversy, with the object of showing how such grasp of rhetorical theory could have made the issues clearer.

For this purpose, it would be hard to think of a better example than the Scopes "evolution" trial of a generation ago. There is no denying that this trial had many aspects of the farcical, and it might seem at first glance not serious enough to warrant this type of examination. Yet at the time it was considered serious enough to draw the most celebrated trial lawyers of the country, as well as some of the most eminent scientists; moreover, after one has cut through the sensationalism with which journalism and a few of the principals clothed the encounter, one finds a unique alignment of dialectical and rhetorical positions.

The background of the trial can be narrated briefly. On March 21, 1925, the state of Tennessee passed a law forbidding the teaching of the theory of evolution in publicly supported schools. The language of the law was as follows:

> Section 1. Be it enacted by the general assembly of the state of Tennessee, that it shall be unlawful for any teacher in any of the universities, normals and all other public schools of the state, which are supported in whole or in part by the public school funds of the state, to teach any theory that denies the story of the Divine creation of man as taught in the Bible, and to teach instead that man has descended from a lower order of animals.

That same spring John T. Scopes, a young instructor in biology in the high school at Dayton, made an agreement with some local citizens to teach such a theory and to cause himself to be indicted therefore with the object of testing the validity of the law. The indictment was duly returned, and the two sides prepared for the contest. The issue excited the nation as a whole; and the trial drew as opposing counsel Clarence Darrow, the celebrated Chicago lawyer, and Williams Jennings Bryan, the former political leader and evangelical lecturer.

The remarkable aspect of this trial was that almost from the first the defense, pleading the cause of science, was forced into the role of rhetorician; whereas the prosecution, pleading the cause of the state, clung stubbornly to a dialectical position. This development occurred because the argument of the defense, once the legal technicalities were got over, was that evolution is "true." The argument of the prosecution was that its teaching was unlawful. These two arguments depend upon rhetoric and dialectic respectively. Because of this circumstance, the famous trial turned into an argument about the orders of knowledge, although this fact was never clearly expressed, if it was ever discerned, by either side, and that is the main subject of our analysis. But before going into the matter of the trial, a slight prologue may be in order.

It is only the first step beyond philosophic naïvete to realize that there are different orders of knowledge, or that not all knowledge is of the same kind of thing. Adler, whose analysis I am satisfied to accept to some extent, distinguishes the orders as follows. First there is the order of facts about existing physical entities. These constitute the simple data of science. Next come the statements which are statements about these facts; these are the propositions or theories of science. Next there come the statements about these statements: "The propositions which these last statements express form a partial universe of discourse which is the body of philosophical opinion."[3]

To illustrate in sequence: the anatomical measurements of *Pithecanthropus erectus* would be knowledge of the first order. A theory based on these measurements which placed him in a certain group of related organisms would be knowledge of the second order. A statement about the value or the implications of the theory of this placement would be knowledge of the third order; it would be the judgment of a scientific theory from a dialectical position.

It is at once apparent that the Tennessee "anti-evolution" law was a statement of the third class. That is to say, it was neither a collection of scientific facts, nor a statement about those facts (*i.e.*, a theory or a generalization); it was a statement about a statement (the scientists' statement) purporting to be based on those facts. It was, to use Adler's phrase, a philosophical opinion, though expressed in the language of law. Now since the body of philosophical opinion is on a level which surmounts the partial universe of science, how is it possible for the latter ever to refute the former? In short, is there any number of facts, together with generalizations based on facts, which would be sufficient to overcome a dialectical position?

Throughout the trial the defense tended to take the view that science could carry the day just by being scientific. But in doing this, one assumes that there are no points outside the empirical realm from which one can form judgments about science. Science, by this conception, must contain not only its facts, but also the means of its own evaluation, so that the statements about the statements of science are science, too.

The published record of the trial runs to approximately three hundred pages, and it would obviously be difficult to present a digest of all that was said. But through a carefully selected series of excerpts, it may be possible to show how blows were traded back and forth from the two positions. The following passages, though not continuous, afford the clearest picture of the dialectical-rhetorical conflict which underlay the entire trial.

The Court (in charging the grand jury)

You will bear in mind that in this investigation you are not interested to inquire into the policy of this legislation.[4]

THE PROSECUTION

THE DEFENSE

Mr. Darrow: I don't suppose the court has considered the question of competency of evidence. My associates and myself have fairly definite ideas as to it, but I don't know how the counsel on the other side feel about it. I think that scientists are competent evidence—or competent witnesses here, to explain what evolution is, and that they are competent on both sides.

Attorney-General Stewart: If the Court please, in this case, as Mr. Darrow stated, the defense is going to insist on introducing scientists and Bible students to give their ideas on certain views of this law, and that, I am frank to state, will be resisted by the state as vigorously as we know how to resist it. We have had a conference or two about the matter, and we think that it isn't competent evidence; that is, it is not competent to bring into this case scientists who testify as to what the theory of evolution is or interpret the Bible or anything of that sort.

Mr. Neal: The defendant moves the court to quash the indictment in this case for the following reasons: In that it violates Sec. 12, Art. XI, of the Constitution of Tennessee: "It shall be the duty of the general assembly in all future periods of the government to cherish literature and science. . . ."

THE PROSECUTION

Mr. McKenzie: Under the law you cannot teach in the common schools the Bible. Why should it be improper to provide that you cannot teach this other theory?

THE DEFENSE

I want to say that our main contention after all, may it please your honor, is that this is not a proper thing for any legislature, the legislature of Tennessee or the legislature of the United States, to attempt to make and assign a rule in regard to. In this law there is an attempt to pronounce a judgment and a conclusion in the realm of science and in the realm of religion.

Mr. Darrow: Can a legislative body say, "You cannot read a book or take a lesson or make a talk on science until you first find out whether you are saying against Genesis"? It can unless that constitutional provision protects me. It can. Can it say to the astronomer, you cannot turn your telescope upon the infinite planets and suns and stars that fill space, lest you find that the earth is not the center of the universe and that there is not any firmament between us and the heaven? Can it? It could—except for the work of Thomas Jefferson, which has been woven into every state constitution in the Union, and has stayed there like a flaming sword to protect the rights of man against ignorance and bigotry, and when it is permitted to overwhelm them then we are taken in a sea of blood and ruin that all the miseries and tortures and carrion of the middle ages would be as nothing. . . If today you can take a thing like evolution and make it a crime to teach it in the public schools, tomorrow you can make it a crime to teach it in the

THE PROSECUTION

THE DEFENSE

private schools, and the next year can make it a crime to teach it to the hustings or in the church. At the next session you may ban books and the newspapers.

Mr. Dudley Field Malone: So that there shall be no misunderstanding and that no one shall be able to misinterpret or misrepresent our position we wish to state at the beginning of the case that the defense believes that there is a direct conflict between the theory of evolution and the theories of creation as set forth in the Book of Genesis.
Neither do we believe that the stories of creation as set forth in the Bible are reconcilable or scientifically correct.

Mr. Arthur Garfield Hays: Our whole case depends upon proving that evolution is a reasonable scientific theory.

Mr. William Jennings Bryan. Jr. (in support of a motion to exclude expert testimony): It is, I think, apparent to all that we have now reached the heart of this case, upon your honor's ruling, as to whether this expert testimony will be admitted largely determines the question of whether this trial from now on will be an orderly effort to try the case upon the issues, raised by the indictment and by the plea or whether it will degenerate into a joint debate upon the merits or demerits of someone's views upon evolution. . . . To permit an expert to testify upon this issue would be to substitute trial by experts for trial by jury. . . .

Mr. Hays: Are we entitled to show what evolution is? We are entitled to show that, if for no other reason than to determine whether the title is germane to the act.

THE PROSECUTION

Mr. Williams Jennings Bryan: An expert cannot be permitted to come in here and try to defeat the enforcement of a law by testifying that it isn't a bad law and it isn't—I mean a bad doctrine—no matter how these people phrase the doctrine—no matter how they eulogize it. This is not the place to prove that the law ought never to have been passed. The place to prove that, or teach that, was to the state legislature. . . . The people of this state passed this law, the people of the state knew what they were doing when they passed the law, and they knew the dangers of the doctrine—that they did not want it taught to their children, and my friends, it isn't—your honor, it isn't proper to bring experts in here and try to defeat the purpose of the people of this state by trying to show that this thing they denounce and outlaw is a beautiful thing that everybody ought to believe in. . . It is this doctrine that gives us Nietzsche, the only great author who tried to carry this to its logical conclusion, and we have the testimony of my distinguished friend from Chicago in his speech in the Loeb and Leopold case that 50,000 volumes have been written about Nietzsche, and he is the greatest philosopher in the last hundred years, and have him pleading that because Leopold read Nietzsche and adopted Nietzsche's philosophy of the superman, that he is not responsible for the taking of human life. We have the doctrine—I should not characterize it as I should like to characterize it—the doctrine that the universities that had it taught, and the professors who taught it, are much

THE DEFENSE

THE PROSECUTION

more responsible for the crime than Leopold himself. That is the doctrine, my friends, that they have tried to bring into existence, they commence in the high schools with their foundation of evolutionary theory, and we have the word of the distinguished lawyer that this is more read than any other in a hundred years, and the statement of that distinguished man that the teachings of Nietzsche made Leopold a murderer. . . (*Mr. Bryan reading from a book by Darrow*) "I will guarantee that you can go to the University of Chicago today—into its big library and find over 1,000 volumes of Nietzsche, and I am sure I speak moderately. If this boy is to blame for this, where did he get it? Is there any blame attached because somebody took Nietzsche's philosophy seriously and fashioned his life on it? And there is no question in this case but what it is true. Then who is to blame? The university would be more to blame than he is. The scholars of the world would be more to blame than he is. The publishers of the world—and Nietzsche's books are published by one of the biggest publishers in the world—are more to blame than he is. Your honor, it is hardly fair to hang a 19-year-old boy for the philosophy that was taught him at the university."
. . . Your honor, we first pointed out that we do not need any experts in science. Here is one plain fact, and the statute defines itself, and it tells the kind of evolution it does not want taught, and the evidence says that this is the kind of evolution that was taught, and no number of scientists could come in here, my friends, and

THE DEFENSE

THE PROSECUTION

override that statute or take from the jury its right to decide this question, so that all the experts they could bring would mean nothing. And when it comes to Bible experts, every member of the jury is as good an expert on the Bible as any man they could bring, or that we could bring.

Mr. Stewart: Now what could these scientists testify to? They could only say as an expert, qualified as an expert upon this subject, I have made a study of these things and from my standpoint as such an expert, I say that this does not deny the story of divine creation. That is what they would testify to, isn't it? That is all they could testify about. Now, then, I say under the correct construction of the act, that they cannot testify as to that. Why? Because in the wording of this act the legislature itself construed the instrument according to their intention. . . . What was the general purpose of the legislature here? It was to prevent teaching in the public schools of any county in Tennessee that theory which says that man is descended from a lower order of animals. That is the intent and nobody can dispute it under the shining sun of this day.

THE DEFENSE

Mr. Malone: Are we to have our children know nothing about science except what the church says they shall know? I have never seen any harm in learning and understanding, in humility and open-mindedness, and I have never seen clearer the need of that learning than when I see the attitude of the prosecution, who attack and refuse to accept the information and intelligence, which expert witnesses will give them.

THE COURT

Now upon these issues as brought up it becomes the duty of the Court to determine the question of the admissibility of this expert testimony offered by the defendant.

It is not within the province of the Court under these issues to decide and determine which is true, the story of divine creation as taught in the Bible, or the story of the creation of man as taught by evolution.

If the state is correct in its insistence, it is immaterial, so far as the results of this case are concerned, as to which theory is true; because it is within the province of the legislative branch, and not the judicial branch of the government to pass upon the policy of a statute; and the policy of this statute having been passed upon by that department of the government, this court is not further concerned as to its policy; but is interested only in its proper interpretation and, if valid, its enforcement.... Therefore the court is content to sustain the motion of the attorney-general to exclude expert testimony.

THE PROSECUTION

Mr. Stewart (during Mr. Darrow's cross-examination of Mr. Bryan): I want to interpose another objection. What is the purpose of this examination?

Mr. Bryan: The purpose is to cast ridicule upon everybody who believes in the Bible, and I am perfectly willing that the world shall know that these gentlemen have no other purpose than ridiculing every Christian who believes in the Bible.

THE DEFENSE

Mr. Darrow: We have the purpose of preventing bigots and ignoramuses from controlling the education of the United States, and you know it, and that is all.

Statements of Noted Scientists as Filed into Record by Defense Counsel

Charles H. Judd, Director of School of Education, University of Chicago: It will be impossible, in my judgment, in the state university, as well as in the normal schools, to teach adequately psychology or the science of education without making

THE PROSECUTION

THE DEFENSE

constant reference to all the facts of mental development which are included in the general doctrine of evolution. . . . Whatever may be the constitutional rights of legislatures to prescribe the general course of study of public schools it will, in my judgment, be a serious national disaster if the attempt is successful to determine the details to be taught in the schools through the vote of legislatures rather than as a result of scientific investigation.

Jacob G. Lipman, Dean of the College of Agriculture, State University of New Jersey: With these facts and interpretations of organic evolution left out, the agricultural colleges and experimental stations could not render effective service to our great agricultural industry.

Wilbur A. Nelson, State Geologist of Tennessee: It, therefore, appears that it would be impossible to study or teach geology in Tennessee or elsewhere, without using the theory of evolution.

Kirtley F. Mather, Chairman of the Department of Geology, Harvard University: Science has not even a guess as to the original source or sources of matter. It deals with immediate causes and effects. . . . Men of science have as their aim the discovery of facts. They seek with open eyes, willing to recognize it, as Huxley said, even if it "sears the eyeballs." After they have discovered truth, and not till then, do they consider what its moral implications may be. Thus far, and presumably always, truth when found is also found to be

THE PROSECUTION

THE DEFENSE

right, in the moral sense of the word. . . . As Henry Ward Beecher said, forty years ago, "If to reject God's revelation in the book is infidelity, what is it to reject God's revelation of himself in the structure of the whole globe?" *Maynard M. Metcalf, Research Specialist in Zoology, Johns Hopkins University:* Intelligent teaching of biology or intelligent approach to any biological science is impossible if the established fact of evolution is omitted.

Horatio Hackett Newman, Professor of Zoology, University of Chicago: Evolution has been tried and tested in every way conceivable for considerably over half a century. Vast numbers of biological facts have been examined in the light of this principle and without a single exception they have been entirely compatible with it. . . . The evolution principle is thus a great unifying and integrating scientific conception. Any conception that is so far-reaching, so consistent, and that has led to so much advance in the understanding of nature, is at least an extremely valuable idea and one not lightly to be cast aside in case it fails to agree with one's prejudices.

Thus the two sides lined up as dialectical truth and empirical fact. The state legislature of Tennessee, acting in its sovereign capacity, had passed a measure which made it unlawful to teach that man is connatural with the animals through asserting that he is descended from a "lower order" of them. (There was some sparring over the meaning of the technical language of the act, but this was the general consensus.) The legal question was whether John T. Scopes had violated the measure. The philosophical question, which was the real focus of interest, was the right of a state to make this prescription.

We have referred to the kind of truth which can be dialectically established, and here we must develop further the dialectical nature of the state's case. As long as it maintained this dialectical position, it did not have to go into the "factual" truth of evolution, despite the outcry from the other side. The following considerations, then, enter into this "dialectical" prosecution.

By definition the legislature is the supreme arbiter of education within the state. It is charged with the duty of promoting enlightenment and morality, and to these ends it may establish common schools, require attendance, and review curricula either by itself or through its agents. The state of Tennessee had exercised this kind of authority when it had forbidden the teaching of the Bible in the public schools. Now if the legislature could take a position that the publicly subsidized teaching of the Bible was socially undesirable, it could, from the same authority, take the same position with regard to a body of science. Some people might feel that the legislature was morally bound to encourage the propagation of the Bible, just as some of those participating in the trial seemed to think that it was morally bound to encourage the propagation of science. But here again the legislature is the highest tribunal, and no body of religious or scientific doctrine comes to it with a compulsive authority. In brief, both the Ten Commandments and the theory of evolution belonged in the class of things which it could elect or reject, depending on the systematic import of propositions underlying the philosophy of the state.

The policy of the anti-evolution law was the same type of policy which Darrow had by inference commended only a year earlier in the famous trial of Loeb and Leopold. This clash is perhaps the most direct one with the Scopes case and deserves pointing out here. Darrow had served as defense counsel for the two brilliant university graduates who had conceived the idea of committing a murder as a kind of intellectual exploit, to prove that their powers of foresight and care could prevent detection. The essence of Darrow's plea at their trial was that the two young men could not be held culpable—at least in the degree the state claimed—because of the influences to which they had been exposed. They had been readers of a system of philosophy of allegedly anti-social tendency, and they were not to be blamed if they translated that philosophy into a sanction of their deed. The effect of this plea obviously was to transfer guilt from the two young men to society as a whole, acting through its laws, its schools, its publications, etc.

Now the key thing to be observed in this plea was that Darrow was not asking the jury to inspect the philosophy of Nietzsche for the purpose either of passing upon its internal consistency or its contact with reality. He was asking precisely what Bryan was asking of the jury at Dayton, namely that they take a strictly dialectical position outside it, viewing it as a partial universe of discourse with consequences which could be adjudged good or bad. The point to be especially noted is that Darrow did not raise the question of whether the philosophy of Nietzsche expresses necessary truth, or whether, let us say, it is essential to an understanding of the world. He was satisfied to point out that the state had not been a sufficiently vigilant guardian of the forces molding the character of its youth.

But the prosecution at Dayton could use this line of argument without change. If the philosophy of Nietzsche were sufficient to instigate young men to criminal actions, it might be claimed with even greater force that the philosophy of evolution, which in the popular mind equated man with the animals, would do

the same. The state's dialectic here simply used one of Darrow's earlier definitions to place the anti-evolution law in a favorable or benevolent category. In
sum: to Darrow's previous position that the doctrine of Nietzsche is capable of
immoral influence, Bryan responded that the doctrine of evolution is likewise
capable of immoral influence, and this of course was the dialectical countering
of the defense's position in the trial.

There remains yet a third dialectical maneuver for the prosecution. On the
second day of the trial Attorney-General Stewart, in reviewing the duties of the
legislature, posed the following problem: "Supposing then that there should come
within the minds of the people a conflict between literature and science. Then
what would the legislature do? Wouldn't they have to interpret?. . . . Wouldn't
they have to interpret their construction of this conflict which one should be
recognized as higher . . . in the public schools?"

This point was not exploited as fully as its importance might seem to warrant;
but what the counsel was here declaring is that the legislature is necessarily the
umpire in all disputes between partial universes. Therefore if literature and
science should fall into a conflict, it would again be up to the legislature to assign
the priority. It is not bound to recognize the claims of either of these exclusively
because, as we saw earlier, it operates in a universe with reference to which these
are partial bodies of discourse. The legislature is the disposer of partial universes.
Accordingly when the Attorney-General took this stand, he came the nearest of
any of the participants in the trial to clarifying the state's position, and by this we
mean to showing that for the state it was a matter of legal dialectic.

There is little evidence to indicate that the defense understood the kind of
case it was up against, though naturally this is said in a philosophical rather than
a legal sense. After the questions of law were settled, its argument assumed the
substance of a plea for the truth of evolution, which subject was not within the
scope of the indictment. We have, for example, the statement of Mr. Hays, already
cited, that the whole case of the defense depended on proving that evolution is a
"reasonable scientific theory." Of those who spoke for the defense, Mr. Dudley
Field Malone seems to have had the poorest conception of the nature of contest.
I must cite further from his plea because it shows most clearly the trap from which
the defense was never able to extricate itself. On the fifth day of the trial Mr.
Malone was chosen to reply to Mr. Bryan, and in the course of his speech he made
the following revealing utterance: "Your honor, there is a difference between
theological and scientific men. Theology deals with something that is established
and revealed; it seeks to gather material which they claim should not be changed.
It is the Word of God and that cannot be changed; it is literal, it is not to be
interpreted. That is the theological mind. It deals with theology. The scientific
mind is a modern thing, your honor. I am not sure Galileo was the one who brought
relief to the scientific mind; because, theretofore, Aristotle and Plato had reached
their conclusions and processes, by metaphysical reasoning, because they had no
telescope and no microscope." The part of this passage which gives his case away
is the distinction made at the end. Mr. Malone was asserting that Aristotle and

Plato got no further than they did because they lacked the telescope and the microscope. To a slight extent perhaps Aristotle was what we would today call a "research scientist," but the conclusions and processes arrived at by the metaphysical reasoning of the two are dialectical, and the test of a dialectical position is logic and not ocular visibility. At the risk of making Mr. Malone a scapegoat we must say that this is an abysmal confusion of two different kinds of inquiry which the Greeks were well cognizant of. But the same confusion, if it did not produce this trial, certainly helped to draw it out to its length of eight days. It is the assumption that human laws stand in wait upon what the scientists see in their telescopes and microscopes. But harking back to Professor Adler: facts are never determinative of dialectic in the sense presumed by this counsel.

Exactly the same confusion appeared in a rhetorical plea for truth which Mr. Malone made shortly later in the same speech. Then he said:

> There is never a duel with truth. The truth always wins and we are not afraid of it. The truth is no coward. The truth does not need the law. The truth does not need the forces of government. The truth does not need Mr. Bryan. The truth is imperishable, eternal and immortal and needs no human agency to support it. We are ready to tell the truth as we understand it and we do not fear all the truth that they can present as facts.

It is instantly apparent that this presents truth in an ambiguous sense. Malone begins with the simplistic assumption that there is a "standard" truth, a kind of universal, objective, operative truth which it is heinous to oppose. That might be well enough if the meaning were highly generic, but before he is through this short passage he has equated truth with facts—the identical confusion which we noted in his utterance about Plato and Aristotle. Now since the truth which dialectic arrives at is not a truth of facts, this peroration either becomes irrelevant, or it lends itself to the other side, where, minus the concluding phrase, it could serve as a eulogium of dialectical truth.

Such was the dilemma by which the defense was impaled from the beginning. To some extent it appears even in the expert testimony. On the day preceding this speech by Malone, Professor Maynard Metcalf had presented testimony in court regarding the theory of evolution (this was on the fourth day of the trial; Judge Raulston did not make his ruling excluding such testimony until the sixth day) in which he made some statements which could have been of curious interest to the prosecution. They are effectually summarized in the following excerpt:

> Evolution and the theories of evolution are fundamentally different things. The fact of evolution is a thing that is perfectly and absolutely clear. . . . The series of evidences is so convincing that I think it would be entirely impossible for any normal human being who was conversant with the phenomena to have even for a moment even the least doubt even for the fact of evolution, but he might have tremendous doubts as to the truth of any hypothesis. . . .

We first notice here a clear recognition of the kinds of truth distinguished by Adler, with the "fact" of evolution belonging to the first order and theories of evolution belonging to the second. The second, which is referred to by the term "hypothesis," consists of facts in an elaboration. We note furthermore that this scientist has called them fundamentally different things—so different that one is entitled to have not merely doubts but "tremendous doubts" about the second. Now let us imagine the dialecticians of the opposite side approaching him with the following. You have said, Professor Metcalf, that the fact of evolution and the various theories of evolution are two quite different things. You have also said that the theories of evolution are so debatable or questionable that you can conceive of much difference of opinion about them. Now if there is an order of knowledge above this order of theories, which order you admit to be somewhat speculative, a further order of knowledge which is philosophical or evaluative, is it not likely that there would be in this realm still more alternative positions, still more room for doubt or difference of opinion? And if all this is so, would you expect people to assent to a proposition of this order in the same way you expect them to assent to, say, the proposition that a monkey has vertebrae? And if you do make these admissions, can you any longer maintain that people of opposite views on the teaching of evolution are simply defiers of truth? This is how the argument might have progressed had some Greek Darwin thrown Athens into an uproar; but this argument was, after all, in an American court of law.

It should now be apparent from these analyses that the defense was never able to meet the state's case on dialectical grounds. Even if it had boldly accepted the contest on this level, it is difficult to see how it could have won, for the dialectic must probably have followed this course:

First Proposition, All teaching of evolution is harmful.

Counter Proposition, No teaching of evolution is harmful.

Resolution, Some teaching of evolution is harmful.

Now the resolution was exactly the position taken by the law, which was that some teaching of evolution (i.e., the teaching of it in state-supported schools) was an anti-social measure. Logically speaking, the proposition that "Some teaching of evolution is harmful," does not exclude the proposition that "Some teaching of evolution is not harmful," but there was the fact that the law permitted some teaching of evolution (e.g., the teaching of it in schools not supported by the public funds). In this situation there seemed nothing for the defense to do but stick by the second proposition and plead for that proposition rhetorically. So science entered the juridical arena and argued for the value of science. In this argument the chief topic was consequence. There was Malone's statement that without the theory of evolution Burbank would not have been able to produce his results. There was Lipman's statement that without an understanding of the theory of evolution the agricultural colleges could not carry on their work. There were the statements of Judd and Nelson that large areas of education depended upon a knowledge of evolution. There was the argument brought out by Professor Mather of Harvard: "When men are offered their choice between science, with its confident and unanimous acceptance of the evolutionary principle, on the one hand, and religion, with its necessary appeal to things unseen and unprovable, on the other, they are much more likely to abandon religion than to abandon science. If such a choice is forced upon us, the churches will lose many of their best educated young people, the very ones upon whom they must depend for leadership in coming years."

We noted at the beginning of this chapter that rhetoric deals with subjects at the point where they touch upon actuality or prudential conduct. Here the defense looks at the policy of teaching evolution and points to beneficial results. The argument then becomes: these important benefits imply an important beneficial cause. This is why we can say that the pleaders for science were forced into the non-scientific role of the rhetorician.

The prosecution incidentally also had an argument from consequences, although it was never employed directly. When Bryan maintained that the philosophy of evolution might lead to the same results as the philosophy of Nietzsche had led with Loeb and Leopold, he was opening a subject which could have supplied such an argument, say in the form of a concrete instance of moral beliefs weakened by someone's having been indoctrinated with evolution. But there was really no need: as we have sought to show all along, the state had an immense strategic advantage in the fact that laws belong to the category of dialectical determinations, and it clung firmly to this advantage.

An irascible exchange which Darrow had with the judge gives an idea of the frustration which the defense felt at this stage. There had been an argument about the propriety of a cross-examination.

The Court: Colonel [Darrow], what is the purpose of cross-examination?

Mr. Darrow: The purpose of cross-examination is to be used on trial.

The Court: Well, isn't that an effort to ascertain the truth?

Mr. Darrow: No, it is an effort to show prejudice. Nothing else. Has there been any effort to ascertain the truth in this case? Why not bring in the jury and let us prove it?

The truth referred to by the judge was whether the action of Scopes fell within the definition of the law; the truth referred to by Darrow was the facts of evolution (not submitted to the jury as evidence); and "prejudice" was a crystallized opinion of the theory of evolution, expressed now as law.

If we have appeared here to assign too complete a forensic victory to the prosecution, let us return, by way of recapitulating the issues, to the relationship between positive science and dialectic. Many people, perhaps a majority in this country, have felt that the position of the State of Tennessee was absurd because they are unable to see how a logical position can be taken without reference to empirical situations. But it is just the nature of logic and dialectic to be a science without any content as it is the nature of biology or any positive science to be a science of empirical content.

We see the nature of this distinction when we realize that there is never an argument, in the true sense of the term, about facts. When facts are disputed, the argument must be suspended until the facts are settled. Not until then may it be resumed, for all true argument is about the meaning of established or admitted facts. And since this meaning is always expressed in propositions, we can say further that all argument is about the systematic import of propositions. While that remains so, the truth of the theory of evolution or of any scientific theory can never be settled in a court of law. The court could admit the facts into the record, but the process of legal determination would deal with the meaning of the facts, and it could not go beyond saying that the facts comport, or do not comport, with the meanings of other propositions. Thus its task is to determine their place in a system of discourse and if possible to effect a resolution in accordance with the movement of dialectic. It is necessary that logic in its position as ultimate arbiter preserve this indifference toward that actuality which is the touchstone of scientific fact.

It is plain that those who either expected or hoped that science would win a sweeping victory in the Tennessee courtroom were the same people who believe that science can take the place of speculative wisdom. The only consolation they had in the course of the trial was the embarrassment to which Darrow brought Bryan in questioning him about the Bible and the theory of evolution (during which Darrow did lead Bryan into some dialectical traps). But in strict consideration all of this was outside the bounds of the case because both the facts of evolution and the facts of the Bible were "items not in discourse," to borrow a phrase employed by Professor Adler. That is to say, their correctness had to be determined by scientific means of investigation, if at all; but the relationship between the law and theories of man's origin could be determined only by legal casuistry, in the non-pejorative sense of that phrase.

As we intimated at the beginning, a sufficient grasp of what the case was about would have resulted in there being no case, or in there being quite a different case. As the events turned out science received, in the popular estimation, a check in the trial but a moral victory, and this only led to more misunderstanding of the province of science in human affairs. The law of the State of Tennessee won a victory which was regarded as pyrrhic because it was generally felt to have made the law and the lawmakers look foolish. This also was a disservice to the common weal. Both of these results could have been prevented if it had been understood that science is one thing and law another. An understanding of that truth would seem to require some general dissemination throughout our educated classes of a *Summa Dialectica*. This means that the educated people of our country would have to be so trained that they could see the dialectical possibility of the opposites of the beliefs they possess. And that is a very large order for education in any age.

Notes

1. Mortimer J. Adler, *Dialectic* (London: Paul, Trench, & Trubner), p. 75.
2. Cf. Adler, *ibid,* pp. 243-44: Dialectic "is a kind of thinking which satisfies these two values: in the essential inconclusiveness of its process, it avoids ever resting in belief, or in the assertion of truth; through its utter restriction to the universe of discourse and its disregard for whatever reference discourse may have toward actuality, it is barren of any practical issue. It can make no difference in the way of conduct."
3. Adler, ibid., p. 224.
4. All quotations are given verbatim from *The World's Most Famous Court Trial* (Cincinnati: National Book Company, 1925), a complete transcript.

The Role of *Pathos* in the Decision-Making Process: A Study in the Rhetoric of Science Policy

By Craig Waddell

Over the course of this century, Western science and technology have developed at a staggering pace, producing both immense benefits—such as antibiotics and increased crop yields—and frightening consequences—such as industrial pollution and the threat of nuclear holocaust. As science and technology have become increasingly sophisticated and increasingly powerful, scientists and engineers have been called upon to accept greater responsibility for the social impact of their work. Thirty years ago, for example, Bertrand Russell argued that, "It is impossible in the modern world for a man of science to say with any honesty, 'My business is to provide knowledge, and what use is made of that knowledge is not my responsibility'" (391-92).

Scientists can exercise social responsibility in at least two ways. First, they can use their *authority* as scientists to lobby for appropriate application of their work. And second, they can use their *knowledge* as scientists to inform the public about the potential risks and benefits of their work.[1] However, scientists themselves are not necessarily the first to recognize the risks inherent in their work, and scientists alone should not make critical public-policy decisions about issues that involve the health and well-being of the community. As the members of the Cambridge Experimentation Review Board said in their 1976 report on recombinant DNA research:

> Knowledge, whether for its own sake or for its potential benefits to humankind, cannot serve as a justification for introducing risks to the public unless an informed citizenry is willing to accept those risks. Decisions regarding the appropriate course between the risks and benefits of a potentially dangerous scientific inquiry must not be adjudicated within the inner circles of the scientific establishment. (n. pag.)

The purpose of this study, then, is to explore the ways in which nonscientists inform themselves and make decisions about complex scientific and technological controversies. Scientists play an important role in this process by way of the two functions mentioned above—lobbying and informing. Hence, this process involves, among other things, the presentation and interpretation of complex technical arguments. Michael Halloran has pointed out that such arguments fall within the province of rhetoric because of "the increasing importance of scientific matters in the arena of public affairs, the traditional realm of rhetoric" (1984, 81; this volume, page 48); Randall Bytwerk contends that public debates about science and technology are "probably among the more significant areas for rhetoric today" (188).[2] Yet while Aristotle describes rhetoric as appealing to the whole person through a complex interplay of *ethos, logos,* and *pathos,* the privileged position enjoyed by *logos* in Western culture has often led to the denial of any appropriate role for *pathos* in science-policy formation.[3]

Faced with the elevated status of *logos* and the degraded status of *pathos,* we would do well to recall that, like appeals to emotion, rational appeals are themselves problematic for at least four reasons. First, as Aristotle points out, like emotional appeals, rational appeals can be inauthentic and deceptive (1400b35). Second, rational appeals can be tautological; too timid to venture a logical leap, they tell us nothing more than the obvious. Third, rational appeals may lead to *agreement,* but not to *conviction;* that is, they may lack the motive force to move us to action when action is called for. Finally, as Sidney Callahan points out, naked reason can lead to morally indefensible conclusions: "A rational argument without any apparent logical flaws may be presented—in, for instance, proposals for using torture, or harvesting neomorts, or refusing to treat AIDS—but our moral emotions prevent us from giving assent" (12).

Furthermore, in denying our emotions, we may actually *increase* the danger of emotional manipulation. As Masanao Toda points out, "[d]ismissing emotions as just noisome irrationality and pretending that we are beyond the sway of emotions are both sure ways of making ourselves susceptible to emotional manipulations" (135). Thus, overcoming our prejudice against emotion should open up the decision-making process and foster the critical skills needed to counteract demagoguery. At the same time, greater acceptance of a legitimate role for emotion should infuse the decision-making process with the "moral emotions" that Callahan argues are essential to humane decisions. To achieve these two goals, we must advocate wider acceptance of the role emotional appeals play in public-policy formation. Moreover, since emotional appeals (e.g., appeals to race hatred) clearly *are* destructive of the community, we must develop a clearer understanding of how we distinguish appropriate from inappropriate emotional appeals.

Toward this end, this paper explores the prejudice against emotion in what has widely been regarded as both a successful and an unprecedented case of public participation in science-policy formation: the Cambridge Experimentation Review Board's (CERB) participation in the 1976-77 moratorium on recombinant DNA research in Cambridge, Massachusetts.

While we can learn much about the decision-making process by examining catastrophic failures, such as the design and launch decisions that led to the explosion of the space shuttle *Challenger*, such failures often obscure more subtle and routine problems with that process. Hence, examining ostensibly successful decisions offers two distinct advantages: first, it provides the opportunity to learn from positive example; second, and more importantly, it allows us to focus our attention on those more subtle and routine problems; in this case, on the prejudice against emotion.

CERB can be faulted in many respects. Goodell, for example, argues that the board "quickly developed a hierarchy of expertise . . . with only a few members participating fully. In particular, the women were at the bottom of the hierarchy" (39). However, the board's decision-making process was successful in the sense that it addressed and allayed community fears and resulted in a decision that allowed potentially valuable research to proceed at what has since proven to be minimal risk to the community. Even some of the leading opponents of the research conceded that the process was successful. For example, Jonathan Beckwith acknowledged that "in the long run, I think the results of the moratorium were . . . pretty reasonable . . . there were problems here or there, but overall, I was not that unhappy with it" (15).

Some might argue that CERB's decision was inherently flawed in that, since the board was restricted by its charge to considering health and safety issues, it did not consider underlying ethical issues, such as those relating to human genetic engineering. Krimsky has argued, however, that the safety of the research was itself an ethical issue (Krimsky, June 1986, 16). Furthermore, as I point out below, in spite of its charge, CERB did hear and consider arguments (e.g., from George Wald and David Nathan) about the pros and cons of genetic engineering.[4]

Focusing on CERB's crucial November 23, 1976 debate, this paper examines the interplay between *logos, pathos,* and *ethos* in the committee's struggle to distinguish appropriate from inappropriate emotional appeals. I will argue that the construction of appropriateness is best understood by considering *logos, pathos,* and *ethos* as complementary appeals that continually blend and interact throughout the rhetorical process. For example, *logos* and *pathos* interact in that emotional appeals are generally built on a rational foundation; conversely, logical appeals generally have an emotional component. *Logos* and *ethos* interact in that a speaker can enhance his or her credibility by appearing rational; conversely, a speaker's arguments are more likely to be judged rational if he or she projects positive character traits. Finally, *ethos* and *pathos* interact in that the audience's judgment of the appropriateness of an emotional appeal may be influenced by their assessment of the speaker's *ethos*; conversely, the audience's assessment of the speaker's *ethos* may be influenced by their assessment of the appropriateness of his or her emotional appeals.[5]

This paper relies heavily on two primary sources: materials collected by MIT's Oral History Program under the direction of Prof. Charles Weiner, which are now housed in the MIT Institute Archives, and approximately thirty interviews I conducted with the former CERB members, with scientists who testified before CERB, and with other participants in this case.

The History of the Cambridge RDNA Moratorium

Early in 1976, Harvard University announced plans to renovate space in its Biological Laboratories to conduct potentially hazardous recombinant DNA experiments. In response to public opposition to this plan and to similar plans at MIT, the Cambridge City Council convened two public hearings during the summer of 1976 to debate the advisability of conducting such experiments in Cambridge. At the second of these hearings, the council passed a resolution calling for a good-faith moratorium on the most troublesome of these experiments, those classified at physical containment levels 3 and 4 (P3 and P4) by the National Institutes of Health (NIH). At this same hearing, the council issued an order establishing the Cambridge Experimentation Review Board and directing the city manager to determine the powers, responsibilities, and membership of the board. In response to this order, City Manager James Sullivan narrowed the responsibility of the board to determining whether the proposed research would have any adverse effects on public health within the city of Cambridge. He selected eight Cambridge residents from diverse ethnic groups, neighborhoods, and professions to serve on the board: as chair, Daniel Hayes, a businessman and former mayor of Cambridge; Sr. Lucille Banach, a nurse and hospital administrator; Dr. John Brusch, a physician specializing in infectious diseases; Constance Hughes, a nurse and social worker; Sheldon Krimsky, a philosopher of science at Tufts University; William LeMessurier, a structural engineer; Mary Nicoloro, a community activist; and Cornelia Wheeler, a community activist and former city councilor.[6]

At the suggestion of CERB member Sheldon Krimsky, the board organized itself on a "citizen-jury" model and began receiving testimony from the leading proponents and opponents of the research. Board members soon became frustrated, however, by their inability to adequately cross-examine the expert witnesses who testified before them. Thus, they agreed to arrange a debate on November 23, 1976 between proponents and opponents of the research; the format, allowed for cross-examination by peers. During this debate, opponents of the proposed research were represented by five men: George Wald, Higgins Professor of Biology at Harvard and Nobel laureate, 1967, in physiology or medicine; Jonathan King, associate professor in the Department of Biology at MIT; Richard Goldstein, assistant professor in the Department of Microbiology at Harvard Medical School; Jonathan Beckwith, professor of microbiology at Harvard Medical School; and Melvin Nemkov, a physician at Central Hospital in Somerville. Speaking in favor of the research were: David Baltimore, American Cancer Society Professor of Microbiology at MIT and Nobel laureate, 1975, in physiology or medicine; Mark Ptashne, professor in the Department of Biochemistry and Molecular Biology at Harvard; Walter Gilbert, American Cancer Society Professor of Microbiology at Harvard who received the Nobel Prize in chemistry in 1980, four years after this debate; Alwin Pappenheimer, professor in the Department of Biology at Harvard; David Nathan, Chief of Hematology and Oncology at Children's Hospital in Boston; and Edward Kass, Director of the Channing Laboratory of Infectious Diseases at Harvard Medical School.

This debate became the turning point in the board's proceedings and led to their unanimous decision to approve the research. In December, the board drafted its report, which it submitted to the city council at a public meeting in January. Following CERB's recommendations, the Cambridge City Council passed an ordinance on February 7, 1977, permitting recombinant DNA research at the P3 level of containment to proceed in Cambridge, but under stricter regulation than that specified in the NIH guidelines. Thus, the Cambridge moratorium ended seven months after it began with the first legislation in the United States to govern recombinant DNA research (Krimsky, "Research" 14).

The Underdetermination of Policy by Logos: CERB's November 23, 1976 Debate

In *Revolutions and Reconstructions in the Philosophy of Science,* Mary Hesse presents the "by now fairly uncontroversial proposition that all scientific theories are *underdetermined* by facts" (187; see also vii-viii). Similarly, from a rhetorical point of view, science policy is underdetermined by *logos.* Despite our cultural bias in favor of reason and against emotion, in practice, appeals to reason are a necessary but seldom, if ever, a sufficient condition for the formation of science policy. CERB's participation in the Cambridge moratorium clearly demonstrates this point.

During my interviews with them, CERB members criticized the opponents of the proposed research for their emotional appeals; in their report, they argue that their decision was "as unemotional and objective" as they were capable of offering. In practice, however, they did not reject emotional appeals as inappropriate, but, instead, they rejected what they saw as inappropriate emotional appeals. Appeals to emotion, then, became *apparent* as "emotional appeals" (something to be scorned) only when committee members deemed them misleading or abusive.

For example, during one interview, LeMessurier criticizes Wald for being "all emotion and all undocumented statements and preconceptions." A moment later, however, he praises Nathan's emotional argument about the life-saving potential of the proposed research. But when questioned about this apparent contradiction, he acknowledges that the proponents "have their emotions too" (LeMessurier 5–6). In fact, perhaps the most interesting and instructive contrast between acceptable and unacceptable emotional appeals in CERB's November 23 debate is that between the opening statements of these two men, David Nathan and George Wald. To a large extent, these two opening statements reflect the entire debate in capsule form. As Nathan and Wald present their cases, each combines appeals to reason, emotion, and character; yet Nathan wins the assent of the committee while Wald does not. Before examining these opening remarks closely, I must briefly establish the context within which they were presented.

Setting the Stage: Format and Introductions

CERB's November 23 debate was held from 5 p.m. until approximately 10 p.m. in the executive dining room at Cambridge City Hospital. The committee selected this relatively small (16' x 24') room for the debate because they did not want to repeat the somewhat chaotic experience of the two public hearings sponsored by the city council the previous summer. In addition to the participants (CERB members, proponents, and opponents), only about thirty members of the public and press attended the debate (Hayes, March 1988, 5).

The debate begins with preliminary remarks by Hayes in which, after outlining the format, he emphasizes that the charge of the committee is to render an opinion as to whether the proposed research "may have any adverse effect on public health within the city" (Debate 2). After Hayes' preliminary remarks, the debate proceeds in the following stages: introductions (Ptashne for the proponents, followed by King for the opponents), thirty minutes of opening remarks for each side (Wald, King, and Goldstein, followed by Ptashne, Baltimore, and Nathan), cross-examination by peers, break, questions from the committee and continued cross-examination by peers, and closing remarks (Wald and King, followed by Baltimore and Nathan). While all of the proponents speak at least once, two of the opponents, Beckwith and Nemkov, remain silent throughout the approximately 55,000-word debate.

The Logos of Pathos:
Saving the Germ Plasm vs. Saving the Children

To understand the interplay between *logos* and *pathos* in this debate, we must first understand the logical foundations that Wald and Nathan provided for their arguments.

When I interviewed Daniel Hayes in 1986, he recalled that while he and his fellow committee members had been presented with many abstract arguments, David Nathan was the first witness to confront the committee with concrete human suffering (March 1986, 7). The contrast between the abstract and the concrete is perhaps most evident in the contrast between Wald's appeal for "the inviolability of the human germ plasm" and Nathan's appeal for compassion for children with cancer and genetic disease.

Wald begins his opening remarks with a detailed analysis of the safety hazards inherent in recombinant DNA research. He describes the danger, in light of microorganisms' propensity for mutation, of inadvertently creating hazardous recombinations; he contends that there is no evidence that transfer of genetic material between microorganisms and higher organisms occurs naturally; then he argues that even desirable genes, such as the gene that produces insulin, could not be introduced into the human body without creating life-threatening problems, such as insulin shock.

The second part of Wald's opening remarks is a brief but impassioned appeal for the inviolability of the human germ plasm. In light of CERB's charge of considering whether the proposed research would have any adverse effects on public health within the city of Cambridge, Wald acknowledges that this appeal may be "beyond the business of this committee." "Nevertheless," he continues, "that's where this thing is heading, and those of us who are concerned have to be concerned about this too" (Debate 9). He argues that this research will lead directly to human genetic engineering and, thus, should raise many ethical questions. He contends that:

> We've got something mortal and something immortal in our bodies that all of us are carrying around. The mortal part of us is our bodies. The immortal part of us is the germ plasm—the egg and the sperm. They represent an unbroken line of life that goes back without a break approximately three billion years now to the first living organism ever to appear on this planet. If that line of life had ever been broken, we could not be here. (Debate 9)

Wald concludes his opening remarks with the following:

> Some years ago, long before recombinant DNA techniques came up . . . I *proposed* that, rather rapidly, one begin to establish a principle in law of the inviolability of the human germ plasm, just as we have a principle in law of inviolability of human life. The inviolability of the human germ plasm. Three billion years of continuous life—we should not be about to turn it over to anybody to toy with. (Debate 9–10)

King, who follows Wald in delivering his opening remarks, develops this theme, arguing that species "don't exchange genes for very deep reasons" (Debate 10). As Baltimore points out in his own opening statement, this argument seems designed to evoke a mystical awe or reverence for biological processes (Debate 29).

As other participants present their opening remarks, approximately forty-five minutes passes before Nathan has an opportunity to respond to Wald. Throughout the controversy, the opponents had repeatedly argued that the advocates of the research were biased because they were doing the research themselves. In response to this argument, Nathan begins his opening statement by suggesting his disinterestedness in the exclusive benefits (e.g., monetary gain or prestige) of the research: "I should explain why I'm here, because I don't *do* recombinant DNA research" (Debate 32). When I interviewed him, Nathan argued that disinterestedness was his primary strength in the debate (Nathan 13–14).

After asserting he will derive no personal benefits from the proposed research, Nathan immediately offers a far more compelling claim of *strong* interest in the potential benefits this research promises for his patients:

> I'm a parasite. I wander around in basic science laboratories to find a way to better deal with the problems that I deal with on a clinical basis. And I deal with children with cancer, and I deal with children with inherited disease. *And I don't share Dr. Wald's love of the germ plasm!* I

must tell you that I've had *rough encounters* with germ plasms all my medical career. If I had a lot of the mothers in here who have children with difficulty in their germ plasm, I think you'd know why I feel the way I feel. (Debate 32)

In this passage, Nathan both challenges Wald's naturalistic fallacy and evokes a concrete image of human suffering. Wald has argued that the human germ plasm is as it ought to be. Nathan objects, pointing out that we have ample reason to tamper with the germ plasm. In recalling this argument almost a decade later, LeMessurier suggested that the potential benefits of the research would be so overwhelming that "it would be almost evil" not to proceed with it (LeMessurier 5–6). Hayes recalled that Nathan's pathetic appeal "really put the pressure to" the committee (Hayes, March 1986, 7).[7] Nathan extends this appeal as he continues his opening statement:

I'm in the business of changing things for particular patients who *need* a change. Now I absolutely *count* on developments in molecular biology to deal with the problems my patients have *right now.* Now, I've got . . . a ward full of them over there. There are twenty with cancer *now* in the ward, and I can *name* them for you. And there are thirty seen a day in the out-patient services, and I've got I don't know how many . . . what percent admissions to the Children's Hospital with genetic diseases. . . . looked at broadly, it's approximately 30 percent of admissions. (Debate 32)

Although Nathan does not detail specific cases, by offering to name the children on his ward, he evokes a concrete image of their suffering. In contrast, Wald's argument about "three billion years of continuous life" and "the inviolability of the human germ plasm" seemed detached and abstract. As Krimsky pointed out, "George Wald's arguments were much more philosophical, and the committee was probably not as comfortable with his arguments. They really didn't get down to concrete situations" (Krimsky, March 1986, 8). Rhetoricians have long pointed out that the concrete tends to have greater persuasive force than the abstract.[8] However, almost anyone would have difficulty forming a concrete conception of three billion years of evolution; certainly, it is far easier to empathize with afflicted children than it is to empathize with germ plasm.

CERB members found Nathan's argument about the potential benefits of this research extremely compelling. Late in the debate, Nicoloro, who was initially opposed to the research, says, "I was impressed by Dr. Nathan . . . he was pale with the passion of the children at the Children's Hospital." She then calls for strengthening the NIH guidelines, "Because DNA research is going on; it's needed!" (Debate 125–26).[9] During a later interview, Hughes, who was initially opposed to the research, argued that "if I had a child who had a genetic problem, I think I would turn heaven and earth to try to find something that was going to help that child" (Hughes 5). Krimsky pointed out that "When you start raising issues of childhood cancer, it has a very persuasive force to it" (Krimsky, March 1986, 9).

Having challenged Wald's argument about the inviolability of the human germ plasm, Nathan defends the proposed research, arguing that before he could accept a ban on such research, he would "have to have a *terribly* good reason presented to [him] by careful people" (Debate 33). He then characterizes the nature of the reasons he has heard from the opposition:

> I've *yet* to hear—and I've *tried* to follow this, because I realize that I'm a pressing parasite; I don't want this work to stop; I want it *desperately* to move forward. I realize I'm prejudiced. But I've tried *very, very* hard to listen to what the hazards really are, and I listened *intently* just now. And all I can say is that from any bit of medical training that I've ever had, *I couldn't understand a word of it—not a single word!* (Debate 33)

Thus, Nathan suggests that the committee has *not* been presented with terribly good reasons by careful people. Furthermore, in acknowledging his failure to understand the opponents' complex arguments, Nathan identifies with the members of the committee, who also had difficulty with these arguments.

Nathan goes on to challenge Wald's arguments about the hazards of this research. He acknowledges that one *can* make a nonpathogenic bacterium pathogenic through recombinant experiments; but he contends that this *cannot* be done with the attenuated strains that will be used in the proposed experiments: "That *defies* the very evolution that Professor Wald is praising. . . . That's alchemy!" (Debate 33). Nathan immediately qualifies this statement, however, acknowledging that "anything is possible." Nevertheless, he contends that he must weigh "the *vague* possibilities" presented by the opponents against the immense potential benefits of this research (Debate 34). Brusch later pointed out that the committee found that proponents enhanced their credibility when they acknowledged the *possibility* of risk, but suggested that it was remote (Brusch 4).

Nathan next contends that recombinant DNA research is essential to efforts to understand the various hemoglobin diseases with which he works: sickle cell anemia and, particularly, thalassemia, "a dreadful disease for those who happen to have it" (Debate 34).[10] Although Nathan did not describe the symptoms of thalassemia, the three medical professionals on the committee (Brusch, Banach, and Hughes) were certainly familiar with these symptoms and could certainly empathize with Nathan's impassioned plea for his patients.

Nathan concludes his opening remarks by urging the committee to try to understand his point of view:

> Now, if they tell me, "Look, Nathan, you're going to ruin the water supply. This is how you're going to do it." And . . . if you've even got a 5 percent chance, I'm not going to take *my family*, live here, and expose them to a terrible risk in order to deal with a few patients. . . . I'd let them

die! Why not let a few thousand patients die to protect a whole city? I'd *certainly* do that. I'm not a madman. But, you've got to understand why I'm so exercised. *I've* got the patients, I can *name* them, and they're not telling me anything that is anything but a bunch of, of, of [odd noise?]. But they have no biological base! And that's got me goofy. And that's why I'm here. (Debate 35)

In describing the conditions under which he *would* oppose recombinant DNA research, Nathan presents himself as a reasonable but unpersuaded by the opponents' arguments. In noting that he is raising a family in Cambridge, he emphasizes that he has a vested interest in community health.

After concluding his opening remarks, Nathan responds to one question from Wald and then remains silent until offering, in his closing remarks, the final words of the debate. Thus, for the 3.5 hours between his response to Wald and his closing remarks, Nathan remains detached from the fray. Wald, on the other hand, continues to participate and continues to take a beating.

The Ethos of Pathos: Discrediting the Opposition

About half-way through the debate, Baltimore characterizes the opponents as representing "one extreme" on an enormous spectrum of opinion (Debate 73). Throughout the debate, the proponents give the opposition, particularly Wald and King, no quarter, repeatedly challenging their credibility and pointing out the extreme nature of their positions.

Krimsky recalls that the committee was particularly sensitive to how well the proponents and opponents were able to defend their claims in the presence of their peers.[11] During the debate, the proponents take advantage of every opportunity to highlight the opponents' inability to support their views. Rather early, for example, the following exchange takes place between Wald and Ptashne:

Wald: I gave the chairman a little packet. You will find in that packet a paper, October 14, 1976, *New England Journal of Medicine,* "New Pathogenic *E. coli* That Produces Infantile Diarrhea." It's not a recombinant DNA job, as far as one knows, but it leads to the next question. . . .
Ptashne: What is that supposed to mean, George!? That's *insulting!*
Wald: It's *not* a. . . Oh, *please!*
Ptashne: It's innuendo; that's what it is! You do it all the time!
Wald: Oh, I withdraw the "so far as one knows" . . . I meant that as a joke. (Debate 46–47)

Later in the debate, Ptashne objects again when Wald seems to suggest that Legionnaires' disease may be the result of a recombinant DNA experiment (Debate 67–68). Still later, Wald clarifies the reason for his reference to Legionnaires' disease. It is *not,* he acknowledges, a "recombinant job, but . . . it provides a *wonderful* model for what this kind of trouble would look like if it ever happened" (Debate 90). Nevertheless, his previous "joke" about the new pathogenic *E. coli* taken together with his observation that "I haven't the *faintest* idea what that Legionnaires' thing is" reinforced Ptashne's claim that Wald frequently engaged in innuendo and elicited a second, there-he-goes-again response from the proponents (Debate 67–68).

In addition to being accused of innuendo, Wald is repeatedly charged with misunderstanding or misrepresenting the facts. For example, when he argues that researchers have produced a new human carcinogen by putting the SV40 protein coat on a mouse cancer virus, Ptashne characterizes this statement as "a complete misunderstanding of *all* of virology." He explains that SV40 was not and, in fact, *could not have been* used in the experiments Wald describes (Debate 98–99). Later, Wald argues that Fort Detrick's record of 423 infections in twenty-five years with only three deaths was a consequence of the fact that "Fort Detrick is an *isolation* situation" (Debate 118).[12]

Having been charged with dealing in innuendo and with misunderstanding or careless mismanagement of the facts, the opponent's suffer a third blow to their credibility when a confession of hazardous laboratory practices casts Jonathan King as both reckless and hypocritical. King argues that the proposed research should be carried out in an isolated research center, such as the Centers for Disease Control, which he contends is "*totally* different than the situation here at MIT" (Debate 130). When Kass asks him how the CDC is "totally different" from MIT, the following exchange ensues:

> King: Do you know what we do with *all* the Salmonella plates produced in the biology department? Salmonella plates that have plasmids in them for antibiotic resistance. They go into the garbage. Last year, when we called on. . . .
> Kass: *You* do that!?
> King: *Everybody* in the biology department.
> Kass: *You* throw Salmonella into the garbage!?
> King: I. . . .
> Kass: You should be put in *jail!* (Debate 130–131)

King's indiscretion is assailed again about fifteen minutes later, immediately before the closing remarks, when Kass expresses his outrage at "*sitting here listening to somebody bleeding for the workers, bleeding for the people of Cambridge, and telling you he put Salmonella in the garbage!!*" (Debate 140). When King tries to clarify his point, Kass adds vehemently, "How can you be trusted with *anything* if you run around and do a thing like this!?" (Debate 140). When I interviewed him, Kass referred to dumping Salmonella as "totally unconscionable" (Kass 13); and Baltimore recalled: "That was one of the stupidest things Jon King said. And Kass was right to rake him over the coals" (Baltimore 14).

Although King's indiscretion reflected most strongly upon himself, by association, it also compromised the credibility of the opposition as a whole.

The Symphony of Rhetorical Appeals

Clearly, the decisions of the Cambridge Experimentation Review Board were shaped by the dynamic interaction between *logos, ethos,* and *pathos* rather than by rational appeals alone. Ultimately, the committee found the *pathos* in Nathan's arguments appropriate, but deemed that in Wald's arguments *inappropriate*. A closer examination of the interaction between these three rhetorical appeals may explain this discrepancy.

Emotion as Judgments

The interaction between *logos* and *pathos* becomes clearer if we think of emotional appeals as enthymemes. Aristotle explains that the enthymeme "must consist of a few propositions, fewer often than those which make up the normal syllogism. For if any of these propositions is a familiar fact, there is no need even to mention it; the hearer adds it himself" (1357a17–19). Bitzer argues that it is this participation of the audience in constructing the argument, not formal deficiency, that is the defining characteristic of the enthymeme (408). John Gage agrees: "[t]he enthymeme cannot be constructed in the absence of a dialectical relationship with an audience, since it is only through what the audience contributes that the enthymeme exists as such" (157).

The power of the enthymeme, then, derives from this: If the audience accepts the premises of the enthymeme, it is drawn into and participates in the construction of the argument; thus, the audience is inclined to persuade itself (Bitzer 408). The *problem* with the enthymeme, however, is that if the speaker or writer misjudges, and the audience does *not* accept his or her premises or share the values inherent in them, the argument will collapse.

The suppressed members of an enthymeme can be made explicit through rational reconstruction.[13] For example, Nathan's emotional appeal rests on a rational foundation which can be rationally reconstructed as follows:[14]

Major premise: Saving human lives is valuable.

Minor premise: This research can save human lives.

Conclusion: Therefore, this research is valuable.

This logical conclusion is complemented by a parallel, pathetic conclusion, compassion for the afflicted children and their families. This conclusion results not from traditional logic, but from deontic logic, a logic of imperatives:[15]

Major premise: If your child were suffering from these diseases, you would want those in a position to help to show compassion.

Minor premise: Other people's children are suffering from these diseases, and you are in a position to help.

Conclusion: Therefore, you should show compassion.

Wald's emotional appeal can also be rationally reconstructed:

Major premise: Violating three billion years of evolution is dangerous.

Minor premise: This research violates three billion years of evolution.

Conclusion: Therefore, this research is dangerous.

Early in the debate, Baltimore suggests that this argument appeals to "mystical" reverence for biological processes (Debate 29). This reverence is complemented by Wald's own *ethos,* which Banach described as follows: "It was just something about the man—I can remember his white hair, and just awe of him when he came into the room" (Banach 22).

Recognizing the mystical element in Wald's argument, Nathan tries to counteract it in his own opening statement. He notes that while he has great respect for his own religious faith, he is "in the business of changing things for patients who *need* a change" (Debate 32). Hence, Nathan's emotional appeal is based on compassion, not mystical reverence. He evokes images, not of germ plasm and three billion years of evolution, but of individual children, suffering from dreadful diseases, diseases that this research offers a hope of curing. The major premise in both arguments was suppressed; but while the committee accepted the suppressed premise in Nathan's argument (Saving human life is valuable), they apparently did not accept that in Wald's argument (Violating three billion years of evolution is dangerous). Hence, Nathan's argument succeeded, while Wald's failed.[16]

Since both of these emotional appeals can be rationally reconstructed, the distinction between them is not that one rests upon a rational foundation while the other does not. Thus, the question still remains: Why was one appeal found to be superior to the other? While the concreteness of Nathan's appeal may have been a contributing factor, emotional appeals based on abstractions are not always ineffective. Witness, for example, the numerous historical instances in which whole populations have been mobilized for war by appeals to abstract national, religious, or racial ideals.[17] In CERB's case, however, this small group of well-informed citizens who were accountable to the community deemed compassion a more appropriate consideration than mystical reverence. This determination was shaped by the contrast between Wald's *ethos* and Nathan's *ethos.* These, in turn, were shaped by the appeals each of these men made to reason and emotion.

Contrasts in Ethos

Nathan's *ethos* is established in at least four ways. First, his appeal to compassion and his affiliation with Children's Hospital help to establish him as a caring, involved person. Second, by acknowledging the possibility of risk ("anything is a possibility"), and by acknowledging his own biases ("I realize that I'm a pressing parasite; I don't want this work to stop"), Nathan establishes himself as reflective and reasonable. Third, in spite of his biases, Nathan makes clear that he is disinterested in the exclusive benefits of the research ("I don't do recombinant DNA research"). Finally, by twice mentioning in his opening remarks that he is raising a family in Cambridge, Nathan indicates his vested interest in community health. Thus, if we assume the unstated premise that we can trust a father not to recommend a course of action that he believes will jeopardize the health and safety of his own children, Nathan asserts his trustworthiness.

The four prominent features of Nathan's *ethos*—compassion, reasonableness, disinterestedness, and trustworthiness—contrast sharply with the *ethos* established by the opponents in general and by George Wald in particular. First, Wald's emotional appeal is based on mystical reverence, not compassion. The general line of argument that he and King advance is characterized as "very poor social policy and almost inhumane" by Baltimore (Debate 31); LeMessurier later argued that it would be "almost evil" not to carry out this research. Second, the proponents repeatedly characterized the opponents as extreme and unreasonable; evidence in later interviews and in the debate itself indicates that committee members agreed with this characterization. For example, after King has argued that sanctions should be put in place before even the most benign, low-risk (P1) experiments proceed in Cambridge, Nicoloro agrees with Ptashne's assertion that "that is not a rational, sensible, scientific position from *any point of view*" (Debate 133).

Third, while Nathan acknowledges his professional biases yet suggests that he is disinterested in the exclusive benefits of the research, the opponents deny that they *have* any biases. For example, when LeMessurier asks what biases the opponents might have, King responds by noting that he has subjected himself to an extremely unpleasant situation and jeopardized his professional status by objecting to this research. He contends that he cannot possibly expect to derive any benefit from his opposition (Debate 106).[18] Ptashne later challenged the sincerity of this protest by arguing that the opponents were using the issue to build their radical political careers and credentials (Ptashne). And in later interviews, several members of the committee indicated that they were aware of the opponents' political biases. For example, Hayes contended that, in addition to their scientific concerns, King and Beckwith "also had some political opinions. They were part of the scientific community that were against a number of things" (Hayes, March 1986, 5).

Finally, the trustworthiness of the opponents is brought into question by three factors described earlier: Wald's repeated mismanagement of facts; his tendency to engage in innuendo; and, perhaps most strikingly, King's admission of having dumped Salmonella into the garbage.

The Social Construction of Appropriateness

According to Aristotle, rhetoric concerns itself with the realm of contingent knowledge. As Carolyn Miller has pointed out, "[o]ver the centuries, we have become less certain than Aristotle was about many things, and what we call the 'new rhetoric' reflects the extension of uncertainty to matters other than Athenian civic affairs" (43).

In the absence of certainty, in the absence of an objective foundation upon which to base belief, we must accept that both *rational* and *appropriate* are social constructs. In light of the rational bias in Western culture, these two constructs are often used almost interchangeably to mean *prudent, judicious,* or *sensible.* One goal of this study, however, is to *broaden* our concept of rationality, with respect to both arguments and our responses to those arguments, to include *emotional* as well as logical appropriateness.[19] More accurately, one goal of this study is to encourage wider acknowledgement and acceptance of the extent to which our concept of rationality is *already* shaped by our sense of emotional as well as logical appropriateness. Thereby, I hope to make the emotional component of the decision-making process more amenable to criticism. From this perspective, the question to ask of a behavior, judgment, decision, appeal, or response is not "Is it rational?" or "Is it emotional?" but "Is it *appropriate?*"

If we accept that there is no Truth, no objective foundation upon which to base belief, then we must simultaneously accept responsibility for *constructing* a foundation, for *constructing* truths.[20] Among the truths we construct are our determinations of what is rational and what is appropriate. A perennial objection to constructivist arguments, however, is that they lead to a radical relativism in which no adjudication between appropriate and inappropriate is possible since no absolute principles apply across contexts.

Gadamer offers a solution to this problem, however, with his notion of a "fusion of horizons" between the past and the present (Gadamer 273). From this perspective, even in the absence of an objective, absolute foundation for belief, we can still define general principles that cut across immediate contexts because these immediate contexts are themselves situated within a larger and more stable, yet still socially constructed, cultural context, what Gadamer calls our "thrownness." That is, we are born (or "thrown") into a historical context, a culture, a tradition in whose values, prejudices, and presuppositions we are steeped. What understanding we have, we gain not by *freeing* ourselves of these prejudices, but by *applying* them; for we understand only by virtue of the questions we ask; those questions are framed by our prejudices. We distinguish appropriate from inappropriate prejudices when we expose our prejudices to the test of experience (Gadamer 236–37).

Hence, the prevailing values and presuppositions of a culture cut across immediate contexts and apply in a wide range of cases much as absolute principles would. However, unlike moral absolutes, values and presuppositions are socially constructed and, hence, mutable. The ideal rhetor both embodies and appeals to what his or her society deems the best and most noble of its sentiments and prejudices.[21] When the times require it, the ideal rhetor helps the society adapt to new and changing circumstances by helping it to define values that are appropriate to the issues raised by these new, often more complex, conditions.[22]

Thus, almost 2,000 years ago, Quintilian made appropriateness, rather than persuasion, the measure of the ideal orator.[23]

The social construction of appropriateness, however, is not simply a process of noble rhetors enlightening their audiences; as the above discussion of the enthymeme suggests, rhetors and audiences together co-create meaning and values through the give-and-take of epistemic rhetoric. Like the ideal rhetor, the ideal audience embodies, responds to, and—when the times call for it—*helps to determine* the society's best and most noble sentiments. With the ideal audience, then, an argument is not appropriate because it is persuasive, it is persuasive because it is appropriate. Hence, both the characteristics of the audience and the nature of the decision-making model adopted by or imposed upon that audience are crucial factors in the determination of appropriateness. In CERB's case, the committee (audience) was composed of citizens who were: (1) representative of the community at risk; (2) disinterested in the exclusive benefits of the research (i.e., benefits not shared by the general population, such as immediate financial rewards and prestige); (3) willing and able to inform themselves adequately about the relevant issues (all of the committee members had some relevant technical or political expertise) and (4) accountable to the community.

Despite these strengths, from a rhetorical perspective, the committee was hampered by its prejudice against emotion and by its tendency to engage in rational rather than rhetorical reconstruction when called upon to defend its decisions. That is, the committee tended to justify its decision in terms of logical appeals rather than in terms of the full range of rhetorical appeals with which it was presented. That the committee, nevertheless, embodied and responded to the best and most noble appeal: "If I had a child who had a genetic problem, I think I would turn heaven and earth to try to find something that was going to help that child" (Hughes 5). This response exemplifies the ancient principle of reciprocity or reversibility embodied in the golden rule and the categorical imperative, a principle tracing back at least as far as Aristotle, who responded when asked how we should behave to our friends: "As we should wish them to behave to us" (Diogenes Laertius V.21).[24]

In their final report, CERB members remark that "it is our sincere belief that a predominantly lay citizen group can face a technical scientific matter of general and deep public concern, educate itself appropriately to the task, and reach a fair decision" (n. pag.). Even the opponents of the research seem to agree with this assessment *with respect to CERB* (Beckwith 15; King reported in Hayes, June 1986, 32). However, since CERB was carefully selected for its task by the Cambridge city manager, the committee's confidence in the ability of *any* lay citizen's group to achieve similar results may be a bit misplaced. For example, while the public hearings sponsored by the city council during the summer of 1976 served a vital function in that they aroused public awareness of and interest in the recombinant DNA controversy, the mass audience at these hearings was anonymous rather than accountable, and it was certainly not as well versed on the issues as CERB came to be by the end of its extensive, four-month investigation. If we consider the city council itself as the primary audience at these hearings, the criterion of accountability is met, but even the councilors were not as well informed as the CERB members came to be. Hence, the audience at these public hearings—whether that audience be the mass in attendance, the city council, or both—played an important part in bringing the controversy to a head, but that audience was not adequately prepared to resolve the controversy by evoking and responding to the best and most noble sentiments of the culture. This role, instead, was filled by CERB.

Conclusions: The Rhetorical Tradition, Pedagogy, and the Democratic Audience

In classical times, rhetoric was intimately associated with pedagogy; in fact, rhetoric was the foundation of the ancient curriculum (Marrou 194). Nevertheless, as Bruce Kimball notes, liberal education in antiquity was confined to the privileged few (13). The goal of rhetorical training was to prepare an elite cadre of civic leaders who could win the hearts and minds of their less sophisticated fellow citizens. Unfortunately, this situation led to the practical necessity of adapting the appeals of the rhetor—who, ideally, embodied the best and most noble of the culture—to the more pedestrian tastes and sentiments of the audience. As Quintilian noted:

> I am surprised that *deliberative* oratory also has been restricted by some authorities to questions of expediency. If it should be necessary to assign one single aim to deliberative I should prefer Cicero's view that this kind of oratory is primarily concerned with what is honourable. I do not doubt that those who maintain the position first mentioned adopt the lofty view that nothing can be expedient which is not good. That opinion is perfectly sound so long as we are fortunate enough to have wise and good men for counsellors. But as we most often express our views before an ignorant audience, and more especially before popular assemblies, of which the majority is usually uneducated, we must distinguish between what is honourable and what is expedient and conform our utterances to suit ordinary understandings. (III.viii.1–2)

For Quintilian, even the ideal orator may deviate from the truth if "he is led to do so by consideration for the public interest" (III.vii.25); that is, if it is expedient to do so when arguing before an uneducated audience. This view of rhetoric and the uneducated audience is not peculiar to Quintilian. In fact, as Perelman and Olbrechts-Tyteca point out, "[i]t is [this] aspect of rhetoric which explains why Plato opposed it so fiercely in his *Gorgias* and which was propitious to its decline in the estimation of philosophers" (7).

The tragic flaw of classical rhetoric, then, is not that it employs emotional appeals, but that it allows for the willful deception of the unsophisticated audience. Such deception can be achieved through *logos* as well as through *pathos* (see, for example, Aristotle 1400b35). In effect, this solution (i.e., willful deception) to the problem of the uneducated audience disenfranchises that audience, for the audience is not a full and active participant in the shaping of public policy. Instead, the audience is a passive mass, which is to be deceived into accepting a solution predetermined by the rhetor. While I agree that it is sometimes necessary to do what is expedient, I also agree with Cicero and Quintilian that rhetoric should primarily be concerned with what is honorable; disenfranchising the audience is an expedient but not an honorable solution to this problem. A better solution, particularly given our expanded, contemporary conception of democracy, is for rhetoric to concern itself not simply with the education of a cadre of rhetors, but with the education of the democratic audience as well. As Thomas Jefferson said:

> I know of no safe depository of the ultimate powers of society but the people themselves; and if we think them not enlightened enough to exercise their control with a wholesome discretion, the remedy is not to take it from them, but to inform their discretion by education. (Letter to W.C. Jarvis, September 28, 1820)

Committees such as CERB provide both an expedient and an honorable solution to the problem of the unsophisticated audience. Encouraging careful selection of the audience which is to make a decision is clearly superior to permitting willful deception of an unsophisticated audience by a rhetor. Ideally, however, in an industrially advanced democracy, we should not need to be as cautious as the Cambridge city manager was in selecting decision makers and in setting up committees. Ideally, the vast majority of citizens would be civic minded, technologically literate, and rhetorically competent. As Frederick Antczak says in *Thought and Character: The Rhetoric of Democratic Education:*

> In democracy, the people rule—that is, they rule insofar as they make their own decisions. But those decisions grow more complex, more intellectually demanding every day. The average citizen . . . is called to deliberate on an increasingly formidable variety of issues, each demanding a different way of knowing. Decisions must be made on, among other things, problems of toxic waste disposal, . . . the reliability of nuclear power plant construction and operation, . . . the proper limits and accountability of recombinant DNA research. . . .

> Life in an increasingly technologized society imposes increasing intellectual demands on society's decision makers. . . .If in such an era democracy's decisions are to be made intelligently and effectively, the public must somehow be reconstituted intellectually. (197–198)

When I interviewed the scientists who testified before CERB, they repeatedly emphasized the importance of education in science to prepare the public for participation in science-policy debates (e.g., Pappenheimer 17, and Beckwith 21). But as C.P. Snow has pointed out, "the intellectual life of the whole of western society is increasingly being split into two polar groups. . . . [l]iterary intellectuals at one pole—at the other scientists" (4–5). If we are to rejoin these "two cultures," we must not simply emphasize the inadequacies of scientific or humanistic education, we must find more and better ways of interrelating the two. Merely increasing the number of science and engineering courses taken by humanities and social science majors and increasing the number of humanities and social science courses taken by science and engineering majors will not solve the problem if these courses do not emphasize the intimate connections between sciences and the humanities in modern life. As Johnston, Shaman, and Zemsky point out in *Unfinished Design: The Humanities and Social Sciences in Undergraduate Engineering Education,* "[m]any technical problems are also social problems—or ethical, or political, or international problems—and some ability to confront them as such is also an increasingly necessary part of the professional equipment of engineers" (8). As the Hastings Center's report *On the Uses of the Humanities: Vision and Application* suggests:

> For all the social impact of the sciences, their meaning and significance cannot be determined until they have been critically interpreted; and the resources of the humanities provide a way of doing so. For all of the potency of the bureaucratic state to shape behavior, it is historical context, evaluative and expressive interpretations, and the restlessness of the willful imagination that will set public goals and give color to its civic flesh. (1)

Much has been written and said over the past three decades about a new rhetoric, a rhetoric that will revive what is best and most pertinent in classical rhetoric and adapt it to the problems of contemporary life. I would argue that principal among the features of classical rhetoric that are most pertinent to the problems of modern life are: (1) its emphasis on *invention*, not just style; (2) its emphasis on *ethos* and *pathos,* not just *logos;* (3) its emphasis on public deliberation about contingent areas of knowledge, not discipline-specific discourse; and (4) its emphasis on application and pedagogy, not just scholastic debate. In adapting the last of these features to contemporary life, we need to foster a rhetoric concerned not only with the education of an elite cadre of rhetors, but also with the education of the democratic audience. The audience (public) must be prepared to participate constructively in the scientific and technological controversies that are becoming increasingly crucial to our nation and to our world.

Acknowledgments

This paper is based upon the author's dissertation research (Rensselaer, 1989), under the direction of Michael Halloran. A previous version of this paper was presented at the

1988 joint meeting of the Society for the Social Studies of Science and the European Association for the Study of Science and Technology in Amsterdam.

Notes

1. They can also, of course, refuse to do work which they consider unethical.
2. Philip Wander describes two forms of rhetoric of science. One concerns itself "with the efforts made by scientists to persuade one another" (227). The other concerns itself with the place of science in the deliberation of public policy (226–27). To clarify this distinction, I refer to the first as rhetoric of science and to the second (as in my title) as rhetoric of science policy.
3. One crude measure of the prejudice against *pathos* is the relative poverty of articles on this topic cited in the Philosopher's Index database. In a recent (1988) search of this 136,000-record database, I found 237 citations for *logos* and 114 citations for *ethos,* but only 20 citation for *pathos.*
4. For a more detailed analysis of CERB's decision-making process, see Waddell, 1989.
5. A distinction should be made here between the appropriateness of an emotional response (anger, resentment, indignation, compassion, empathy, etc.) and the appropriateness of basing a decision or action upon that response. A similar distinction should be made between the appropriateness of an emotional *response* and the appropriateness of an emotional *appeal.*
6. To this point, I have used the terms *public* and *nonscientist* almost interchangeably. It should be clear now, however, that many of the members of CERB have technical backgrounds in medicine, physics, and engineering. Nevertheless, they qualify as nonscientists in that they are not practicing, experimental scientists; they are certainly nonscientists in the more restricted sense that they are not expert in the particular fields of science under consideration (i.e., microbiology, molecular biology, and pathology.)
   CERB members qualify as members of the public in that they are representatives of the community at risk. Furthermore, *public,* like *civilian* or *laity,* is a relative, community-defining term. *Civilian,* for example, describes a different set of people when used, respectively, by a soldier, a police officer, and a prison inmate. In relation to the government, those members of society who are not part of the government constitute the public. Hence, the overlap between my use of the terms *public* and *nonscientist:* CERB members are both members of the public and nonscientists in that they are not part of the relevant scientific community. While the term *nonexpert* might serve in place of either *public* or *nonscientist,* this term lacks the connotation of community representativeness and the association with science carried, respectively, by the latter two terms.
7. As Thomas Farrell has pointed out, "[s]ome knowledge *demands* that a decision be made. It forces our options, insofar as the very apprehension and comprehension of such knowledge requires that some action be taken" (10). The committee seems to have found the knowledge Nathan provided to be of this sort. My task will be to attempt to explain why.
8. Bitzer, for example, suggests that "[Charles] Baldwin's statement that enthymemes are concrete means that such arguments, when successful, always require human commitment or action." He goes on to say that although concreteness is not an essential feature of the enthymeme, "most enthymemes probably are concrete" (403).
9. Since CERB was charged with determining whether the proposed rDNA research would have any adverse effects on public health within the city of Cambridge, it is significant here that Nicoloro concludes that rDNA research is *needed,* not that it is *safe.*
10. Thalassemia major is a severe, hereditary anemia that appears during childhood. It is characterized by fatigue, enlargement of the heart and spleen, jaundice, leg ulcers, formation of gallstones, and thickening of cranial and facial bones. Although the prognosis varies, the younger the child when the disease appears, the more unfavorable the outcome.
11. "If we found duplicity—if we found that people were just saying things for effect, but amongst their colleagues they couldn't defend it—that would be a sign of which direction our confidence would run" (Krimsky, March 1988, 10).
12. Fort Detrick is located in Frederick, Maryland (population, 27,000). It is the former site of a U.S. Army chemical and germ warfare center.
13. To rationally reconstruct an *argument* is to restate a loosely woven argument such that its logical foundation is made explicit. In so doing, hidden assumptions are converted to formal, logical relationships; missing or implied premises are made explicit; and the entire argument is presented in syllogistic form such that the conclusion follows from the premises.

To rationally reconstruct a *decision* is to emphasize, in defending or justifying the decision, the logical appeals upon which the decision was founded (or, in their absence, to *construct* such appeals), and to *deemphasize* or ignore entirely the irrational or nonrational influences on the decision.

14. In arguing that emotional appeals rest on a rational foundation, I do not mean to suggest that emotion is *controlled* by reason, but only that there is a dialectic between the two.

15. Traditional logic is concerned with *logical* modalities, with questions of what is *necessarily*, or *possibly*, or *certainly not* the case. Deontic logic, on the other hand, is concerned with *deontic* modalities, with questions of what *should, ought,* or *must* be done. Deontic logic, then, clearly falls within the province of rhetoric, which Isocrates defined as the art of determining "what we should do or what we should say" (*Antidosis* 270).

16. The choice between compassion and mystical awe or reverence might also be viewed in terms of Kohlberg's moral-dilemma model and Festinger's cognitive dissonance theory. In this view, the choice is not between appropriate and inappropriate, but between more appropriate and less appropriate. Hence, forced to choose between two, dissonant moral beliefs (that the germ plasm should be inviolable and that in some cases, the germ plasm is defective and needs to be "violated" for humane reasons), the committee may have subsequently denigrated the rejected option, and, in effect, converted it from *less* appropriate to *inappropriate*.

17. It may be, however, that the most successful abstract appeals, as Strunk and White suggest, "furnish particular instances of their application" (17).

18. King later acknowledged that "the issue of genetic engineering can be seen as a vehicle for promoting the democratization of science" (634).

19. This goal, I believe, is consistent with Perelman and Olbrechts-Tyteca's goal of "broadening the concept of proof" (510); with Walter Fisher's goal of "broadening the concept of 'good reasons'" (377); and with Evelyn Fox Keller's goal of "reconceptualizing objectivity" (114).

20. Thus, having asked where, in the absence of absolute knowledge, we get the universal premises that are essential to rhetorical arguments, Charles Mudd contends that, "[w]e create them, we find them, we *invent* them" (414).

21. Halloran argues that Quintilian's description of the ideal orator as "a good man skilled in speaking" suggests that "the model the orator aspires toward is a person who has interiorized all that is best in the culture, and who can apply this wisdom in the public forum, influencing his fellow citizens to think and act on particular issues in accord with their common heritage" (1976, 235).

22. This is not to suggest that individuals, groups, or even nations do not violate long-standing and deep-rooted cultural values. It is only to suggest that within the context of the larger (in some cases, transnational) culture, these values tend to reassert themselves. Within the context of the larger Western culture, for example, the "thousand-year Reich" lasted all of a dozen years. Thus, values serve as cultural ballast, and while our ship may be capsized, it has a strong tendency to right itself again.

23. "Too much insistence cannot be laid upon the point that no one can be said to speak appropriately who has not considered not merely what it is expedient, but also what it is becoming to say. . . . these two considerations generally go hand in hand. . . . Sometimes, however, the two are at variance. Now, whenever this occurs, expedience must yield to the demands of what is becoming. . . . the end which the orator must keep in view is not persuasion, but speaking well, since there are occasions when to persuade would be a blot on his honor" (Quintilian XI.I.8–11).

24. In some instances, reciprocity may be based upon narrow self-interest—on a calculated, tit-for-tat exchange; but this is not the sense in which I am using the term here. Instead, the principle of reciprocity I am concerned with is based on empathy and the role-playing ability expressed in such questions as "How would I want to be treated (or how would I want my loved ones to be treated) if I (or they) were in this situation?"

Works Cited

Aristotle. *The Rhetoric*. Trans. W. Rhys Roberts. New York: Random House, 1954.

Antczak, Frederick J. *Thought and Character: The Rhetoric of Democratic Education*. Ames: Iowa State University Press, 1985.

Banach, Sr. Mary Lucille. Transcript of an interview by Craig Waddell, June 17, 1986. Recombinant DNA Oral History Collection, Institute Archives and Special Collections, MIT Libraries.

Baltimore, David. Transcript of an interview by Craig Waddell, March 31, 1988. Recombinant DNA Oral History Collection, Institute Archives and Special Collections, MIT Libraries.

Beckwith, Jonathan. Transcript of an interview by Craig Waddell, April 1, 1988. Recombinant DNA Oral History Collection, Institute Archives and Special Collections, MIT Libraries.

Bitzer, Lloyd F. "Aristotle's Enthymeme Revisited." *Quarterly Journal of Speech* 45 (1959): 399-408.

Brusch, John L. Transcript of an interview by Craig Waddell, March 20, 1986. Recombinant DNA Oral History Collection, Institute Archives and Special Collections, MIT Libraries.

Bytwerk, Randall L. "The SST Controversy: A Case Study of the Rhetoric of Technology." *Central States Speech Journal* 30 (1979): 187-98.

Callahan, Sidney. "The Role of Emotion in Ethical Decisionmaking." *Hastings Center Report* 18.3 (1988):9-14.

Cambridge Experimentation Review Board. "Guidelines for the Use of Recombinant DNA Molecule Technology in the City of Cambridge." December 21, 1976.

Cambridge Experimentation Review Board. Transcript of the Cambridge Experimentation Review Board Debate on the Hazards of Recombinant DNA Experimentation, November 23, 1976, transcribed and annotated by Craig Waddell. Recombinant DNA Oral History Collection, Institute Archives and Special Collections, MIT Libraries.

Diogenes Laertius. *Lives of Eminent Philosophers.* Vol. I. New York: G.P. Putnam's Sons, 1925.

Farrell, Thomas B. "Knowledge, Consensus, and Rhetorical Theory." *Quarterly Journal of Speech* 62 (1976): 1-14.

Farrell, Thomas B., and Thomas Goodnight. "Accidental Rhetoric: The Root Metaphors of Three Mile Island." *Communication Monographs* 48 (1981): 271-300.

Festinger, Leon. *A Theory of Cognitive Dissonance.* Stanford: Stanford University Press, 1957.

Fisher, Walter R. "Toward a Logic of Good Reasons." *Quarterly Journal of Speech* 64 (1978): 376-84.

Gadamer, Hans-Georg. *Truth and Method.* New York: Crossroads Publishing Co., 1982.

Gage, John. "An Adequate Epistemology for Composition: Classical and Modern Perspectives." In *Essays on Classical Rhetoric and Modern Discourse.* Ed. Robert Connors, Lisa Ede, and Andrea Lunsford. Carbondale: Southern Illinois University Press, 1984. 152-69.

Goodell, Rae S. "Public Involvement in the DNA Controversy: The Case of Cambridge, Massachusetts." *Science, Technology & Human Values* 4.27 (Spring 1979): 36-43.

Gross, Alan G. "Public Debates as Failed Social Dramas: The Recombinant DNA Controversy." *Quarterly Journal of Speech* 70 (1984): 397-409.

Halloran, S. Michael. "Tradition and Theory in Rhetoric." *Quarterly Journal of Speech* 62 (October 1976): 234-240.

Halloran, S. Michael. "The Birth of Molecular Biology: An Essay in the Rhetorical Criticism of Scientific Discourse." *Rhetoric Review* 3 (1984): 70-83. In this volume, pp. 39–50.

Hastings Center, The. *On the Uses of the Humanities: Vision and Application.* Hastings-on-Hudson: The Hastings Center, 1984.

Hayes, Daniel. Transcript of an interview by Craig Waddell, March 19, 1986. Recombinant DNA Oral History Collection, Institute Archives and Special Collections, MIT Libraries.

Hayes, Daniel. Transcript of an interview by Craig Waddell, June 18, 1986. Recombinant DNA Oral History Collection, Institute Archives and Special Collections, MIT Libraries.

Hayes, Daniel. Transcript of an interview by Craig Waddell, March 29, 1988. Recombinant DNA Oral History Collection, Institute Archives and Special Collections, MIT Libraries.

Hesse, Mary. *Revolutions and Reconstructions in the Philosophy of Science.* Bloomington: Indiana University Press, 1980.

Hughes, Constance. Transcript of an interview by Craig Waddell, June 18, 1986. Recombinant DNA Oral History Collection, Institute Archives and Special Collections, MIT Libraries.

Isocrates. "Antidosis." Trans. George Norlin. In *Isocrates,* Volume II. Cambridge: Harvard University Press, 1956.

Johnston, Joseph S., Jr., Susan Shaman, and Robert Zemsky. *Unfinished Design: The Humanities and Social Sciences in Engineering Education.* Washington, D.C.: Association of American Colleges, 1988.

Kass, Edward. Transcript of an interview by Craig Waddell, March 30, 1988. Recombinant DNA Oral History Collection, Institute Archives and Special Collections, MIT Libraries.

Keller, Evelyn Fox. "Feminism and Science." In *The Signs Reader: Women, Gender & Scholarship.* Ed. Elizabeth Abel and Emily K. Abel. Chicago: University of Chicago Press, 1983. 109-122.

Kimball, Bruce A. *Orators & Philosophers: A History of the Idea of Liberal Education.* New York: Teachers College Press, 1986.

King, Jonathan. "A Science for the People." *New Scientist* 74 (June 16, 1977): 634-636.

Kohlberg, Lawrence. *The Philosophy of Moral Development.* New York: Harper & Row, Publishers, 1981.

Krimsky, Sheldon. "Research Under Community Standards: Three Case Studies." *Science, Technology, & Human Values* 11 (1986): 14-33.

Krimsky, Sheldon. Transcript of an interview by Craig Waddell, March 19, 1986. Recombinant DNA Oral History Collection, Institute Archives and Special Collections, MIT Libraries.

Krimsky, Sheldon. Transcript of an interview by Craig Waddell, June 16, 1986. Recombinant DNA Oral History Collection, Institute Archives and Special Collections, MIT Libraries.

Krimsky, Sheldon. Transcript of an interview by Craig Waddell, March 30, 1988.

LeMessurier, William J. Transcript of an interview by Craig Waddell, March 21, 1986.

Marrou, H.I. *A History of Education in Antiquity*. 1948. Madison: University of Wisconsin Press, 1982.

Miller, Carolyn R. "The Rhetoric of Decision Science, or Herbert A. Simon Says." *Science, Technology, & Human Values* 14 (1989): 43-46.

Mudd, Charles S. "The Enthymeme and Logical Validity." *Quarterly Journal of Speech* 45 (1959): 409-14.

Nathan, David. Transcript of an interview by Craig Waddell, March 28, 1988. Recombinant DNA Oral History Collection, Institute Archives and Special Collections, MIT Libraries.

Pappenheimer, Alwin. Transcript of an interview by Craig Waddell, March 28, 1988. Recombinant DNA Oral History Collection, Institute Archives and Special Collections, MIT Libraries.

Perelman, Chaim and L. Olbrechts-Tyteca. *The New Rhetoric: A Treatise on Argumentation*. Notre Dame: University of Notre Dame Press, 1969.

Ptashne, Mark. Personal interview, May 12, 1988.

Quintilian. *The Institutio oratoria*. Trans. H.E. Butler. Cambridge: Harvard University Press, 1980.

Russell, Bertrand. "The Social Responsibilities of Scientists." *Science* 131 (1960): 391-92.

Snow, C.P. *The Two Cultures and A Second Look*. Expansion of the 1959 version. Cambridge: Cambridge University Press, 1964.

Strunk, William, and E.B. White. *The Elements of Style*. 2nd ed. New York: Macmillian Company, 1972.

Toda, Masanao. "Emotion and Decision Making." *Acta Psychologica* 45 (1980): 133-55.

United States Department of Health, Education, and Welfare. "Recombinant DNA Research Guidelines." *Federal Register* July 7, 1976.

United States. "Recombinant DNA Research Guidelines: Draft Environmental Impact Statement." *Federal Register* September 9, 1976.

Waddell, Craig. "The Fusion of Horizons: A Dialectical Response to the Problems of Self-Exempting Fallacy in Contemporary Constructivist Arguments." *Philosophy and Rhetoric* 21 (1988): 103-115.

Waddell, Craig. "Reasonableness vs. Rationality in the Construction and Justification of Science Policy Decisions: The Case of the Cambridge Experimentation Review Board." *Science, Technology, & Human Values* 14 (1989): 7-25.

Wander, Philip C. "The Rhetoric of Science." *Western Speech Communication* 50 (1976): 226-35.

Owning a Virus:
The Rhetoric of Scientific
Discovery Accounts

by Carol Reeves

On April 23, 1984, Margaret Heckler, at that time the secretary of Health and Human services, announced at a press conference that "the probable cause of AIDS has been found." She went on to explain that the cause was a new virus and its discoverer was "our eminent Dr. Robert Gallo," chief of the National Cancer Institute (Crewdson 2). Gallo was known for his discovery of a family of human retroviruses[1]—Human T-cell Leukemia Virus (HTLV)—that caused a rare form of cancer, and had received the coveted Lasker prize, the top honor in American biomedicine which often precedes the Nobel Prize. In several papers published in 1984, Gallo claimed to have found antibodies to what he believed was a new member of the Leukemia virus family—HTLV-III—in the blood serum of AIDS patients and that he had several isolates of the virus growing in his lab. He had also published electron micrographs of virus particles. On the day of Heckler's announcement, the American biomedical community was ready to accept Gallo as the first to solve the mystery of AIDS etiology, and was ready to view him as "one of the paradigmatic figures of the 20th century", as Samuel Broder described him (in Crewsdon).

However, the Gallo lab was not alone in searching for the cause of AIDS and developing a blood test. At the time of Heckler's announcement, the American patent office had two applications for a patent to the new AIDS blood test, the ELISSA (Enzyme-Linked Immuno-Absorbent Assay), one from the American Government and one from the Pasteur Institute in Paris. Gallo's chief rivals in the search for the cause of AIDS, a team headed by Luc Montagnier of the Pasteur Institute, had also applied for a patent on their ELISSA. They had argued at various conferences and in publications in 1983 and 1984 that they had found no antibodies to Gallo's leukemia virus in the blood of their AIDS patients, that the virus they had identified, Lymphadenopathy-Associated Virus, was unrelated to HTLV, and that they had discovered it many months before Gallo detected any retrovirus activity in blood samples from AIDS patients. Moreover, they were convinced that the American blood test had been made from the virus isolate the French had sent Gallo in the spirit of scientific collaboration. By 1985 a storm was brewing over who had first isolated the AIDS virus and who deserved the patent rights to the AIDS blood test. In May 1985 the US patent office issued a patent to the US government on the ELISSA made with Gallo's AIDS virus. From May to December 1985, representatives from the Pasteur Institute confronted senior officials at the Department of Health and Human Services (HHS) with evidence that they, not Gallo, had first isolated the virus and developed the ELISSA and that the American patent was invalid. The HHS rejected their claim. On December 12, 1985, the French filed suit against the US government challenging the patent.[2]

From 1985 to 1986, while the controversy heated up surrounding the question of who first discovered the AIDS virus and developed the ELISSA, Gallo published a flurry of review articles about research linking retroviruses to AIDS. Three of these five papers (in *Cancer*, *Cancer Research*, and *Scientific American*) were written by Robert C. Gallo alone; two of them (in *The New England Journal of Medicine* and *Annals of Internal Medicine*) were written by Gallo and a cowriter from his laboratory. In these accounts, Gallo and his cowriters use the review or discovery account genre to promote Gallo's role in AIDS research and his virus—HTLV-III—as the cause of AIDS. These five discovery accounts represent an intriguing opportunity to examine how one scientist and two members of his laboratory team demonstrate and attempt to maintain rhetorical authority during a periord that represents a significant shift in knowledge about a devastating human problem.

The necessity for rhetorical discourse during a shift in knowledge has been noted by several theorists. Sociologist Robert Merton notes that there is a rhetorical dimension involved when collectives compete to be the first to produce a public interpretation of a phenomenon and naturally want their interpretations to prevail over other interpretations (110-11). Likewise, Thomas Kuhn identifies persuasion as essential to scientific revolutions: "To discover how scientific revolutions are effected, we shall therefore have to examine not only the impact of nature and logic, but also the techniques of persuasive argumentation effective within the quite special groups that constitute the community of scientists" (*Structure* 94). It is impossible to determine here whether the discovery of retroviruses or the cause of AIDS had anything like the revolutionary impact upon whole scientific paradigms as did the identification of the double helix model of DNA. Nevertheless, the fact that these papers are directed at several different communities during a time when there was considerable competition for what Joseph Gusfield calls "ownership" of a public problem necessitates a close examination of the strategies Gallo and his cowriters use to maintain that ownership.

Gusfield explains that ownership "indicates the power to define and describe the problem" (13). People or institutions who "own" a public problem "can make claims and assertions. They are looked at and reported to by others anxious for definitions and solutions to the problem" (10). These papers demonstrate Gallo's determination to maintain the privileges of ownership—to be viewed as the first to discover the cause of AIDS and to have the authority to name, define, and describe the phenomenon and control its public interpretation. These accounts serve not only to disseminate knowledge but also to defend Gallo's primary role in generating that knowledge.

Gallo claims "ownership" in AIDS research by crediting the divergent course his research has taken and his team's contributions to what is now routine laboratory practice, and by discrediting the early publications of the French team. Another way that Gallo gains rhetorical authority in these papers is to construct contexts that accomodate his intended audience's interests and beliefs. Gallo's readers range from biomedical specialists who read *Cancer, Cancer Research,* and *Annals of Internal Medicine* to medical practitioners who read *The New England Journal of Medicine* to a wider, less specialized, audience for the more popular publication, *Scientific American.* While there are obvious differences in the style and format of these papers due to editorial policy and convention, there are some contrasting features that I believe go beyond the necessities of convention and serve as flexible argumentative resources in Gallo's attempt to gain rhetorical ownership in AIDS research. Gallo demonstrates an acute understanding that different audiences have different epistemological assumptions and therefore need different conceptual constructions of the scientific enterprise as well as the phenomena under investigation.

Steve Woolgar describes discovery accounts as reviews or "popular articles" that "present summary accounts of research" on a particular phenomenon (396). Woolgar finds a discrepancy between formal, published accounts and informal accounts gleaned from conversations and interviews. Published accounts "tended to give the impression of a relatively straightforward progression through a series of logical steps . . . leading to the discovery" (415) whereas personal interviews revealed that the discovery process is often marked by uncertainty, error, confusion, and surprise. In formal accounts, moreover, the debatable issues—such as the date of discovery, for example—are assumed to be closed and consensus achieved. Woolgar's findings imply that the valedictory occasion for the formal accounts may require writers to give the impression that their discovery process was unproblematic and that the issues have been black-boxed. I am concerned here neither with comparing a scientist's formal and informal accounts nor with discovering whether Gallo's accounts are scientifically or historically accurate. My program is simply to identify the ways that Gallo uses the valedictory occasion as an active resource in an open battle for ownership in AIDS research.

Crediting the Gallo Lab

One way that Gallo establishes his "ownership" in AIDS research is to credit himself and his lab team and to discredit the French team. In two papers for which he was the sole writer, (*Cancer* and *Cancer Research*), Gallo emphasizes the role his lab played in retrovirus AIDS research in the introductions. In the *Scientific American, Annals,* and *New England Journal of Medicine* papers, discussion of the primary role of the Gallo team appears in subsequent body paragraphs. Given the nature of the introductions to *Scientific American* and *The New England*

Journal of Medicine, which emphasize the disease itself rather than the Gallo team's role in investigating the problem, it could be argued that staking territory in AIDS research for the benefit of his specialist peers in biomedicine is a high rhetorical priority for Gallo. Highlighting his lab's investigative role in AIDS research in the introduction to papers for specialists could be his way of tacitly reminding them of his ownership.

In the introductions to the *Cancer* and *Cancer Research* papers, Gallo discusses how the work done in his lab was the result of a readiness to travel outside mainstream cancer research and take an approach that differed from routine methodology. In *Cancer,* after he briefly summarizes the mainstream work on cellular onc genes, he writes:

> However, these studies do not generally deal with primary causes. My co-workers and I are also interested in the molecular mechanisms involved in the pathogenesis of neoplasms, but our approach has been different. (2317)

He then explains that his group has "focused on particular types of cells and particular forms of growth abnormalities" and that they "are also interested in the primary causes of these diseases" (2317). Also, he insists that he and his coworkers used the discoveries of others in ways that mainstream investigation did not. In both *Cancer* and *Cancer Research,* he describes how his lab creatively utilized the discovery of reverse transcriptase in ways the rest of the community did not. In *Cancer,* he writes that

> [w]hat stimulated us to work, rather than just to think about these questions, was the discovery of reverse transciptase in retroviruses, which opened up new areas to molecular biology and paved the way to the understanding of the life cycle of these viruses. To us it also offered something more. (2317)

In *Cancer Research,* he writes that "reverse transcriptase paved the way for the beginning of biochemical understanding of retrovirus replication . . . ; for me it was also a powerful and sensitive new tool for detecting low levels of retroviruses in human leukemias" (4524). Gallo knows the scientific community values the kind of flexible, divergent thinking that leads to the solution of major technical problems. Kuhn argues that the sciences "do demand just that flexibility and open-mindedness that characterize, or indeed define, the divergent thinker. . . . Unless many scientists possessed them to a marked degree, there would be no scientific revolutions and very little scientific advance" (*Tension* 227). Whether or not Gallo's work was flexible and divergent is a question for sociologists. I am concerned here, with how Gallo's introductions go further than the perfunctory assertion that the work about to be described represents a "significant advance" in knowledge. Gallo is doing more than building a context for the information he is about to provide. He is reminding his audience that his work represents a creative and divergent approach.

Gallo also reminds his readers of the skepticism and dogma that his team opposed. In *Cancer Research* he notes in the introduction that

> [o]ur idea that retroviruses might be present in small amounts in some human leukemias received considerable resistance. . . . Our premise was therefore contrary to the experience and dogma current at that time, and our attempts were considered futile. (4524)

Likewise, in the introduction to the *Cancer* paper, he writes that "when we set out to do these experiments, there were very strong doubts expressed by many investigators about the existence of any human retroviruses." Because these viruses had not been detected by electron microscopy, "it was assumed that no human retroviruses existed" (2317).

In the midsection of the *Scientific American* paper, Gallo asserts that "the prologue to the discovery of the first human retrovirus is a history of skepticism" (88), and later, he places himself in the position of countering common wisdom:

> Under the influence of Termin's ideas I decided to search for reverse transcriptase in human leukemic cells, hoping to find a retrovirus there. In so doing, I was gainsaying acceptable wisdom. (91)

The premise evoked in these examples is that it is so often necessary to consider the impossible, to go outside mainstream assumptions and dogma in order to conduct research having an impact on the growth of scientific knowledge. He is implying that without his willingness to consider the possibility of a human retrovirus, science would be without the benefits resulting from such efforts, namely the information about the relationship among human retroviruses, T-cell leukemia, and acquired immunodeficiency. Though it is uncertain whether retrovirology had a revolutionary impact on whole paradigms within several scientific communities, it is clear that Gallo wants his audience to think of his work as having a singularly important impact.

While he refers to other laboratory teams doing research on retroviruses and AIDS, Gallo insists that by being the first to ask that all-important initial question about the role of retroviruses in AIDS and by developing technology with which to isolate the virus, his lab actually reached "the top" first. Though he gives credit to the collaborative efforts of labs across the country, his references to his chief competitors, the Pasteur Institute team, who first published findings of the isolation of retroviruses from AIDS patients, traverses a fine line between credit and discredit: He concedes that they were the first to publish findings of retrovirus activity in AIDS patients, but dismisses those findings as "inconclusive." In so doing, Gallo discredits the work of the French team and their claim that the AIDS virus is a newly discovered virus unrelated to Gallo's HTLV.

According to Gallo, the obstacles involved in growing and propagating the virus prevented the French team from definitively characterizing the virus, and, in effect, from actually being "the first" to make the discovery. In the *Annals* paper, just after they establish that the problem of losing virus isolates "initially prevented characterization of the virus, development of specific reagents and clear-cut linkage of the virus to the cause of AIDS," Gallo and Wong-Staal write,

> However, in 1983, the group at the Pasteur Institute headed by Luc Montagnier reported their first desciption of the virus associated with AIDS. . . . These investigators also had difficulty in getting adequate amounts of this virus to grow; they could not keep the primary cells in culture indefinitely. . . . Therefore, the virus was not characterized, nor clearly linked to the cause of AIDS in 1983, and specific reagents for the virus could not be developed (683).

In *Cancer Research,* he explains that the French team could "not report evidence indicating that any two isolates [of the virus] were the same" but published their findings anyway because they were "sufficiently convinced of its interest to publish the first identification of the virus" (4528). However, as Gallo points out, his team had also isolated the virus, but rather than publish prematurely, they decided to wait because they "did not feel these results were sufficiently clear to publish at that time" (4528). In explaining that his team did not publish their findings until they had found a way to grow the virus and thus definitively characterize it, Gallo obviously wants to demonstrate that although the other team might have reached the top first, they took a short cut. Gallo implies that the other team lacked the patience and tenacity that contributed to the conclusive work done in his lab.

In addition to reminding his audience that the French team's first published data were inconclusive, Gallo also reminds his readers that he and his colleagues first proposed the notion that AIDS was caused by a retrovirus and thereby initiated investigation of that hypothesis. In the *Cancer Research* paper, he states, "I proposed that AIDS was likely to be caused by a human T-lymphotropic retrovirus in February, 1982 at a Cold Springs Harbor Conference on AIDS" (4527). Likewise, in *Cancer,* he writes that "[w]e reasoned that AIDS was a viral disease and proposed that it was due to a human T-lymphotropic retrovirus in February 1982 at a Cold Springs Harbor Meeting on AIDS" (2321). He and Wong-Staal remind their readers in the *Annals* paper, that "the genes of HTLV-III were first molecularly cloned in our laboratory" (684). Like the explorers who mark their territory with a flag and want the world to acknowledge their symbolic ownership of a place, Gallo wants the scientific community to acknowledge his ownership of the territory of AIDS research.

Gallo also promotes his work as resulting in the kind of new knowledge and technology which initiates major shifts in methodology. In all his accounts, Gallo describes the technology and methods developed in his lab that enabled investigators to first identify human retroviruses and then to isolate the retrovirus causing AIDS. He also reminds readers that these methods have become indispensable to routine practice. In the introduction to the *Cancer* paper, for example, he points out that his discovery of T-cell growth factor or interleukin-2 and the techniques he and his lab team developed to use interleukin-2 to stimulate T-cell growth *in vitro* "became the approach to isolate human retroviruses." Moreover, he reminds his readers that "selection of the right patient, use of sensitive assays for virus, and proper growth of the cells *in vitro* has made isolation of these viruses now a routine" (2318). It is important to keep in mind that there is no clinical necessity for descriptions of these methods since they have become so much a part of routine; however, rhetorical necessity requires that Gallo remind his audience of what team is responsible for developing these methods.

Gallo's emphasis on the role his work played in the discovery of the cause of AIDS likely undergirded his claim that his virus, HTLV-III, was the cause of AIDS and helped to market him as an "owner" in AIDS research. In all the papers, the claim that HTLV-III causes AIDS and that it is the same virus isolated by other researchers seems unproblematic, black-boxed. In *The New England Journal of Medicine,* for example, while he concedes that "there is a noteworthy diversity in the . . . patterns seen among HTLV-III isolates from different patients," he still claims that the French LAV is not another virus but one of the forms of HTLV-III: ". . . we believe that HTLV-III really represents a set of closely related but varying genetic forms, and that lymphadenopathy-associated virus [LAV] is one of these forms" (1295).

In the *Annals* paper, Gallo and Wong-Staal enumerate the evidence pointing to HTLV-III as "clearly" the cause of AIDS:

> There are now more than 100 isolates of HTLV-III in our laboratory. The virus was isolated from all risk groups: homosexuals, drug abusers, blood product recipients, mothers and fathers of infected children, promiscuous heterosexual men, prostitutes, Haitians, and the wives of men with AIDS. We have never been able to detect the virus in a normal donor. *These findings alone clearly link the virus to the cause of AIDS.* (684, emphasis mine)

In their classification of statement types, Latour and Woolgar define a claim as a statement which contains "modalities which draw attention to the generality of available evidence (or lack of it)" (79). The modality "clearly" draws attention away from what could be problematic in Gallo's claim. Other labs had reported isolations of retroviruses from AIDS patients, but no one had confirmed that these isolates represented the same virus or that they were all isolates of Gallo's retrovirus. In fact, a research report written by Luc Montagnier follows the Gallo and Wong-Staal paper in *Annals*. Montagnier claims that the primary causative agent of the acquired immunodeficiency syndrome "is a type of retrovirus that has not been previously recognized" (689). At least as far as the French were concerned, consensus had not reached the point that a definitive statement about the link between HTLV and AIDS was even possible. Yet Gallo's ostensibly unproblematic statement is made after he has narrated the history of research on retroviruses and AIDS and made himself and his lab team chief protagonists in that story. By establishing the primary role of his lab in AIDS research, Gallo lends credibility to his claim that *his* virus is *the* virus causing AIDS.

Building Contexts

Another way that Gallo gains rhetorical authority in these papers is to construct contexts that accomodate his intended audience's interests and beliefs. In his introductions and conclusions, he focuses on issues indicative of his intended audience's assumptions about such matters as AIDS, science, and knowledge.

The main contextual difference between Gallo's introductions in the *Cancer Research* and *Cancer* and his introductions in *The New England Journal of Medicine* and *Scientific American* centers around the issue of *agency*. For specialists, Gallo emphasizes his laboratory team's agency in AIDS research, an emphasis that accomodates the communal ritual of identifying the discoverer and the belief that such an endeavor is even possible given the collaboration that occurs in science. For readers of *The New England Journal of Medicine* and *Scientific American,* however, Gallo emphasizes the disease itself as an agent in causing human suffering and public catastrophe and then counters the grim outlook with the consolation of new knowledge.

In the *New England Journal of Medicine* paper, Gallo and Broder demonstrate a clear intent to communicate with those who will actually treat AIDS patients in their offices or emergency rooms. Throughout the paper the authors counter the bad news—the devastating and frustrating course of the new disease—with good news—the growing knowledge. In the introduction they claim that "one of the major frustrations for practicing physicians and the public alike has been the inability of the scientific community to define the origin of this disease" (1292). They liken this frustration to the sentiments expressed by John Donne: "Perhaps John Donne, in 1624, in reference to his own life-threatening illness, was able to capture some of the frustrating emotions felt by patients with AIDS, persons thought to be at risk of AIDS, and their physicians" (1292). They then quote Donne:

> 'I observe the Phisician, with the same diligence, as hee the disease; I see hee feares, and I feare with him, I overrun him in his feare, and I go the faster, because he makes his path slow; I feare the more, because he disguises his fear, and I see it with more sharpnesse, because hee would not have me see it.' (1292)

The reference to the frustrations felt by patients and doctors alike and the Donne quote embody the text with a highly personal tone. The writers establish an

intimate relationship with their readers as members of the practitioner community—a membership of individuals who have witnessed the AIDS epidemic
first hand. Gallo and Broder want their readers to know they understand the
plight of the practitioner facing patients who have a debilitating and overwhelming disease. But they also want to console their readers with the hope
of new knowledge about the cause of the disease, as they say in their
introduction:

> But recently, several converging lines of research have linked a human
> T-cell lymphotropic retrovirus (HTLV-III) to the pathogenesis of AIDS,
> and this knowledge has brought us closer to understanding the disease.
> (1292)

In the introduction to the paper in *Scientific American*, Gallo employs
metaphors that Susan Sontag has identified in the language of AIDS, the
metaphors of plague and apocalypse. Gallo's first sentence contains the plague
metaphor: "It is a modern plague: the first great pandemic of the second half
of the 20th Century" (47). Sontag says that plague "has long been used
metaphorically as the highest standard of collective calamity, evil, scourge"
(89). Gallo goes on to add that "by now as many as two million people in the
U.S. may be infected" and that "in some areas it may be too late to prevent a
disturbingly high number of people from dying" (47). The "bad news" outlined in the first paragraph is followed by the "good news" outlined in the
second:

> In sharp contrast to the bleak epidemiological picture of AIDS, the
> accumulation of knowledge about its cause has been remarkably quick.
> Only three years after the disease was described its cause was conclu-
> sively shown to be the third human retrovirus (47).

In this introduction Gallo accomodates his readers by evoking the metaphors
they likely associate with AIDS and by reassuring them that though infection has
reached catastrophic proportions, the disease is no longer shrouded in mystery.
Gallo allows himself some dramatic touches that would not be appropriate for
more specialized audiences.

In the conclusions to the papers in *Annals, Cancer, Cancer Research,* and
The New England Journal of Medicine, Gallo appeals to the premise that
scientific explanation of puzzling phenomena is always a desirable end in itself.
Though, as Gallo points out, the retrovirus may pose serious psychological and
physical problems for its victims, it is a "remarkable" biological entity for
scientists to study. Moreover, he explains that not only has research identified the
human retrovirus causing AIDS but also promises to curtail its destructive
capacity. The consolation of new knowledge *about* AIDS is placed against the
morbidity and mortality caused *by* AIDS.

He concludes the *Cancer* and *Cancer Research* papers by reminding his readers of the optimistic results of the contributions he outlined in the introductions and throughout the papers. In conluding the *Cancer* paper, he avoids mentioning the clinical difficulties involved in disease management and vaccine development for a virus that has a great capacity to mutate and states that "preventive measures to control the spread of AIDS with the help of chemotherapeutic agents or a vaccine are currently being actively pursued" (2321). He ends the paper by reminding his readers of the rapid progress in research:

> Thus, in the past 5 years, the first human retroviruses have been discovered and two types have been linked to the cause of two different human diseases. (2321)

He also reminds readers of the interesting knowledge that has been discovered:

> Remarkably, one disease involves an over-proliferation of the T4 lymphocyte, while the other involves premature death of the same cell. (2321)

Finally, he alludes to the technological innovations developed in his laboratory and their effects on basic research:

> The capacity to mimic the disease *in vitro* by infecting target normal T4 cells provides the most powerful system available for the study of altered growth of a human cell. (2321)

In concluding the *Cancer Research* paper, Gallo says:

> The past 5 years have witnessed the beginning of a new era in retrovirology, the era of human retroviruses. Within this short period the causes of two human fatal diseases have been worked out. In both instances in vitro systems have developed which mimic the diseases so that a reasonable amount of information on the involved molecular mechanisms has become available. The future promises much more such information, but what is much less certain is how soon we will learn to correct and/or prevent these diseases. (4530)

Though this statement is not wholly optimistic, it is important to note that Gallo does not ask *whether* we will learn to prevent these diseases, but *how soon*.

In *The New England Journal of Medicine,* after detailing the new knowledge about retroviruses and about the history of their discovery, the writers conclude their paper by enumerating what they believe is optimistic about their findings. Knowledge of the virus "may affect many phases of basic research with clinical implications" (1295): "a stimulus to re-examine endemic forms of cancer that pose special public-health problems for some nations in the Third World" and the possibility of developing "a vaccine for persons who belong to certain risk groups and thereby prevent the disease" (1296). Whereas he emphasizes the implications for future research at the end of the specialized research journals, here Gallo and Broder emphasize the implications for treatment of serious medical problems caused by the virus.

In *Scientific American*, he concludes by painting a very different picture of the scientific community. Gallo warns his readers of the "hubris" of scientists in thinking that infectious disease has been conquered. In the final paragraph, Gallo discusses the "moral" of "this terrible tale." He says that the discovery of retroviruses and their "capacity to cause extraordinarily complex and devastating disease has exposed the claim [that science has conquered infectious disease] for what it was: hubris. Nature is never truly conquered" (65). It is interesting that he ventures a moral attack on the "hubris" of modern science in this paper while in the other papers he applauds the capacity of basic science to explain puzzling phenomena and develop new technology.

In light of the recent negative publicity surrounding Gallo, it would seem prima facie that Gallo was using base rhetoric, was playing to the gallery. However, a more benign interpretation is that Gallo was using two contrasting conceptions of science as flexible tools for arguing in different contexts. Karl Popper suggests that the primary aim of science is "to find *satisfactory explanations* of whatever strikes us as being in need of explanation" and that ". . . scientific explanation, whenever it is a discovery, will be the explanation of the known by the unknown" (191). For scientific culture, *explanation* of puzzling phenomena, rather than *control* of phenomena, is a desirable end in itself. In popular culture, the Frankenstein or mad scientist myth, as well as such highly publicized cases as the development of the atom bomb and human genetic engineering, combine to build a popular image of science as aiming to control nature with often disastrous consequences. The discrepancy between Gallo's celebration of scientific explanation in *Cancer* and *Cancer Research* and his admonishment of scientific hubris in *Scientific American* would seem hypocritical to a positivist but merely ironic to anyone who understands the relationship between rhetoric, culture, and reality.

Will the Real Virus Please Stand Up?

Description of the phenomenon itself offers another flexible argumentative resource for Gallo. Rhetorical ownership implies the authority to name, define, and describe the phenomenon and control its public interpretation. Gallo's descriptions of the AIDS virus itself indicate his sensitivity to the needs and expectations of his specialized and lay audiences as well as his understanding that his descriptions influence their perceptions of the virus and his role in its discovery. In these descriptions two metaphorical conceptions of the virus construct two very different public interpretations of the disease.

Jeanne Fahnestock has traced the "rhetorical life"of scientific facts (1986) as they move from specialized to lay publications. Fahnestock illuminates the ways journalists inscribe scientific findings in language that is more certain and authoritative than the language in scientific research reports and how the translation from scientific to lay communication belies the different epistemological assumptions and rhetorical practices of both communities. But what happens when *one* writer or one team of writers disseminates new knowledge and the story behind its discovery to different audiences, to specialized and unspecialized readers?

The star of the show, the AIDS virus itself, offers an interesting view of the way one writer or a team of writers may very subtly "change" the phenomenon at hand to accomodate different audiences. In both the *Annals* and the *Scientific American* papers, the virus and its cytopathic effects are described in detail, but both descriptions offer different conceptions of the virus. In the *Annals* paper, the human retrovirus causing AIDS is a "remarkable" natural phenomenon with "unique" biological capacities. In the *Scientific American* article, the virus becomes an "invader," a "culprit" that violently attacks its host. Although Gallo cowrote the *Annals* paper with Wong-Staal and although it is impossible to tell which writer composed the descriptions of the virus, the difference in the language used to describe retrovirus activity in these two papers indicates the different needs and attitudes of the two audiences. The biomedical specialist is interested in the virus as a clinical entity and needs to know how the virus differs from other viruses. The *Scientific American* readers likely knew the virus not as a clinical entity but as a killer; Gallo's descriptions of the virus accomodate their expectations and allow them to visualize intercellular activities.

For readers of *Annals,* Gallo and Wong-Staal choose terms which depict the virus as "agent" and the host cell as "scene," but do not necessarily invoke visual images of aggression:

> This enzyme [reverse transcriptase] *converts* viral RNA into double stranded DNA. For virtually all retroviruses, the DNA form *moves* from the cytoplasm to the nucleus where it *integrates* into the host cell DNA. (679, emphasis mine)

In contrast, the description of the same activity in *Scientific American* uses stronger verbs connoting aggressive agency:

> When the virus enters its host cell, a viral enzyme called reverse transcriptase *exploits* the viral RNA as a template to *assemble* a corresponding molecule of DNA. The DNA *travels* to the cell nucleus and *inserts itself* among the host's chromosomes, where it provides the basis for viral replication. (47, emphasis mine)

In describing this activity, Gallo is sensitive to his nonspecialized readers' need for more graphic visual images to help them *imagine* what they will probably never get the chance to *see*. The key differences are in the verbs: "converts" becomes "exploits," "moves" becomes "travels," and "integrates" becomes "inserts itself." The first set of verbs implies that what is occurring is a process in nature, a complex relationship between viruses and human T-cells. But the second set of terms builds a metaphoric theme centered on the retrovirus as an aggressive invader, a model likely to be recognized by his readers.

Mary Hesse explains how metaphor works in scientific knowledge:

> . . . metaphor works by transforming the associated ideas and implications of the secondary [which involves language] to the primary [involving observation] system. These select, emphasize, or suppress features of the primary [which] . . . is 'seen through' the frame of the secondary. (114)

In other words, through language we provide versions of the empirical world. The version of retrovirus activity in the paper for virologists suppresses features of aggression while his version prepared for unspecialized readers emphasizes these features.

Descriptions of how the virus transforms cells and infiltrates the bloodstream are also different. In *Annals* the writers explain that once infection occurs,

> integrated viral genes are duplicated with the normal cellular genes so all progeny of the originally infected cell will contain the viral genes. The cell may become a cancer cell by the expression of one or more viral genes. In other instances, integration of the viral DNA form without expression of the viral genes may lead to a change in the expression of nearby cellular genes, leading to the pathologic effects of the virus. (679)

In *Scientific American* the story goes like this: Once infection occurs, and the T4 cells gets activated,

> instead of yielding 1,000 progeny [as in a healthy system] the infected T cell proliferates into a stunted clone with perhaps as few as 10 members. When those 10 reach the blood stream and are stimulated by antigen, they begin producing virus and die. (50)

Obviously, Gallo constructed the second account so that visual imagery might accomodate the unspecialized reader and the context of a more "popular" publication. We can "see" the "stunted clone" and imagine the pathetic army of only ten fighters entering the blood stream, unaware that they carry within themselves the very enemy they seek.

In the *Annals* paper, Gallo seems intent on bringing the uniqueness of the HTLV family of viruses to light. This uniqueness is defined in terms of such things as the "indirect association with B-cell leukemia and lymphoma," and "a *trans*-acting transcriptional activation of viral long terminal repeat sequences" (686).

Gallo explicitly labels these and other features as "unique family traits." Thus, in this paper, Gallo is concerned with showing how the human retroviruses differ from other viruses in their biological, biochemical, and epidemiologic uniqueness.

The retrovirus' uniqueness in comparison with other viruses plays a lesser role in the *Scientific American* paper. It is more important to gain the readers' understanding by helping them visualize through language of destruction and violence. Gallo depicts the retrovirus as a furious, violent agent that plays havoc with T4 cells:

> The virus bursts into action, reproducing itself so furiously that the new virus particles escaping from the cell riddle the cellular membrane with holes and the lymphocyte dies. (49)

We may well imagine Audie Murphy bursting from a blown-up tank and riddling the German invaders with bullet holes.

Gallo proposes a hypothetical model, as he calls it, of the process of cell death:

> [the cell's] death may depend on an interaction between the viral envelope and the cell membrane. Perhaps that interaction . . . punches a hole in the membrane. Because the virus buds in a mass of particles, the cell cannot repair the holes as fast as they are made; its contents leak out and it dies. (52)

The above description, "for the moment only a model," employs an analogy from our everyday experience. Terms such as "punch out" and "leaks out" suggest a visual model of puncturing a hole in a bag full of water. This modeling is actually both an hypothesis and a tool for explanation.

It is physically impossible for Gallo to directly observe the mechanisms contributing to cell death; thus Gallo employs analogy as a heuristic device for his own problem solving. This analogy also serves rhetorical purposes. The vivid and graphic qualities of his description enable the reader not only to visualize a biochemical process but also to picture a battleground which scientists like Gallo have spent countless hours investigating. Gallo wants the readers of *Scientific American* to know the formidable odds in fighting such an enemy and to understand that "the progress made in only three years" (55) represents the fruits of the valid scientific investigation done in his lab.

Given the exigencies that could have contributed to the rhetorical situation—the lawsuit against Gallo's patent claim as well as the laboratory evidence that the virus isolated by the French was unrelated to HTLV—we can assume that these papers represent Gallo's attempt to reclaim rhetorical ownership in AIDS research, to reassert the primary role of his laboratory in discovering human retroviruses and linking them to AIDS. However, I believe these accounts represent something more than one scientist's attempts to gain rhetorical ownership. These accounts actually tell us more about the social forces surrounding the acceptance of a theory than about the actual real world of the laboratory. Steve

Woolgar has argued that "the practical expression of, or reference to, a phenomenon both recreates and establishes anew the existence of the phenomenon. In describing a phenomenon, participants [that is, writers and readers] simultaneously render its out-there-ness" (246). Gallo's depiction of the virus itself recreates the phenomenon, the retrovirus, which is a different critter for different audiences. And his depiction of his lab's role in human retrovirology and AIDS research not only recreates that role but also rekindles communal expectations about the kind of laboratory investigation that leads to significant discoveries. Though certainly more research of the discovery account genre is needed before we can draw definitive conclusions about them, I suggest that discovery accounts, a recognized form of valedictory address which usually occurs after a community has settled issues and reached consensus, may actually be employed by writers to settle those issues and establish their rhetorical ownership in the field. Gallo adopts the discourse appropriate for a closed debate as an active argumentative resource in an open battle.

Notes

1. According to Gallo and Wong-Staal, "A retrovirus carries within its core an enzyme that synthesizes DNA, reverse transcriptase. This enzyme converts viral RNA into double stranded DNA. For virtually all retroviruses, the DNA form moves from the cytoplasm to the nucleus where it integrates into the host cell DNA" *(Annals* 679)
2. For a thorough account of the controversy between Gallo and the French, see "Science Under the Microscope" by John Crewdson. *The Chicago Tribune,* November 18, 1989.

Works Cited

Broder, Samuel, and Robert C. Gallo. "A Pathogenic Retrovirus (HTLV-III) Linked to AIDS." *New England Journal of Medicine* 311 (1984): 1292–97.
Crewdson, John. "Science Under the Microscope." *The Chicago Tribune,* 18 November 1989.
Fahnestock, Jeanne. " Accommodating Science: The Rhetorical Life of Scientific Facts." *Written Communication* 3 (1986): 275–96.
Gallo, Robert C. "Human T-Cell Leukemia (Lymphotropic) Retroviruses and their Causative Role in T-Cell Malignancies and Acquired Immune Deficiency Syndrome." *Cancer* 55 (1985): 2317–23.
—."The AIDS Virus." *Scientific American* 256 (January 1987): 47–56.
—."The Human T-Cell Leukemia/Lymphotropic Retrovirus (HTLV) Family: Past, Present and Future." *Cancer Research* 45 (1985): 4524–32.
—.and Flossie Wong-Staal. "A Human T- Lymphotropic Retrovirus (HTLV-III) as the Cause of Acquired Immunodeficiency Syndrome." *Annals of Internal Medicine* 103.5 (1985): 679–89.
Gusfield, Joseph R., *The Culture of Public Problems: Drinking-Driving and the Symbolic Order.* Chicago: U of Chicago P, 1981.
Hesse, Mary. *Revolutions and Reconstructions in the Philosophy of Science.* Bloomington: Indiana UP, 1980.
Kuhn, Thomas. *The Structure of Scientific Revolutions.* 2nd ed. Chicago: U of Chicago P, 1970.
—. *The Essential Tension: Selected Studies in Scientific Tradition and Change. Chicago: University of Chicago,* 1977.
Latour, Bruno, and Steve Woolgar. *Laboratory Life: The Social Construction of Scientific Facts.* London: Sage, 1979.
Merton, Robert. "Three Fragments from a Sociologist's Notebooks: Establishing the Phenomenon, Specified Ignorance, and Strategic Research Materials." *Annual Review of Sociology* 13 (1987): 1–28.
Popper, Karl R. *Objective Knowledge: An Evolutionary Approach.* Oxford: Clarendon, 1986.
Sontag, Susan. *AIDS and its Metaphors.* New York: Farrar, Strauss and Giroux, 1989.
Woolgar, Steve. "Writing an Intellectual History of Scientific Development: The Use of Discovery Acccounts." *Social Studies of Science* 6 (1976): 395–422.

Writing Science

Reporting the Experiment:
The Changing Account of
Scientific Doings in the
Philosophical Transactions
of the Royal Society, 1665–1800

by Charles Bazerman

Experimental reports tell a special kind of story, of an event created so that it might be told. The story creates pictures of the immediate laboratory world in which the experiment takes place, of the happenings of the experiment, and of the larger, structured world of which the experimental events are exemplary. The story must wend its way through the existing knowledge and critical attitude of its readers in order to say something new and persuasive, yet can excite imaginations to see new possibilities in the smaller world of the laboratory and the larger world of nature. And these stories are avidly sought by every research scientist who must constantly keep up with the literature.

If each individual writer does not think originally and creatively about how to master recalcitrant language in order to create such powerful stories, it is only because the genre already embodies the linguistic achievement of the three hundred years since the invention of the scientific journal necessitated the invention of the scientific article.[1] The experimental report, as any other literary genre, was invented in response to a literary situation and evolved through the needs, conceptions, and creativity of the many authors who took it up. The corpus of the genre is not only immense, it is rich and varied, synchronically and diachronically. Despite familiar pedagogical prescriptions, the experimental report is no single narrow form.

Fiction, Nonfiction, and Accountabilities

The extent of literary construction is not diminished for the genre's being nonfiction. Nonfiction—a concept defined only negatively, for its not being the regular meat of literary investigation—presents serious literary questions of the representations of worlds in words. Given modern critical understanding and modern epistemology, the traditional distinction between that which is made up (and therefore of literary interest) and that which reflects the world (and therefore trivial linguistically), obscures rather than illuminates. Few today would contend that signs are unmistakable and predetermined reflections of things.

Some contemporary theorists would in fact reduce all texts to fiction, claiming reference itself a fiction.[2] While much may be said for this position, nonfiction creation incorporates procedures tying texts to various realities. An introspective phenomonology of religious experience or a political speech or an annual report is no less nonfiction than an account of doings in a room at a physics laboratory. Differences of nonfictions hang on differences of accountabilities (of both degree and kind) that connect texts to the various worlds they represent and act on.

The concept of accountabilities will run throughout this chapter, as we look at how the various writers and readers, situated in certain communities, following the habits and procedures of observation and representation, are restricted in what they say, do, and think by empirical experience. Many mechanisms (of training, argument, criticism, normative behavior, application, sanction, and reward) realize and elaborate this fundamental commitment of the discourse. This chapter examines a series of accounts of empirical experience, those defining experiences in science of events created so they might be told, experiments. It examines the emergence of standards and procedures for those accounts, means for reconciling accounts and developing more generalized accounts consistent with more specific accounts, situations where discrepancies or uncertainties within or between accounts call for further accounts. The scientific enterprise is built on accounts of nature, and the development of scientific discourse can be seen as the development of ways of presenting accounts.

Other types of communities may have other fundamental accountabilities and means of enforcing and elaborating these accountabilities. Sacred texts, for example, provide the constant ground, pattern, and reference point for communication in some religious communities; all discourse is held accountable to the sacred text by means of discourse style, conceptual assumptions, overt quotation and paraphrase, psychological rewards of certainty, social rewards for piety, and ostracism for blasphemy. Legal discourse is held accountable on one hand to a hierarchically arranged series of court decisions, laws, and constitutions, and on the other to evidence gathered through procedures defined by the system and represented in a manner established by tradition and explicit rule. In certain types of literary critical discourse, the fundamental reference point is a subjective experience of the text, which the critic identifies and attempts to transfer to the reader. The whole enterprise rests on that experience and is elaborated through the socially recognized means of developing such accounts.

As developed here, the concept of accountabilities is closely related to Ludwik Fleck's definition of a fact as a "stylized signal of resistance in thinking" within a thought collective (98). That is, following the thought style (including styles of perception, cognition, and representation) of a group of people engaged in intellectual interchange, certain statements limit what can be appropriately said and thought within the collective. Certain of the constraints are what Fleck calls active elements, actively produced by the thought style; others are passive, where the discourse system so to speak bumps into objects outside itself, which by the thought style must be respected by the thought collective. Facts are perceived and represented through the actively constructed thought style, but reflect the passive constraint imposed by external conditions. (A more complete discussion of Fleck's analysis of facts is presented in Bazerman 1988: 312–313.)

These facts accepted by the community form the basis for the accountability, as I use the term. These facts, outside the immediate active elements of discourse, must be brought into the discourse and accounted for. The process of holding the text accountable to these facts serves to shape the discourse. The mechanisms of accountability permeate the creation, reception, and textual form of statements in the collectives holding themselves accountable in this way.

Fleck goes on to characterize the thought style of contemporary science as actively seeking to include a maximum of passive elements despite their tendency to disrupt other accepted active elements. Put more simply, the fundamental commitment is to empirical experience. Scientific discourse, therefore, is built on accountability to empirical fact (as of course characterized within the thought style of science) over all other possible accountabilities (such as to ancient texts, theory, social networks, grant-giving agencies), and must subordinate other forms of accountability (that is, those other forms which do form part of the scientific thought style) to the empirical accountability.

The Experimental Report as a Historical Creation

Although many kinds of communication pass within scientific communities, experimental reports are close to the heart of the accountability process, for experimental reports present primary accounts of empirical experience. Experimental reports attach themselves to the nature that surrounds the text through the representation of the doings, or experiment. How does the world of events get reduced to the virtual world of words? How did conventions and procedures for this reduction develop? What are the motives and assumptions implicit in the rhetoric and procedure? And what are the accountabilities that limit statements, ensuring the influence of the evidence of the world on human conception? These are equally questions of literary theory and rhetoric as of philosophy of science, for what appears to philosophy of science as the problem of empiricism, appears to rhetoric as the problem of persuasive evidence, and to literary theory as the problem of representation.

One place to turn for answers to these questions is the early history of the experimental report, for the formation of a genre reveals the forces to which textual features respond. A genre consists of something beyond simple similarity of formal characteristics among a number of texts. A genre is a socially recognized, repeated strategy for achieving similar goals in situations socially perceived as being similar (Miller). A genre provides a writer with a way of formulating responses in certain circumstances and a reader a way of recognizing the kind of message being transmitted. A genre is a social construct that regularizes communication, interaction, and relations. Thus the formal features that are shared by the corpus of texts in a genre and by which we usually recognize a text's inclusion in a genre, are the linguistic/symbolic solution to a problem in social interaction.

That a well-established, successful genre is usually realized in relatively static formal features should not hide the social meaning and dynamics of a genre, no more than the active reality of a performed Beethoven quartet should be obscured by the sheet music. By examining the emergence of a genre we can identify the kinds of problems the genre was attempting to solve and how it went about solving them. The history of the experimental report shows how a certain kind of detailed picture of a laboratory event became the standard and how particular information became essential to a successful telling. We can also see forming, as the genre takes shape, a particular literary community with certain critical expectations.

The *Philosophic Transactions of the Royal Society of London,* the first scientific journal in English, carries the main line of the development of scientific journal writing in English through the nineteenth century. Here I follow the development of the genre of experimental report in the pages of the *Transactions* from its founding in 1665 until 1800, when a number of familiar features of the experimental report were firmly in place.

I focus here entirely on the internal development of the genre. Although the genre of experimental article has origins in essay, epistolary, and journalistic writing of the seventeenth century (Frank; Houghton; Kronick; Paradis; Sutherland), the internal dynamics of scientific communication within a journal forum reshape the initial sources to create a new communicative form, powerful enough to influence other forms of communication and the social structure of the community which uses it.

Method

This study is based on an examination of all articles (about 1000 altogether, over 7000 pages) in volumes 1, 5, 10, 15, 20, 25, 30, 35, 40, 50, 60, 70, 80, and 90 of the *Transactions*. From these volumes, all articles using the word "experiment" either in the title or in running text were then selected for closer examination. Then those articles providing only secondary accounts of experiments were eliminated, leaving only articles written by the experimenter reporting on new experiments. This procedure left a remainder of about 100 articles to be analyzed.

Because of the changing character of the writing in the articles and because of the individual character of each separate article, no quantitative comparisons appeared useful, so I resorted to the traditional method of literary criticism, descriptive analysis of each of the separate articles. This method did allow me to explore the varying features of writing as they presented themselves. However, such individual description makes generalization difficult. In order to facilitate the comparison and continuity among the many cases, I narrowed my descriptive analysis to a set of specific questions.

1. To what kind of event does the term "experiment" refer?
2. How fully and in what manner are experimental events described?
3. How fully are apparatus and methodology described? How fully and in what way are methodological concerns discussed?
4. How precisely and completely are results presented? What criteria of selectivity are used? How much and what kind of discussion and interpretation are present?
5. Is the experiment presented as a single event or as part of a series of experiments? In a series, what is the principle of continuity?
6. How is the account of the experiment organized? How are series of experiments organized? Where does the account of experiment or experiments fit within the organization of the entire article?
7. What is the rhetorical function of the experiment within the article?

To facilitate the organization of the material of these separate analyses, particularly with the intention of clarifying historical trends, I then synthesized the analyses from each volume examined, thus forming generalizations about the character of the experimental reporting in each time period. From this collection of chronologically arranged syntheses I extracted the major themes and trends as presented below. The story I will be presenting thus has been filtered several times through my own personal interpretive, selective, and synthetic judgments. I will present detailed evidence from the texts to illustrate and support the story I present, but I will not be presenting all the trees in the forest. If I were to tell more I would risk the reader losing sight of the shape of the forest I believe I have found. On the other hand, I have no more impersonal way of either reconnoitering the shape of the forest or communicating and demonstrating that shape. This is always the dilemma of attempting to make sense of historical and literary material which incorporates the complex actions of many individuals. In terms of persuasion, this essay must rest in the short term only on the impression it gives of a plausible story and in the long term only on whether others crossing the same terrain find the shapes presented here recognizable and useful.

Another consequence of working from individual accounts of the products of many individuals, each reacting to specifics of individual situations, is that the overall trends are likely to wash out many individual variations as well as to appear more uniform than they in fact are. When looking at all the trees in the forest, I find a somewhat more ragged shape than will emerge here, although I will attempt to indicate where the raggednesses are.

The Changing Experiment

Despite our current belief in experiment as one of the foundations of science, only a small part of the volumes examined up to 1800 were devoted to reporting on experiments. Both in terms of the percentage of total articles and percentage of pages, experimental articles accounted for only 5 to 20 percent of each volume through volume 80. Only in volume 90, opening the nineteenth century, did the percentages rise substantially to 39 percent of the articles and 38 percent of the pages.

Until 1800, however, it is clear that experiments were only one of many types of information to be transmitted among those interested in science. The most articles and pages were devoted to observations and reports of natural events, ranging from remarkable fetuses and earthquakes, through astronomical sightings, anatomical dissections, and microscopical observations. Human accomplishments received attention with accounts of technological and medical advances, travelogues of journeys to China and Japan, and an interview with the prodigy Mozart. The reportable business of natural philosophers was hardly restricted to experimenting or even theorizing, which received even less space than experiments.

The relative paucity of experimental accounts should remind us how much the importance we attach to experiments is a function of the rise of the experimental article as a favored way of formulating and discussing science. Although experiments may have their ancient precursors and early books may have experimental accounts within them, the creation of the experimental article has helped create our modern concept of experiment.

Those reported events identified as experiments change in character over the period 1665–1800. The definition of experiment moves from any made or done thing, to an intentional investigation, to a test of a theory, to finally a proof of, or evidence for, a claim. The early definitions seem to include any disturbance or manipulation of nature, not necessarily focused on demonstration of any stated preexisting belief, nor even with the intention of discovery. With time, experiments are represented as more clearly investigative, corroborative, and argumentative.

In the first volume of the *Transactions*, a number of experiments reported are simply cookbook recipes for creating marvellous effects or effects of practical use, such as the directions for coloring marble internally "of use to artisans" (1:125).3 Elsewhere experiments are a method of investigating nature, treated on a par with observations, as in the formula often appearing in the pages of the journals: "experiments and observations." Observations were made upon undisturbed or unmanipulated nature while experiments involved human intervention. That intervention need not imply intention of investigation; for example, one series of experiments grew out of a cook's pickling of mackerels. Only after several days, when the cook noticed that the broth had turned luminescent, did the master of the house identify the phenomenon as something worth observing (1:226–28).

By volumes 5 and 10 the definition of experiment had narrowed in most cases to a conscious investigation of phenomena involving some doings or manipulations, even though cookbook novelties appeared as late as volume 20 (20:42–44, 87–90, 363–65). These experiments, however, are presented as simply allowing the conditions for brute nature to reveal itself. The meaning of the experiment is simply what is observed upon its occurrence. For example, in volume 10 Christiaan Huygens and Denis Papin report on a series of experiments to "know, whether the Vacuum would be of use to the Preservation of Bodies" so they placed various flowers, fruits, and other comestibles in vacuums for various periods of time and observed (10:492–95). Retrospectively, such experiments seem part of a broader investigation of the atmosphere, but nowhere do the reports of these or similar vacuum experiments suggest that questions, theories, problems, or hypotheses were being explicitly explored or tested.

Only in cases of overt controversy are assumptions or hypotheses explicitly set out to be tested, for then the experiment becomes a means of adjudicating between two or more proposed views. Again in volume 10, as part of a report of another series of vacuum experiments, Huygens and Papin, prompted by comments by another investigator, present their alternative view of the reasons for collapse of lungs and then describe a specific experiment that led them to their conclusion.

By volume 20 several experiments have clear hypothesis-testing or debate-solving functions. Experiments are being recognized as created events designed with specific claims about nature in mind. In volume 25, for example, Francis Hauksbee[4] comments, with some pleasure, at the use of experiment as a way to test hypotheses: "... the greatest Satisfaction and Demonstration that can be given for the Credit of any Hypothesis, is, That the Experiments made to prove the same, agree with it in all Respects, without force" (25:2415–17).

In volume 30 five articles place their experiments in the context of extensive discussions of debates which the experiments are set to resolve, such as whether a vacuum is truly empty. In this particular case, the experimenter Jean T. Desaguliers spends a full page reporting on a previous vacuum experiment he had made and the particular objections a group of plenists had to his procedure, against which he sets his current work (30:717–18). In most cases the experiments provide rather direct observations concerning the issue at hand, as in the preceding example where pairs of different objects were dropped in an evacuated column to see whether the time of fall were the same for each. But at least in one case the experiments were at some remove from the issue of contention, indicating that experiments were now accepted within the context of a complex of accepted knowledge rather than simply as brute demonstrations. Desaguliers, in order to dispute Leibniz's explanation of barometric measurements during rain, enters into a theoretical discussion of the weights of bodies falling through a medium, which discussion he then supports through a series of ingenious experiments employing neither barometers nor atmosphere. The experiment stands on established background knowledge for its construction and interpretation (30:570–79).

At this point the experiment's role of adjudicating disputes as to the brute truth of nature starts to shift toward establishing the truth of general propositions that are not necessarily disputed by anyone. Experiments stop being a clear window to a self-revealing nature, but become a way of tying down uncertain claims about an opaque and uncertain nature. The meaning of an experiment is no longer the simple observation of what happens. An experiment is to be understood only in terms of the ideas that motivate it, for nature is no longer considered to be so easy to find. Through volumes 50 and 60 the experiments become increasingly couched in terms of problems—things that despite our familiarity with phenomena we do not understand. In volume 60, Joseph Priestley describes his puzzlement concerning the nature of the electrical phenomenon he calls "lateral explosion" which did not behave the expected way in a series of simple exploratory experiments: "I do not remember that I was ever more puzzled with any appearance in nature than I was with this; and, in the night following these experiments, endless were the schemes that occurred to me, of accounting for them, and the methods with which I proposed to diversify them the next morning, in order to find out the cause of this strange phaenomenon" (60:195). A series of experiments follow, logically solving the puzzle, part by part. Experiments are now clearly represented as part of a process of coming to conclusions. Priestley in another article on electricity comments on the personal intellectual consequences of an experiment: "With respect to the main object of my inquiry, I presently satisfied myself, that the conducting power of charcoal . . ." (60:214).

By volume 80, experiments are subordinated to the conclusions the authors have come to; that is, the experiments are ways of proving or supporting general claims. Hypotheses are presented up front and the series of experiments follow. Priestley, for example, now adopts language of proof rather than of discovery: "That my former supposition . . . is true, will appear, I presume, from the experiments which I shall presently recite" (80:107).

By volume 90 authors talk about the necessity of establishing general knowledge, and the role of experiment in testing our beliefs, as well as filling out knowledge. William Henry comments at the beginning of his report of experiments analyzing muriatic acid, "The theory of the formation of acids . . . must be regarded as incomplete, and liable to subversion, till the individual acids now alluded to have been resolved into their constituent principles" (90:188). Experiments test and justify the general claim, which is part of a larger system of general claims. The language of general proof holds sway: "The above facts prove, that the combination of oxygen and muriatic acid. . ." (90:194).

Methodological Concern

As experiments gain an argumentative function, the reports explain more fully how the experiment was done and why the particular methods were chosen. How nature is prodded is recognized as affecting nature's response. Debates over differing results focus attention on differences in experimental methods and conditions. Methodological care enables experiments to be used as investigative tools and then as proofs. Investigators, in order to satisfy their own problems, make subtle methodological distinctions among different experimenters within the same series. Then toward the end of the period, when experimenters start arguing general propositions, the meaning and validity of the experiment depends on proper methodology.

In the early volumes, how an experiment was performed was generally mentioned in passing, simply to let the reader know what kind of experiment was done. In volume 1, issue 3, for example, the editor, Henry Oldenburg, describes a series of observations and experiments made by Thomas Henshaw on the putrefaction of May-Dew. In each of the series, the procedure is described in an introductory clause or modifying phrase only, such as "Dew newly gathered and filtered through a clean Linnen cloth, though it be not very clear, is of yellowish color . . ." (1:34). The procedure only serves to identify the dew. Only when directions are for practical use (and not, I emphasize, replication) are more detailed instructions given, though these are still vague by modern cookbook standards. Robert Boyle, for example, in volume 1, issue 15, appearing in mid-July, explains, for the benefit of sweltering Londoners, his new method for producing cold, useful for chilling drinks: "Take one pound of Sal Armoniack and about three Pints (or pounds) of Water, put the Salt into the liquors, and stir altogether, if your design be to produce an intense, though but a short coldness; or at two, three, or four several times, if you desire, that the produced coldness should rather last somewhat longer . . ." (1:256–57).

Even as early as the fifth volume, challenged by disagreements, authors demonstrate their experimental care and account for differences in results by describing in greater detail their experimental procedures and the conditions. Disagreements over experimental results on sap flow in sycamores lead Willoughby to consider both the date and weather conditions when the trees were bled. In volume 10, fear of challenge leads Robert Boyle to report that he took great care that a copper mixture was not shaken in the course of the experiment, although he himself does not believe that disturbance of the mixture to be of any consequence (10:468). The most explicit presentation of technique results from Francis Line's challenge of Isaac Newton's results. Newton in response lays out in much greater detail the method of his earlier experiment and the conditions under which the experiment occurred. He further suggests additional experiments and challenges Line to replicate them all. In presenting the method in such great detail, Newton insinuates that Line in doing his first set of experiments got things wrong. Since the debate is over whether such things as reported in Newton's account happen, the method and conditions to make them happen are crucial to the argument. (This incident and the surrounding story are examined more fully in Bazerman 1988:80–127.)

By volume 30 authors claim they design experiments to meet specific objections of opponents. Desaguliers, for example, attempts to answer objections of the plenists that earlier experiments concerning bodies falling in vacuo were done over too short a distance. Desaguliers reports: "To obviate this I contriv'd a machine to this purpose, which consisted of a strong wooden frame . . ."

(30:718). Variations in apparatus that might cause variations in results are also noted, to indicate that the author is not misled or confused by such variations (for example, 30:1078). Delicate parts of the procedures are noted, so as to distinguish the author's proper procedures from his opponent's less careful ones and to indicate specific points where the opposition may have erred. Desaguliers, for example, defending Newtonian optics at length against an extensive attack by John Rizzetti, points out many places where Rizzetti's procedures were misguided and where his judgment may have failed. In one instance Desaguliers comments, "This must have been Signior Rizzetti's mistake . . . for several of the Persons present at my Experiments made the same Mistake at first before they could perform the Experiment in manner above-mentioned; which they at last did. . . . This mistaking a Reflecting for a Refraction has been the Occasion of several more Errors, and Difficulties to be met with in Signior Rizzetti's Book" (35:610).

Articles not engaged in overt contention continue to discuss method only sketchily. However, as experiments become incorporated into stories of discovery, the distinctions between trials become important as events in consciousness, so at least the crucial differences between trials become defined. Richard Watson, for example, introduces the sixth experiment of his series on the solution of salts with the comment: "Thinking that the difference in the bulks of the water before and after solution might be owing to the separation and escape of some volatile principle; I took care to balance as accurately as I could, water and sal gemmae, water, and the salt of tartar, water and vitriolated tartar & c . . ." (60:335). Since persuasion comes through the audience's willingness to accept the experimenter's experience of discovery, detailed accounts of method indicate both the experimenter's care and that he was convinced of his discoveries for good reasons.

Finally, in volumes 80 and 90, as articles present proofs of general hypotheses, the details of the experiments demonstrate care, exactness of results, relevance to thesis, and the elimination of alternatives. Henry, for example, gives a complex rationale for a particular method on the bases of precision and clarity of results:

> I employed the electric fluid, as an agent much preferable to artificial heat. This mode of operating enables us to confine accurately the gases submitted to experiment; the phaenomena that occur during the process may be distinctly observed; and the comparison of the products with the original gases, may be instituted with great exactness. The action of the electric fluid itself, as a decomponent, is extremely powerful; for it is capable of separating from each other, the constituent parts of water, of the nitric and sulfuric acids, of the volatile alkali, of nitrous gas, and of other several bodies, whose components are strongly united. (90:189)

William Herschel, in his experiments on the distinction between the visible and radiant spectrum, takes his measurements in several different configurations to prove that his results are caused by the principle he is trying to prove. Not only that, he rotates the position of the thermometers to ensure the results are not artifacts of faulty measuring devices. In such duplication and varying of measurements to ensure validity of results and to eliminate all other possible variables, Herschel presents his work in a way that approaches the modern concept of controls (90:255–326).

Indeed, throughout the period, the increasingly expressed awareness of possible variables seems to reach toward an unexpressed concept of controls. In recognizing differences of conditions or execution of the experiment that might affect results, the reports started comparing results from different situations. As more experiments report multiple trials with only slight variation of experiment, crucial factors are isolated. Then, as we have seen, multiple trials are explicitly designed to establish distinctions between two sets of conditions. The practice of experimental controls—running an experiment twice, identically except for an isolated crucial variable—is only the next step in argumentative clarity through the representation of method.

The changes in illustrations through the period also express the growing importance of methods. The early issues of the journal frequently illustrate the phenomena being reported on or new technological marvels, but rarely is the apparatus used for an experiment considered worth a picture. However, as experiments become more ingenious, elaborate, or just simply careful, illustrations follow. The first apparatus illustrations I found were in volume 25, showing the brushes and vacuum devices used by Hauksbee to generate static electricity in vacuum. Although not all experiments are illustrated with apparatus diagrams, they do become a prominent feature, as in Desaguliers's answer to Rizzetti, allowing the reader to visualize the experimental procedures and the results (35:575). Herschel's articles in volume 90, punctuated by a number of quite realistic apparatus illustrations, give a concrete feel of what was done. The realism of illustration becomes particularly important as the account or story of the experiment becomes the reader's vicarious surrogate for the actual experiment, as will be discussed below.

Precision and Completeness of Results

As with method, results of the experiments are reported with increasing detail, care, and quantitativeness as the experiment bears more and more weight of argument, persuasion, and then proof. Early results are described vaguely and qualitatively, as though the phenomena of nature were robust, uniform, and self-evident. As disputes arise over reported results, writers become more careful about reporting what they see, and measurement takes a greater role. With the proliferation of quantitatively comparable results, experimenters being puzzling over subtle variations in results; detailed results become a means of figuring out exactly what is going on. Finally, detailed quantitative experimental results, fitting quantitative theoretical results, form the empirical proof of general hypotheses.

In the early volumes, those experiments that provide directions for achieving certain wondrous effects have no explicit results at all, for it is simply assumed that following the recipe will lead to the desired effect. Where results are given they are in the form of general qualitative observations, such as in the example of luminous mackerel broth: "As soon as the Cooks hand was thrust in to the water, it began to have a glimmering. . . . they who look'd on it at some distance, from the further end of another room, thought verily, it was the shining of the Moon through a Window upon a Vessel of Milk; and by brisker Circulation it seem'd to flame" (1:227). Even where quantification of results seems a rather simple matter, as in two experiments in volume 5 concerning expansion of a freezing solution and the timing of respiration, the results were given in purely qualitative terms.

Again, debate and conflict push results to greater detail and precision in exactly the same articles with more detailed accounts of method. Newton, for example, in answering Line, spends a lengthy paragraph describing the three different images cast by a prism, distinguishing the character of these different images, so that anyone repeating the experiment can find the oblong image which was Newton's particular concern and which Line disputed (10:503).

By volumes 15 and 20 quantitative results in measurements of the speed of sound, barometric air pressure, and specific gravity enable comparisons. With the increase of multiple trials, distinctions among results of various trials become a practical expository tool. In volume 50, for example, William Lewis, in investigating mixtures of platinum and gold, creates nine different mixtures of different proportions from 1:1 to 1:95 in order to compare both qualitative and quantitative properties (50:148–55).

By volume 60 the results sought and reported have specific relation to the hypotheses being investigated and tested. James Johnstone reports a series of experiments designed to test hypotheses concerning the function of nerve ganglions; not only are the results found consistent with the hypotheses, but he adds results from experiments reported by other authors. These additional results also support the hypotheses, even though the experimenters did not have the same questions in mind; Johnstone was already aware of the potential for bias in experimental design (60:30–35).

Near the end of the eighteenth century, as arguments move toward proof, the precise reporting of results enable them to be compared to quantitative predictions of hypotheses and thus to serve as direct evidence. Herschel, in volume 90, in order to prove that radiant heat is distinct from the visible spectrum, provides extensive quantitative results, to the point of inductive tedium (90:255–322).

The Ocular Proof and Communal Validation

The increasing precision and detail of method and result accompanies a major change in how accessible the experimental demonstrations are for the readers of the journal. As the actual experiment becomes more of a private affair for the investigator and close associates, verisimilitude of the report reassures the readers that the events happened in the way reported.

In the early years many of the experiments reported in the *Transactions* were demonstrated before the assembled body of the Royal Society at regular meetings. The demonstration is its own meaning, for all to witness and agree it did take place. The report of the experiment is little more than a news report that such an event took place and was witnessed by the assembled body. The validity of events rests on the communal witness and not the story told.

The communal witness remains important validation of the experimental events for much of the earlier part of the period, but as experiments gain subtlety and face conflicting results, experimenters need to control the particular conditions of demonstration. Experiments stay in the laboratory, remote from the lecture hall. Designated competent witnesses travel to the experiment to represent the general membership and a prestigious list of witnesses becomes an important feature of the report. Thus in volume 30 Desaguliers's witnesses include the king and queen as well as the chief members of the Royal Society. Witnesses, however, no matter how prestigious, can always be opposed by equally impressive witnesses who attest to conflicting results, so a precise account of methodology with detailed results, allowing critique, comparison, and replication become part of the argument.

The next change occurs when the problem shifts from the simple existence of phenomena to the meaning of baffling, troublesome phenomena. The experiment, no longer an end in itself, certainly no longer performed in public, becomes a private affair, an event in the individual intellectual journey of the investigator. In the volumes 40 onward there is almost no direct conflict over results, but rather only over theories, and even the theories are presented more as the results of individual research programs rather than highly combative claims and counterclaims. Conflicts and comparisons of results are more likely to occur within the series of experiments of a single scientist trying to work out the subtleties of a complex phenomenon. The series of experiments are not presented as being likely to be replicated. For example, Tiberius Cavallo in volume 70 reports his earlier experiments as part of his puzzling through a problem in electricity, not giving adequate instructions for replication, but at the end he does give detailed replication instructions for one final, contrived experiment so others can convince themselves of his conclusions and can explore the phenomenon further (70:15–29).

Nonetheless, specificity, detail, and plausibility of the experiments are important as part of the story of the intellectual journey of the investigator. Since neither the reader nor any surrogates or representatives, except for the author himself, has witnessed the series of experiments, the account must stand in place of the witness. The reader in order to understand the experimental argument must vicariously witness the experiment through the account. In order to earn the trust of the reader, the story of the experiments must be told plausibly if not persuasively, and the events reported on must provide sufficiently good cause for the investigator to come to the conclusions he reports.

When finally the structure of the series of experiments turns from a representative personal journey to a retrospective guided tour of conclusions and experimental evidence, the account of the experiment has come, at least for the time being, to stand as the proof. In the long run, the experiment or series of experiments may be replicated, but in the persuasive experience of the reading of the argument, the story of the experiment must serve as a surrogate for the actual experiment.[5] By this time papers were read to the Royal Society, but experiments were conducted in private, simply to be reported on.

Organization of the Articles

The organization of the experimental articles serves as an outward manifestation of all the trends discussed to this point. Articles tend to grow longer throughout the period as the argument surrounding the experiment grows and individual investigations rely more and more on series of logically connected experiments rather than single events.

In the first issues most of the information passes through the voice of the editor who simply reports on things he has found out about from a variety of sources. Typically, Oldenburg announces that, "The Ingenious Mr. Hook, did, some months since, intimate to a friend of his, that he had . . ." (1:3). By the end of the first volume, authored articles appear, with much the format that, would maintain through volume 25. The article opens with a short statement of what was done, followed by a narrative of results. Often articles end there, although some discussion of cause or meaning may follow. Articles tend to be short, often only a page or two, and usually discuss only a single experiment or trial.

In those articles reporting on a series of experiments, however, continuity does increase over this early period. At first experiments in a series are only loosely connected, being concerned with the same general phenomenon, as in the many series of experiments putting different objects in vacuum reported in volumes 5 and 10 or as called for in William Petty's "Miscellaneous Catalogue of Mean, vulgar, cheap and simple Experiments," all loosely related to weights and specific gravities (15:849–53). Occasionally a rationale explains specific trials, as when Boyle provides reasons for choosing particular animals to deprive of air: the duck, which breathes in air and dives; the viper which has lungs but is coldblooded; a newborn kitten, recently in the womb without access to the atmosphere, etc. (5:2011–31, 2035–56). A result in an early experiment in the series may lead to new questions to be explored in later trials, such as when a Captain Hall notes that a rattlesnake is less lethal on successive bites; unfortunately, the experimental program was cut short when neighbors began complaining about missing dogs (35:309–15).

As experiments begin to respond to conflicts, their reports focus on the issue in contention. Typically, the report starts with a statement of the phenomenon in dispute and then a discussion of the opponent's work or position. The author's own position with consequent experimental method and supporting results follow, with perhaps some general conclusions, as in Desaguliers's article arguing with Leibniz's explanation of barometric fall in wet weather (30:570–79). By volume 40, the hypothesis or meaning of an experiment often precedes the account of the experiment, even where no particular issue is in contention. Desaguliers, for example, not only begins a paper on statics with a general proposition, but he promises to provide elsewhere a more general theory (40:62–69).

As phenomena are treated as more problematic, articles take on a different organization, opening with an introduction to the problematic phenomenon, often substantiated with the story of an experiment that did not go as expected. With the problem established, the article chronologically describes a series of experiments aimed at getting to the bottom of the mystery. Transitions between pairs of experiments draw conclusions from the previous experiment and point to the rationale or need for the consequent one. In the process of continuous reasoning, the experimenter gradually comes to an adequate understanding of the phenomenon, which is pulled together in a concluding synthesis or explanation of the phenomenon, as in William Hewson's investigations into the nature of blood (60:368–83).

At the end of the period, articles using experiments to prove general claims often begin with philosophic statements about general knowledge. Then a problem is presented, either through a surprising experimental result or through the exposition of a gap in current knowledge. Then a series of claims resolve the problem, followed by supporting experiments. Although a subseries of experiments may be presented chronologically, the larger structure of the article is based on the logical order of the claims to be proved. The conclusion may discuss consequences of these claims, but no synthetic set of conclusions is needed because the claims have already been presented at the beginning. Henry's investigation of muriatic acid (90:188–203) and Herschel's investigations of radiant heat (90:239–326) conform to this general pattern.

Forging Persuasive Forms and a Collective Literature

One early article (15:856–59)—that in many respects resembles articles from a century later—reveals the rhetorical function of the features of the experimental article that emerge by 1800. The anonymous article starts with a general proposition concerning the ease with which larger wheels may be drawn over an obstacle. The experiment, clearly designed as a demonstration of that proposition, is presented in great quantitative detail, as are the results. Moreover, many different trials are set out, isolating variables and allowing exact comparisons proving the general proposition. However, the theoretical point was already well established in the literature (it is attributed to three authors, both ancient and contemporary) and this experiment and the article reporting it are only to convince practical people—wagonmakers and wagon purchasers—of the advantages of what was already known theoretically. The point is not to prove the truth of the statement, but to persuade recalcitrant craftsmen to use a well-established truth.

In those early years, argumentative persuasion could be used for the ignorant artisan, but for those actively pursuing nature, nature was portrayed as speaking for herself. The scientific report was simply a matter of news. Just as an earthquake or passage of a meteor needed to be reported, so did experiments. Not until nature was treated as a matter of contention and then a puzzle could the experiment become part of an argument and could theory or claims hierarchically and intellectually dominate experiments.

With the journal serving as a forum, contention grows. This contention pushes the individual author into recognizing that he is not simply reporting the self-evident truth of events, but rather is telling a story that can be questioned and that has a meaning which itself can be mooted. The most significant task becomes to present that meaning and persuade others of it. Persuasion of claims then lies in a story of personal discovery, supported by good reasons and careful work. Since all people, however, have good reasons, the persuasive story must shift to more universal grounds: the proof of a claim transcending the particulars of an investigation.

To draw the historical lines even more sharply, we observe four loose and overlapping stages in the development of the experimental report. In the first stage, most evident through volume 20 (c. 1665–1700), articles consisted of uncontested reports of events. In the second stage, most evident from volumes 20 through 50 (c. 1700–1760), experimental articles tended to argue over results. Beginning in about volume 50 through volume 70 (c. 1760–1780), articles explored the meaning of unusual events through discovery accounts. Finally, in volumes 80 and 90 (1790 and 1800), experimental articles offered claims and experimental proofs.

In this process we find the beginnings of something like Karl Popper's third world of claims, separate from both nature and the individuals who perceive it. The earliest reports—accounts of what happened, as witnessed by many—recognize only the first world of nature. Contention draws attention to the second world of human perception and consciousness, throwing the authors back on to their own experience and thought (although hedged with the proper respect for nature and empirical methods) as the essence of their reports. Finally the claim or conclusion—Popper's third world—becomes the central item to be constructed within the article, to be supported by empirical evidence from the first world and proper method and reasoning from the second world.

Yet to the end of the period, experimenters present their claims as purely products of their individual interactions with nature, not explicitly recognizing the communal project of constructing a world of claims. In most of the articles the literature is still not treated in any explicitly codified way, as we have all become familiar with in the twentieth century. The experiment still appears solely the result of the individual's invention and understanding. Although the individual scientist has an interest in convincing readers of a particular set of claims, he does not yet explicitly acknowledge the exact placement of the claims in a larger framework of claims representing the shared knowledge of the discipline. Herschel does not relate his theories and findings to a large body of knowledge other than his own, except in the most general way. He presents himself as the only explorer of his terrain, and the experiments are thus the confirmation of the general truths he has discovered in his particular travels. The only consistent use of other literature occurs in debates where discussion of the literature serves to draw lines and marshal forces rather than construct an edifice beyond the immediate claims.

Although the collective intelligence of the scientific community before 1800 is not regularly displayed in explicit codifications of the literature, the collective intelligence is embodied in the way the members of the community have chosen to communicate with one another. Whether the emergence of an argumentative community necessitated a conventional genre in which to carry on that argument or whether the clarification of forms of argument allowed a coherent community to coalesce in discussion is an unanswerable dialectical conundrum. A more exact formulation might be that a community constitutes itself in developing its modes of regular discourse.

In this particular case, the kind of argument the community engaged in, over the regular appearances of natural phenomena, seemed best pursued by increasing descriptive detail and precision, re-creating events increasingly designed to display particular features of that nature. But regularity and particularity proved at odds, creating new problems in symbolizing nature. Particular and general formulations did not always fit together easily, so new modes of discourse were needed to expose the regularities hidden in the anomalous particulars and to demonstrate that general formulations offered precise representations of particulars. The emerging form of experimental report offered a way to harness stories of the smaller world of the laboratory to general claims about the regularities of the larger world of nature. In the attempt to satisfy the objections and desires of the growing scientific community, the experimental report kept changing in form, as it continues to do today—for objections and desires grow with the ability to formulate them. And what is science without objections and desires?

Notes

1. For a comprehensive view of the rise of scientific journals see David A. Kronick, *A History of Scientific and Technical Periodicals.* A. J. Meadows, *Development of Science Publishing in Europe* suggests some of the historical variety of scientific publication. A. J. Meadows, *Communication in Science,* and William Garvey, *Communication: The Essence of Science,* describe some features of the current system of journal communication.

2. Core documents for this position are Jacques Derrida, *Of Grammatology,* and Michel Foucault, *The Order of Things.*

3. This and similar articles are part of a regular program of reporting on the trades, according to Kathleen H. Ochs, "The Failed Revolution in Applied Science," and Marie Boas Hall, "Oldenburg, the *Philosophical Transactions,* and Technology."

4. Francis Hauksbee and John Desaguliers, as frequent contributors to the *Transactions* throughout the early part of the eighteenth century, seem to have had significant influence on the development of the

experimental article. Studies of their development as writers of experimental science would help fill out the story of the evolution of the experimental report. Similarly, a study of the innovations and influence of William Herschel as a scientific writer ought to reveal significant trends in the latter half of the eighteenth century.

5. Steven Shapin, "Pump and Circumstance: Robert Boyle's Literary Technology," reports that a number of the features of virtual representation that did not appear regularly in journals until the second half of the eighteenth century were mobilized in books by Robert Boyle almost a century earlier. This uncoordinated development of book and journal publication raises two questions. First, are the dynamics, constraints, form, and literary situation of book publication significantly different than that of journal publication, so as to encourage the emergence of different textual features at any particular time or to cause any one common feature to emerge at different times? Second, what are the formal interplay and mutual influence of journals and books? Bazerman (1988: 80–127) begins an investikgation of such questions.

Chapter 2 of Shapin and Schaffer, *The Leviathan and the Air-Pump,* offers a more complete view of the proper form of communal knowledge and the importance of empirical experience, direct and virtual, for successful public debate and evaluation of knowledge claims. Much of Boyle's attitude toward empiricism and public debate have been carried out in the history of the rhetoric of the *Transactions* analyzed in this chapter.

Work Cited

Bazerman, Charles. *Shaping Written Knowledge: The Genre and Activity of the Experimental Article in Science.* Madison: University of Wisconsin Press, 1988. [The above essay is Chapter Three of *Shaping Written Knowledge.*]

Derrida, Jacques. *Of Grammatology.* Trans. Gayatri Chakravorty Spivak. Baltimore: The Johns Hopkins University Press, 1977.

Fleck, Ludwik. *Genesis and Development of a Scientific Fact.* Chicago: University of Chicago Press, 1979.

Foucault, Michel. *The Order of Things: An Archeology of the Human Sciences.* Trans. Alan Sheridan. New York: Pantheon, 1970.

Frank, Joseph. *Beginnings of the English Newspaper, 1620-1660.* Cambridge: Harvard University Press, 1961.

Garvey, William D. *Communication: The Essence of Science.* Oxford: Pergamon Press, 1979.

Hall, Marie Boas. "Oldenburg, the *Philosophical Transactions,* and Technology". *The Uses of Science in the Age of Newton.* Ed. John G. Burke. Berkeley: University of California Press, 1983: 21–48.

Houghton, Bernard. *Scientific Periodicals.* London: Clive Bingley, 1975.

Kronick, David A. *A History of Scientific and Technical Periodicals: The Origin and Development of the Scientific and Technical Press, 1665-1790.* 2nd ed. Metuchen, NJ: Scarecrow Press, 1976.

Meadow, A.J. *Communication in Science.* London: Butterworth, 1974.

____, ed. *Development of Science Publishing in Europe.* Amsterdam: Elsevier Science Publishers, 1980.

Miller, Carolyn R. "Genre as Social Action". *Quarterly Journal of Speech* 70 (1984); 151–67.

Ochs, Kathleen H. "The Failed Revolution in Applied Science: Studies of Industry by Members of the Royal Society of London, 1660-1688". University of Toronto, 1981.

Paradis, James. "Montaigne, Boyle, and the Essay of Experience". *One Culture.* Ed. George Levine. Madison: University of Wisconsin Press, 1987: 59–91.

Shapin, Steven. "Pump and Circumstance: Robert Boyle's Literary Technology". *Social Studies of Science* 14 (1984): 481–520.

____, and Simon Schaffer. *The Leviathan and the Air-Pump: Hobbes, Boyle, and the Experimental Life.* Princeton: Princeton University Press, 1985.

Sutherland, James. *The Restoration Newspaper and Its Development.* Cambridge: Cambridge University Press, 1986.

Text as Knowledge Claims:
The Social Construction
of Two Biology Articles

by Greg Myers

Almost every scientific researcher I have interviewed has an anecdote about a referee who reviewed an article of his or hers unfairly, or who required alterations that, in the writer's view, diminished the value of the article. I would like to look at the processes of review and revision from a perspective different from that of the researcher, to see these processes as part of the functioning of a scientific community. I will suggest that the procedures of review and revision of the text can be seen as the negotiation of the status that the scientific community will assign to the text's knowledge claim. This negotiation may not directly address the claim itself and the evidence for it, but may instead focus on the form of the text. Thus, a close study of these texts may help us see one part of what Bruno Latour and Steve Woolgar call 'The Social Construction of a Scientific Fact'.[1] I present two articles as cases of such negotiation, showing the range of possible claims, interpreting the formal changes in the manuscripts as they affect the status of the claim, and accounting for these changes in terms of the social context.[2] I make several kinds of comparisons: earlier articles published by the same authors serve as a background of acceptable form; successive versions of the articles show the process of negotiation; and the differences between the views of editors and reviewers on the one hand, and the views of the writers on the other, show the kinds of tensions on which the negotiation is based.

The pre-publication histories of two recently published articles by biologists raise the issue of social construction with particular clarity. 'tRNA-rRNA Sequence Homologies: Evidence for a Common Evolutionary Origin?' is by David Bloch, a University of Texas Professor of Botany.[3] 'Gamete Production, Sex Hormone Secretion, and Mating Behaviour Uncoupled' is by David Crews, an Associate Professor of Zoology and Psychology, also at the University of Texas.[4] Each article had been through four reviews before it was finally accepted, Dr Bloch sending his to *Nature* (twice) and *Science* before it was accepted in a revised version by the *Journal of Molecular Evolution*, and Dr Crews sending his to *Science* (twice), *Nature*, and *Proceedings of the National Academy of Sciences (PNAS)* before it was finally accepted by *Hormones and Behavior*. The authors rewrote the articles each time, so that the published versions are hardly recognizable as related to the first submissions. I chose these texts because they

are atypical: they mark departures in the careers of two well-established researchers, so they show the tensions operating in the process of publication. Also,
the two authors are different enough from each other to be compared in social
terms—one entering a new field, the other well-established in a network of
researchers. And the long, drawn-out review, though unusual, helps us see the
texts in evolutionary terms: we see in detail responses and decisions that are
usually compressed and unnoticed, even by the participants.

I collected the various manuscripts of these articles that were submitted for
publication, and a few of the many intermediate drafts, together with comments
of the authors' colleagues and of the reviewers, and the authors' responses to these
comments, in correspondence with the editors and in interviews with me. I
marked changes, however trivial, and made some guesses about why these
changes were made. Then I interviewed the authors, asking them for their own
interpretations of the changes and submitted a draft of this paper to them for
comments. Thus my conclusions draw on three kinds of interpretations of the
texts and revisions: my own reading, in complete ignorance of biology; the
comments of authors' colleagues and then of anonymous reviewers; and the
explanations given by the authors themselves. I was interested in the differences
between these readings, not in determining which one was 'correct'. My assumption is that none of them is privileged, so I have relied neither on the authors'
claims for the importance of the articles, nor on the reviewers' doubts. I have tried
also to avoid privileging my own outsider's perspective, but that perspective is
the basis for my narrative.

Why did these articles take so long to get published? The explanations for
the delays depend on whether one focuses on the individual researcher or on the
structure of the research community. A rhetorician might say that the authors had
to invent by trial and error the arguments by which they could persuade their
audience to assent to their claims. A sociologist interested in how social factors
distort objective scientific research might say that the authors' research programmes conflicted with the individual interests of reviewers, who had their own
established research programmes; before they could publish they had to find
journals of sub-specialities in which their work was not a threat. Both these
approaches are useful, but they both underestimate the social nature of the
publication process. The rhetorical approach, in treating the problem as a matter
of strategy, accords the writer more conscious control and detachment from the
audience than I can see: we must remember that Dr Crews and Dr Bloch frame
their ideas, however unorthodox, within disciplinary assumptions. The interests
approach, while it points out some kinds of social influences, overlooks the way
the very form and language of the article tend to create consensus. I argue that
the writing process is social from the beginning, and that there is a tension
inherent in the publication of any scientific article that makes negotiation between
the writer and the potential audience essential. On the one hand, the researcher
tries to show that he or she deserves credit for something new, while, on the other,
the editors and reviewers try to relate the claim to the body of knowledge produced

by the community. But the claim must be both new and relevant to existing research programmes to be worth publishing; the writer cannot please the audience just by being self-effacing. The result of this negotiation is that the literature of a scientific field reproduces itself even in the contributions of those who challenge some of its assumptions.

This tension is brought out in the negotiations over the published form of the text, so we can see, in arguments over the tiniest textual details, the workings of social construction. I will argue that the claims in these articles can be read on several levels. For instance, the claim may be allowed as a description of one species, or as an interpretation of a process applicable to all species, or as an argument about how this process evolved. The claims that are restricted to descriptions of the data are not inherently more scientific, or even more publishable; they are just one level of a hierarchy in which the place of the article is being negotiated. The same claim may be considered 'speculative' or 'well-defined', a 'highly significant' advance or a 'well-known' observation, depending on the body of literature into which it is placed and the audience which is to read it. We can see in the referees' comments negotiations over how the claim is to be placed. In general, the authors start by making high-level claims for the importance of their findings, while the reviewers demand that they stick to the low-level claims that take their findings as part of the existing structure of knowledge.

Much of this negotiation of the status of the claim concerns the 'appropriateness' of a paper to the journal to which it has been submitted. Of course, the authors want to see their papers published in prestigious journals that testify to the importance of their claims. They have practical reasons, because of the interdisciplinary nature of their work, to want to appeal to a broad range of researchers in other specialties who might, if interested, provide data to support their claims. And, like most authors, they want speedy publication, especially because the articles would then back up their related proposals for funding. The referees, on the other hand, see their function as that of sorting papers by levels of importance, by sub-specialty and by genre. This sorting is not automatic, as their use of the word 'appropriate' might suggest: here, too, we see a tension finally resolved in the compromises that allow an article to appear in print.

There is a similar tension over the form of the article. These authors have some difficulty fitting their new interpretations into the form of the research report or the review article, because these forms demand that the claims fit closely into the structure created by other scientific articles. The authors try to bend the constraints of form to fit their ideas, to tell their stories in the context of their own concepts and their own courses of work, while the reviewers try to use the form to make the ideas fit into the literature as a whole. Again, they are arguing not over the writers' failure to use the correct format but over the type of the claim and the importance to be accorded to it.

The Writers

Some readers might assume that Dr Bloch's and Dr Crews's difficulties in getting these articles published could be traced to their own eccentricities or scientific skills, rather than to their claims in relation to the literatures of their fields. But both authors have had successful careers, and they are both familiar with the literature of their fields and with the processes of publication. The *Science Citation Index* shows that most of the cited publications of both authors are reports of experimental findings or field observations published in the core journals of their disciplines; it also shows an important difference in the authors' positions in the sub-specialties to which these new articles matter. Dr Bloch receives about forty citations a year for articles as far back as 1954; his most cited paper is a 1969 *Genetics* article that still gets ten to fifteen citations a year, an unusual number for a fourteen-year-old article in most fields. He seems to have written several cited articles a year through the 1960s (with fewer in the 1970s, as his interests changed), all in journals of cell biology or genetics, in general journals like *PNAS*, or in specialized handbooks. But the article considered here is his first publication in the very competitive field of nucleic acid sequencing. Dr Crews has been first author of five or six cited articles a year since 1974; with the work on which post-doctoral fellows, graduate students and undergraduates were first authors, his laboratory produces about fifteen articles a year. These papers fall into several categories: articles in journals of zoology and endocrinology; those in more general journals, such as *Science* and *PNAS*; chapters in books; and popularizations in *Scientific American* or *BioScience*. Currently, his most cited entries are articles in *Science* (1975) and *Hormones and Behavior* (1976), though a controversial report in *PNAS* (1980) received a number of citations and some news stories soon after its publication. We will see this last article mentioned in reviews of his current work.

The articles I have chosen to discuss differ from these earlier publications in complex ways, which I can summarize by saying that the writers each had a big idea. Most of their earlier articles, whether experimental or theoretical, had answered questions posed in the literature. In these articles, though, Dr Bloch is answering a question that had not been asked, and Dr Crews is giving a new answer to a question that had already been answered. The articles are especially suited to illustrate the tensions I have outlined because they are interpretations of published data in new terms, rather than reports of experiments continuing an established research programme. Thus, if the writers are to have any effect, they need to claim significantly new interpretations and reach a broad audience, for they cannot just contribute data to, or get additional data from, the researchers in their immediate sub-specialty. While the two authors have a similar problem, their processes of publishing these claims are different, partly because of their differing positions in their research communities.

For Dr Bloch, the big idea came when he started work in an area entirely different from the cell biology studies he had pursued for twenty-five years. He traces his interest in evolutionary questions to his reading of Schrödinger's *What is Life?* in graduate school, to his having to present basic concepts in teaching introductory biology courses, to a period of free writing forced on him by a back injury ('I was prone to write,' he says), and to a graduate seminar that allowed him to follow up some of the questions raised during this free period. He has since become familiar with the current work on evolution of nucleic acids, but the fact that the idea did not first arise in response to the current literature helps explain why it proved so difficult for him to fit it back into that literature. In 1981 he wrote, but did not try to publish, a paper on 'The Evolution of Evolution', and several times since then he has applied for support for this line of research, with proposals in which he uses many passages from the article manuscripts I am studying here. But so far these proposals have been unsuccessful so he is, as he puts it, 'moonlighting', working on nucleic acid sequences while his laboratory is funded for his cell biology work, and he continues to publish some articles on his cell biology findings. Dr Bloch, then, is an unusual example of a writer completely new to a specialized field who is (as he must be for our purposes) quite familiar with the way biology journals in general review articles for publication.

In his acknowledgements, Dr Crews traces his big idea to discussions with colleagues whom he cites as 'innocent bystanders' and to mentors who taught him 'the value of a comparative approach'. He also pointed out in conversation how these teachers made him sceptical of 'deterministic models of behaviour'. He sees himself as carrying on, in his experimental studies, a form of the nature versus nurture argument. Several factors in his career might tend to make him receptive to unorthodox ideas. His training in psychology and then in evolutionary biology, before he turned to herpetology, gives him an unusual background for a researcher whose approach is physiological; he sees himself, as does Dr Bloch, as working across disciplinary lines. His popularizing articles require him to explain, and to rethink, basic principles, as Dr Bloch must do for his freshman classes. The many grant proposals needed to support his large laboratory, most of them to health agencies, require him constantly to justify his work with reptiles in terms of its significance for humans, so he must consider its ultimate relevance. Whatever the reason, he, like Dr Bloch, proposed a claim that was at variance with the current literature in his field, that needed support from researchers in several fields, and that led him to try a different form from that of his earlier reports and reviews of research. He, like Dr Bloch, based a proposal for funding on this claim, and his proposal and article evolved together. Meanwhile, he and his colleagues in the laboratory continue to publish other articles reporting new data from their experiments and field observations.

The differences I have suggested in the positions of the two writers may or may not affect the judgement of editors and referees. But I will suggest that these differences are felt before the articles are submitted, in the writing and revision of the papers. Dr Bloch has no research network:[5] few, if any, of his graduate school friends, colleagues and students work on this aspect of nucleic acid research, and the leaders in the field do not know of his work. In this sense, he, as a full professor, is more isolated than a new graduate student, who has, at least, a sponsor. He values the help of students a great deal, and gives them co-author status; and he remains in close contact with R. Guimares, one of his co-authors, who visited Texas from Brazil for five months in 1981. But he does not, or did not then, have constant informal contact with co-workers who are expert in this area. So he does not hear arguments against his claim before he submits a paper, and he has no day-to-day sources for new arguments to support his work, or new data that could be relevant. After years of collaborative work in cell biology, he is acutely aware of what he is missing; he would like to get the project funded, not primarily for the equipment and time it would give him, but to have a post-doctoral fellow with whom he could 'bounce around some ideas.' His co-author on a more recent manuscript that grew out of this project is a physicist specializing in statistical mechanics; his quest for collaborators has taken him outside of the discipline of biology.

Dr Crew's laboratory, on the other hand, is an important node in his network of researchers. His graduate school training and two post-doctoral research positions related to his current work, his teaching of dozens of undergraduates and graduates who are now themselves teaching at other schools, and his dozens of conference papers and lectures, all give him daily formal and informal contacts with many researchers in both zoology and psychology. A recent article of which he is first author lists five co-authors at three other universities. Any article he submits for publication has been criticized by many readers; it is already the product of a community. And to a large degree, he has internalized this community; he could easily predict the contents of negative reviews. Dr Crews's position in a network does not mean his article or proposals are always accepted but, as we will see, it enables him to be considerably more flexible in the negotiation over the status of his claim, in the revision of his article, and in his choice of outlets and audiences.

Determining the Claim

Despite these differences in the positions of the authors in their fields, their two articles go through roughly similar stages on the way to publication. I will illustrate the negotiations over the published form of each article with six texts. 1) First, each author wrote a wide-ranging draft he did not submit for publication. 2) Then they wrote more limited and conventional manuscripts for submission to major interdisciplinary journals: *Nature* (in Dr Bloch's case) and *Science* (in Dr Crews's case). Dr Bloch's was rejected without review, while Dr Crews's was reviewed by two referees who split their decisions; it was also rejected. 3) Each author then revised and resubmitted the manuscript to the same journal, with a covering letter asking for reconsideration; both were reviewed and again rejected. 4) Still confident that their manuscripts were important, they resubmitted to other prestigious interdisciplinary journals, Dr Bloch revising somewhat for *Science*, Dr Crews revising drastically for *Nature*. This time Dr Bloch's article got an ambivalent but generally favourable review, but was still rejected, while Dr Crews's article was returned without review. 5) After these rejections by *Science* and *Nature*, both authors submitted to journals with more limited audiences that seemed more likely to accept the articles. Dr Bloch sent a revised version to a journal recommended by one of the referees at *Nature*, the *Journal of Molecular Evolution*. It was accepted on the condition that certain changes suggested by the referees and editor be made. Dr Crews submitted the unrevised *Nature* manuscript to *PNAS*, where the referees were generally favourable, but still recommended rejection. 6) Finally, both articles were published. Dr Bloch's manuscript was accepted in its revised form in the *Journal of Molecular Evolution*, where it appeared in December 1983. Dr Crews's unrevised manuscript was accepted at *Hormones and Behavior* on the basis of its previous reviews, and appeared in March 1984. The revisions between each of these stages are extremely complex, ranging from massive cuts and additions to shifting of an adjective or a comma. I will focus on changes that seem to affect the scope or the form of the article, for these are the features that seem most crucial in the negotiation of the status of the claim.

A citation such as '(Watson and Crick, 1953)' refers to a single knowledge claim an article makes—in this case, for instance, the claim that the structure of the DNA molecule is a double helix with chains of phosphates on the outside and particular pairs of bases connecting them. Nigel Gilbert has shown how one published article may contain a number of possible knowledge claims, from which the authors and readers select the claim relevant to the model by which they are interpreting the article.[6] Latour and Woolgar have arranged the interpretations of the claims in an article in a five-level scale of statements from 'fact-like status' to 'artefact-like status'. They show how statements can be transformed from one type of statement to another by addition or deletion of 'modalities', statements about the statements, as in 'The Structure of GH.RH was *reported* to be X'.[7] Trevor Pinch also proposes a hierarchy of claims: his is arranged in terms of what he calls increasing 'externality', from the lowest level—statements about the observing apparatus ('Splodges on a graph were observed')—to the highest level—statements about phenomena at several removes from the observing apparatus ('Solar neutrinos were observed').[8] He points out the similarities between these poles and the philosophers' opposition of claims with high veracity to claims with greater theoretical significance. I see a hierarchy similar to these in the two articles I am describing, but I prefer to define it in terms of the distance between the author's claims and the claims of the particular part of the scientific literature in which they are to be placed. The issue is the way the claim fits in what Pinch calls the 'evidential context'. The higher-level claims, in each case, involve contradiction of large bodies of the literature, of claims that underlie many research programmes or claims that are particularly well entrenched. The lowest-level claims contradict nothing, but neither do they add anything to what has been accepted. Like Pinch, I see this hierarchy as determining not just the degree of acceptance or rejection of a particular claim but which claim is accepted or rejected. Like Latour and Woolgar, I am trying to base this hierarchy in terms of the language of the claims rather than in terms of the experiment.

As Pinch points out, higher-level claims are likely to be profound but risky, while lower-level claims are likely to be taken as correct, but are also likely to be trivial. Both the biologists I am studying try to make the highest-level claim the editors and reviewers will allow (Figure 1). Dr Bloch's highest-level claim appears only in the early draft he did not submit for publication; he identifies a fundamental concept he says links several kinds of evolution. 'Transfer of control...given the name "surrogation", marks the appearance of new kinds of behaviour at every level of organization and process, including evolution itself.' His first manuscript submitted for publication just presents the model and makes the claim that a 'primordial tRNA produces through successive rounds of elongation a molecule with multiple functions of gene, message, and scaffolding, and which serves as the source of the original tRNAs and rRNAs'. Supporting this model, in the same manuscript, is a more limited claim, an interpretation of data: 'The patterns and distributions of homologies make phylogenetic relatedness a more plausible explanation than evolutionary convergence.' This is the major claim that remains in his first revised version of the article after the reviewers'

FIGURE 1

Scope of Claims
Dr Bloch

I. 'Transfer of control...given the name "surrogation", marks the appearance of new kinds of behaviour at every level of organization and process, including evolution itself.'

II. 'A primordial tRNA produces through successive rounds of elongation a molecule with multiple functions of gene, message, and scaffolding, and which serves as the source of the original tRNAs and rRNAs.'

III. 'The patterns and distributions of homologies make phylogenetic relatedness a more plausible explanation then evolutionary convergence.'

IV. 'The existence of homologous sequences among tRNAs and 16S rRNA is demonstrated.'

V. The sequence one tRNA is...

Dr Crews

I. Environmental factors may influence the evolution and development of three aspects of reproduction: (i) The functional association among gamete production, sex hormone secretion, and mating behaviour, (ii) the functional association between gonadal sex...and behavioural sex, (iii) the functional association among the components of sexuality.'

II. Environmental factors may cause gamete production, sex hormone secretion, and mating behaviour to be dissociated.

III. Gamete production, sex hormone secretion, and mating behaviour are dissociated in some species of each class of vertebrates.

IV. Gamete production, sex hormone secretion, and mating behaviour are dissociated in the red-sided garter snake.

V. The red-sided garter snake mates at the beginning of warm weather, when sex hormone levels are low.

criticisms. A still more limited claim shows the relation on which the interpretation of a common origin for the two molecules is based: 'The existence of homologous sequences among tRNAs and 16S rRNA is demonstrated.' This is the claim that most interests the reviewers, who generally agree in finding his data on matching sequences intriguing.

It would be possible for Dr Bloch to make an even more limited claim, stating the sequences of the RNAs without insisting on the homologies, but, since he is using already-published data rather than doing his own sequencing, such a very limited claim would not be publishable. Even the observation of homologies is trivial, in his own view, without some explanation of why this pattern should be noticed. So we cannot simply say that Dr Bloch should avoid speculation; he has to try to make one of his higher-level claims stick. If the model for the evolution of RNA is accepted, he will have one piece of his larger argument in place. He may have selected this particular piece because he can define the claim for the homologies clearly and design a research programme to support it using computers (to which he has access), but requiring no new (and expensive) equipment. This awareness of what constitutes a 'well-defined' claim and a practical research design are part of what he brings with him from his cell biology work; he doesn't have the contacts, but he does know the conventions.

Dr Crews, like Dr Bloch, makes higher claims for the implications of his work that are supported by lower-level claims interpreting his observations, and still lower claims showing what he has observed. His highest claim (the claim that relates to nature versus nurture arguments) is that environmental factors may influence the evolution and development of three aspects of reproduction: '(i) the functional association among gamete production, sex hormone secretion, and mating behavior, (ii) the functional association between gonadal sex (=male and female individuals) and behavioural sex, (iii) the functional association among the components of sexuality.' His first submitted manuscript limits the claim somewhat by focusing on the first of these aspects—that, contrary to what he calls the prevailing paradigm, these processes can be dissociated. Supporting this claim is the more limited claim that there exist many species in which gamete production, sex hormone secretion, and mating behaviour *are* dissociated, and that these species need to be studied further. This is the claim that the more favourable reviewers emphasize as an addition to the structure of claims in the literature. When Dr Crews revises his claim to this level, he loses some of the argument for ecological approaches he was making in his first draft; the claim for dissociation can be made at this level without any reference to the environmental factors leading to such dissociation. Supporting this claim is the still more limited claim that these processes of reproduction are certainly dissociated in at least one species, the red-sided garter snake. This last claim is accepted even by hostile reviewers, but Dr Crews has already published his studies of garter snakes, so to make that claim alone would be trivial. It belongs not in *Science* but in his popular article for *Scientific American.*[9] Dr Crews, like Dr Bloch, must limit the scope of his claims. But, as we will see, he has the advantage that his claims, however contrary to established research programmes, emerged from those programmes and can be related back to them.

The hierarchy of claims has some relation to the hierarchy of journals to which Dr Bloch and Dr Crews submitted their articles—at least in the case of some of the most prestigious journals, which insist that the claims in articles they publish be of interest beyond any one sub-specialty.[10] Dr Bloch's decision to send his article first to *Nature* and Dr Crews's decision to send his first to *Science* indicate how important they considered their claims, since the editors of these journals say in their instructions to contributors that they select 'items that seem to be of general significance' (*Science*) or 'reports whose conclusions are of general interest or which represent substantial advances of understanding' (*Nature*). Neither publication is limited to biology articles, but since there is no biological equivalent of *Physical Review Letters*, they fill the role of rapid publication, prestigious journals. And they offer the access to the broad audience both authors need if they are to find a wide range of data to support their hypotheses.

The articles were rejected, originally, on the grounds of their 'appropriateness' to these journals, rather than on the grounds of faulty interpretation of data. The editor of *Nature* returned Dr Bloch's article without review, saying that the journal was unable to publish manuscripts that, 'like yours, are very long and speculative,' and suggesting that he send it to the *Journal of Theoretical Biology*. The words 'long' and 'speculative', and the alternative journal suggested, place the article in the hierarchy of claims: the editor did not accord the claim in this form the status that would justify such broad implications, so much space, or such a broad audience. Dr Bloch's response, in a covering letter with a revised manuscript, shows he reads the editor's criticisms as part of a negotiation, not just as a formal criticism of the length of the article. But his response also shows that he was not yet willing to give much up by making major revisions. He argued that though the article seemed speculative to the editor, it was in fact 'an analysis of hard data' that 'makes predictions' on the basis of the model, and 'these predictions were fulfilled'. He argued that the speculation at the beginning 'is appropriate' and 'would be conspicuous by its absence'. If it seems to belong in a more specialized, more purely theoretical journal, then the editor has overlooked 'a completely new slant on the origins of RNAs and of coding mechanisms...a "Rosetta Stone" for the origin of life'. He apologized for seeming to make 'extravagant' assertions of his claim, but such assertions were the only way he could press a claim not grounded in the literature.

This letter from Dr Bloch asking for reconsideration suggests that he saw the rejection by *Nature* as an oversight on the part of one editor, not as part of the community's assessment of his claim. But when his revised version was given to referees (three rather than two, suggesting that the editor tried to resolve some ambivalence), they made comments similar to those given by the editor, focusing on the status of the claim rather than the evidence or argument for it. As a high-level claim about the origin of RNA it lacks rigour; as a low-level claim making some observations about homologies in RNAs it lacks general interest beyond the sub-specialty concerned with molecular evolution. One referee suggested that the claim could not be supported by the literature of molecular biology, and thus belonged in what he saw as a less rigorous sub-specialty: 'The manuscript drifts into unsubstantiated speculation; this, however, is common in evolutionary papers.' Another shared these doubts about the 'highly speculative evolutionary model'. Two of the reviewers suggested it be sent to 'a more appropriate journal' or 'a more specialized journal'. The other suggested that *Nature* could 'publish a much briefer account of the homologies together with a brief possible interpretation of it'. While two of these readers raised statistical questions, and one referred to an earlier article, famous in the field, that puts forth a similar idea, none of them attacked the evidence as much as they questioned the status of the claim itself. The reviews at *Science*, while much more favourable, also split the broad claims from the narrower claims: 'It is not clear that the empirical observations of homologies and the discussions of pmfRNA [that is, the model for common origins] can both be adequately presented in a single paper which meets *Science* page limitations.' The editor apparently took this tension between claims as unresolvable, for she rejected the article without suggesting any rewriting.

The comments on the claim by referees at the more specialized journal that accepted Dr Bloch's paper, the *Journal of Molecular Evolution*, were actually quite similar to those at *Nature* and *Science*, but here these comments did not indicate that the paper was inappropriate to the journal. 'The hypothesis of this paper is of interest to evolutionists', one began, suggesting that Dr Bloch had found his niche in the hierarchy. While, by the *Nature* referees' standards, the hypothesis was quite far from the data, here the referee found claims that were, in Popperian terms, falsifiable: 'Both the hypothesis and the data are clear-cut enough so that if the authors are wrong they will hear about it quickly from other scientists.' But even the most enthusiastic reviewer was concerned that the article, while significant, did not entirely fit in the current structure of claims. If the statistics check out, he said, 'This is an important finding that needs to be explained', and he 'strongly recommends' publication. But before Dr Bloch can make higher-level claims, his lower-level claims must be accepted by the rest of the research community of the sub-specialty. 'The essence of an initial paper should be to document the reality of the homologies rather than extensive studies of their origin. If such discussion is to be included at all, it should be far more balanced and less speculative.' In these terms, his discussion was 'speculative', not so much because it ran ahead of the data, but because it ran ahead of the literature.

The criticism of Dr Crews's first manuscript by the referees at *Science* also focused on the placement of his claim. One saw it as placed too low on the hierarchy, in relation to the accepted knowledge of the field: 'I learned very little...the model is really very simple.' But he could imagine a more important article on the same topic and by the same author that would be appropriate to this journal. 'I am ambivalent. *Science* needs some articles in important areas such as this. I think the author, who has done some very important work, can do a better job of putting things together...*Science* is read by such a wide audience that this article will certainly reach the audience that needs it most.' This ambivalence makes sense if we see not just an article to accept or reject but a knowledge claim, the status of which is indicated by a number of formal features that can be negotiated separately. The referee can choose to accept the author's claim without accepting some aspects of its form. The other, entirely negative, review also placed the status of Dr Crews's claims. The referee located three claims Dr Crews was making: one was 'an accepted fact' (that is, a claim at the lowest, trivial level); another was 'not a new or startling observation' (also at a low level); and the third was 'a quantum leap from faulty premises' (a higher level than the referee will grant). He saw, as did the more positive reviewer, that the negotiations here concerned the status that this journal could confer upon the claim: publication of this material in a journal as prestigious as *Science* 'could set the field back by providing a straw man for those that feel it necessary to refute the thesis'. In effect, he didn't say, 'it's wrong', but said, 'it doesn't belong'.

Dr Crews's letter to the editor in response to the *Science* criticisms shows that he, like Dr Bloch, was aware that formal points are part of a negotiation of the status of his claim. He too defended his claim in a language quite different from that of the article itself. But, unlike Dr Bloch, who could only point out to the editor the importance of his claim, Dr Crews is enough a part of the network of researchers to be able to fit his claim into its structure of knowledge, or at least to try. His originality, he said in his letter, was not in the observation but in recognizing its larger implications. Even if 'this observation has been around for at least forty years, its significance at the conceptual level has been unappreciated...'. While the reviewer said his claim was 'an accepted fact', he could show that the opposite view is held by the standard textbook (from which he quotes) and in two recent *Science* articles. He implied that his refutation of this view was appropriate to the same journal and audience.

But in a second review at *Science*, the referees were even further apart than before. One referee said again that the claims were either well-known already ('Only the naive who had done no reading would suggest that...no one has ever claimed that...'), or out of touch with current knowledge ('it is entirely different from the well-documented findings...'). Finally, he or she questioned whether the paper said anything definite: 'I was unable to find this experimentally testable hypothesis, as were two of my colleagues who I asked to read the paper.' The other referee seemed to have read a different article: 'The author of this paper provides a valuable service...a needed jolt...another important contribution...a clever and reasonable hypothesis.' Although the referee said 'any biologist with even a passing interest' in the topic was aware of some of the specific instances Dr Crews cites, he or she saw the usefulness of a list showing 'a large number of such exceptions', and of his evolutionary hypothesis, which 'will certainly generate debate and further research'. In a sense, the two referees *did* read different articles. Dr Crews suggests that the first reviewer was a classical neuroendocrinologist, and the second a comparative zoologist. In that case, they were placing the claims of the article in different hierarchies—so different that where one reader found a clever and reasonable hypothesis, the other found no hypothesis at all.

The comments on the shortened manuscript Dr Crews submitted to *PNAS* show that it is not necessarily enough that the claim of the article be significant; it must have the right sort of significance. While the referees granted the importance of the article, they both said that it did not make the kind of claims appropriate to this journal. *PNAS* is certainly a prestigious journal, but the check-list it sends to referees to get their comments suggests that, unlike *Nature* and *Science*, it selects claims I have put lower on the hierarchy, claims that are basically presentations of new data that fit an already established conceptual framework. So one referee said that 'this article breaks new ground', but decided that, 'While I found the hypothesis as presented quite interesting and worthy of serious experimental attention, I do not think this idea merits a separate *PNAS* article' The other review showed clearly how the status of the claim may be separated from the question of appropriateness of a specific audience and journal, so I will quote it at length.

> I have little problem with my recommendation regarding this paper: it does not belong in *PNAS*. I think that the points raised are of great importance, that the scholarship is genuinely profound, that the conceptualization is original, that the presentation is crystal clear and not obfuscated by unnecessary information and argumentation. I would strongly urge its publication in a more general journal (obviously, I would think first of *Science* or *Nature*) where it will receive the attention it deserves...It is because I consider this survey/thesis to be highly significant that I do not think it belongs in a journal that publishes 'data' papers.

Though it may have struck Dr Crews as ironic that his paper would be rejected for being 'highly significant', and that he would be referred back to the journals he had spent months trying to satisfy, this reviewer's decision was consistent with earlier reports in focusing on the 'issue of appropriateness', and on determining just what kind of claim was being made, rather than evaluating the evidence for the claim. So a 'highly significant' theoretical formulation was much out of place in *PNAS*, for the reviewer, as would have been an article with weak data or unimportant claims. This interpretation of the reports is confirmed by the decision at *Hormones and Behavior* (which often publishes his group's work—three articles in that issue alone), for Dr Crews simply sent that journal the reports he had received from other journals, and the article was accepted without further review or revision. The editors seem to have accepted the favourable reviews I have quoted, and to have discounted the unfavourable comments, which dealt with the article's relation to other specialties or its appropriateness for more general journals. So the same placement of claims that provided grounds for rejection at *Science* and *PNAS* was taken, at a more specialized journal, as grounds for acceptance.

What has changed in the course of these negotiations? Both Dr Bloch and Dr Crews have altered their claims, choosing a somewhat more limited claim before submitting the manuscripts for publication, and then cutting their more controversial, higher-level claims in their revisions. In exchange for publication, they accepted a different level in the hierarchy of claims. They have also settled for less prestigious journals and more specialized audiences—accepting, then, a somewhat different status than they had first proposed for these claims.

Determining the Form

So far I have described only those referees' comments concerning the statement of the claim itself and its appropriateness to a specific journal. But referees' comments about such matters as length, organization and style are not just matters of taste; they, too, help define the status of the claim. As there is a tension in determining the appropriateness of the claim for a particular journal, between assertions of originality and participation in an established structure of knowledge, there is a tension in determining the form of the article, between the construction of the idea as the author tells it and the conventional formats of the report or review article, both of which emphasize the placement of the article within a body of literature. These formats, though flexible within limits, embody the attitudes of a sub-specialty towards claims, methods and use of the existing literature.[11] And the conventional tone of scientific articles carries assumptions about the appropriate persona for the researcher.[12] The author has a story which he can neither tell as his own, ignoring the literature, nor fit completely into the format without distorting the shape of his idea.

A number of writers have commented on the difference between narratives of the actual experience of science, with all their odd sources of ideas, wrong turnings, and unexpected discoveries, and the presentation of science in journal articles, the form of which suggests a method of pure inductive logic.[13] Thus it may seem strange for me to speak of the authors' stories in describing the forms of these articles, as if they had presented their ideas in autobiographical fashion. But Latour has shown how some narratives can remain within the form of a research report, latent in the methods section or in descriptions of physiological processes. And Mayr and others have commented on the particular importance of narratives to biological argument which, unlike the physical sciences, must often deal with unique events in time: 'Explanations in biology are not provided by theories but by "historical narratives".'[14] These narratives need only be implied in most articles, so that, for instance, observations of successive generations of Dr Crews's *Cnemidophorus* can make sense without a retelling of the narrative of genetics. Perhaps Dr Bloch and Dr Crews use somewhat unconventional forms for these articles because they find that they need to retell a whole narrative from the beginning, rather than dealing with just one incident within the narrative given by the scientific literature. In these terms, each deviation from what the editors expect may be not an error but an assertion of the status of the claim. In the simplest example, an editor will not allow an unusually long article unless he or she considers it unusually significant. The reviewers' comments suggest that a similar kind of evaluation is made whenever the organization or tone of an article departs from the conventions. As the authors gradually move from the somewhat unconventional forms of their earlier manuscripts to the more conventional versions that are finally published, they are accepting the status these referees accord their claims.

The earliest draft by each author reflects the route he took to this research programme. Though Dr Bloch's early draft, 'The Evolution of Control Systems: The Evolution of Evolution', apparently follows the format of a review article, with an abstract, introduction, definitions, examples and copious citations, the style is personal and exploratory, allowing for digressions (labelled as such), asides, suggestions of possible lines of thought left unexplored, and references to a wide range of authors outside the sub-specialty, from Darwin to Delbrück to Prigogine. The evolution of RNA, the topic of the article he eventually submitted for publication, is here just a one-page example of the genetic code. Only at this early stage do we see in the text the relation of this model to the ideas of code, information, control and culture with which Dr Bloch began his thinking. Only a reader of this draft, or someone who had had a chance to hear Dr Bloch talking informally or at a poster session, would suspect that his real goal was an explanation of the origin of life.

Dr Crews's early draft, titled 'New Concepts in Behavioral Endocrinology', also shows more of the relation of this research to his larger thinking than does the first version submitted for publication. The paper seems to have been written for people who are already receptive to his ideas, terminology, and criticisms of current concepts; the eight names in the acknowledgements suggest that only a close group of colleagues had read it at that stage. For this audience, he could safely follow an organization that is more exploratory than argumentative, opening with broad questions, making a leisurely review of his recent work, and presenting his alternatives only in the last pages. Here, one can still see the relation of his research on dissociation of gamete production and hormones to his larger assertion of the importance of environmental factors in all aspects of evolution and development of reproduction. Dr Crews's first draft, like Dr Bloch's, is closer to the form he uses for oral presentations than to that of his other articles; it lacks only the slides with cartoons of lizards.

The manuscript Dr Bloch sent to *Nature* is much more conventional than this early draft. But by comparing it to one of his recent articles on cell biology, we can see in it a tension between the form of the report on research and the more exploratory form he had given up.[15] In the *Nature* submission, six pages of introduction provide the reasoning behind the model, and then just two pages of methods and results describe the work, before nine pages of discussion and two pages on 'Further Evolution'. So the article is about 32 percent introduction, 10 percent methods and results, and 58 percent discussion. In contrast, the cell biology article is about 11 percent introduction, 47 percent results and discussion, and 42 percent discussion. (The version that was finally published was revised to be somewhat more conventional in its structure, with just 4 percent introduction, 41 percent methods and results, and 55 percent discussion.) If Dr Bloch's problem is that he is answering a question that has not yet been asked, his solution here is to start in his introduction with the most fundamental questions—the conditions for the first protein synthesis—and work towards his interpretation. In cutting the methods and result sections so drastically, he may be assuming that his extensive tables (with forty of his sixty references) are striking enough in themselves to attract attention, and too straightforward to need explanation. The bulk of the paper is in a discussion titled, 'Common Descent, Evolutionary Convergence, or Coincidence', in which he gives his interpretation of these results. The last section, 'Further Evolution', does not correspond even roughly to a section of a conventional article, but crowds in some of the ideas from the early draft, relating all this back to his broader claims. The form still looks like a personal essay embodying the researcher's thought, rather than a research report embodying the discipline's criteria for judgement.

While the organization of Dr Bloch's first submitted version suggests big ideas, his tone is as cautious as he can make it. 'A panoramic view of evolution offers clues that can serve as a guide in ordering the early stages.' The tentativeness of the diction balances the enormous claim; he finds 'clues' and a 'guide', not a demonstration. When he describes the model in his introduction, the verbs are almost all conditional ('could provide a configuration') and the claims tentative ('is envisioned as a hairpin structure'). His characteristic method of argument, here and elsewhere, is to survey a broad question, suggest possible answers and argue against the alternatives until only his own view is left. He tries to give the impression of a balanced approach, but the responses of the referees indicate that he does not successfully avoid giving the impression that he has a prior commitment to one interpretation, that of common origins.

The mixture of boldness and caution in Dr Bloch's tone is apparent in his presentation of what he tells me is 'gratuitous but suggestive evidence', a ratio, which he saves for last, involving the information content possible with the number of RNA nucleotides. 'This is a tantalizing bit of numerology that evokes no ready explanation from current views of RNA functions or relationships.' On the one hand, he is claiming to introduce a new view of the evolution of life; on the other, he injects his characteristically self-mocking tone with 'tantalizing' and 'numerology'. (This last word is particularly charged with connotations.)[16] A conclusion Dr Bloch adds in the version after this one can serve as an example of the style of much of his writing. 'The scattered homologies are likened to the shards with which the archaeologist reconstructs pottery of ancient civilizations.' The awkward sentence structure shows how hard it is to work this simile, which Dr Bloch uses often in his oral presentations of his work, into the passive constructions of the scientific article. We could treat this mixed tone, like the exploratory organization and metaphorical conclusion, as a tactical error on Dr Bloch's part. But since we know he can write straightforward research reports, it seems reasonable to take these departures from form as assertions that his claim is important enough to justify some background for the argument, some speculation in the conclusion, and some personal style in the presentation.

While Dr Bloch's departures from the form of the research report can be seen most clearly in the structure of his manuscript, Dr Crews's departures from the form of the review article are largely a matter of tone and emphasis. And he is aware of the effect of these departures, again because of his immersion in a network. As I have suggested, Dr Crews was guided in his reframing of the draft for publication by the marginal comments, sometimes quite acerbic, of a number of his colleagues. For instance, one reader points out, 'It takes a long time (many paragraphs) before you get to the *new concepts*', and responds to an 'indirectness' in the argument by proposing 'a different strategy of organization' which he describes clearly, and which Dr Crews adopts. Another raises potentially troublesome questions about the physiology of a particular species, giving the kind of detailed argument one seldom sees in a referee's report. A graduate student working at another laboratory where Dr Crews has contacts compiles a three-page list of ambiguous phrasing and terminology. In each case, the reader defines part of the potential response of the zoological and endocrinological communities, before Dr Crews submits the manuscript for judgement. All these different styles of handwriting in the margins of various drafts are the visible sign of the invisible college.

Despite these suggestions, Dr Crews's first submitted version shows some tension between what he wants to say and the review article form in which he must say it. We can see these tensions by comparing the tone of various passages of this manuscript with similar passages in a review article Dr Crews published in *Science*, the same journal to which this manuscript was submitted, in 1975, when he was a post-doctoral fellow, and was perhaps more cautious.[17] A review article presumably summarizes the recent work of a research programme, drawing on a broad survey of the literature, tactfully and impersonally presented. But the title of Dr Crews's 1983 manuscript says he will give 'New Concepts in Behavioral Neuroendocrinology', challenging the work of this research programme. The 1975 article has the unthreatening textbook-like title, 'Psychobiology of Reptilian Reproduction'. This earlier article begins with what might be considered the stereotypical opening sentence of a scientific review article: 'The interaction of behavioral, endocrinological, and environmental factors regulating reproduction has been the subject of intensive investigation in recent years.' The diction of the first sentence of the new article is provocative and even combative: 'Much of the information on the causal mechanisms of vertebrate reproductive behavior has been gathered on *highly inbred stocks* of rodents and birds living under *artificial conditions* . . . Some of the organismal level concepts that have emerged are *overly narrow* and *sometimes unrealistic*' (emphasis added). I assume that the phrases I have emphasized would be red flags to other naturalists: he is saying that they are studying something unnatural. He uses a vocabulary with contrasting connotations to describe his own work; he proposes to investigate 'species diversity under naturalistic conditions' and quotes comparative biologists who say such an approach leads to 'new insights' and 'new paradigms of thought'. He is particularly bold in attacking the most commonly studied species as well as the most commonly held ideas; psychologists have money, time, prestige and egos invested in their laboratory animals, and might respond more fiercely to attacks on their mice than to attacks on their minds.

A similar sharpness of tone is apparent in a comparison of the conclusion of the 1983 manuscript with that of the 1975 article. The earlier article ends with a concession to the competing research programme in a subordinate clause and a conventional reference to the continuing advances of the field: 'Thus, while the utilization of inbred species contributes greatly to our understanding of the factors regulating reproduction, the integration of these factors can only be appreciated fully in an ecological context where the adaptive significance of such interactions become apparent.' The 1983 article ends with a statement of a similar idea, but frames it in terms of a call for more work on the whiptail lizard (*his* species), so that 'it becomes possible to apply evolutionary theory to gain insight into the evolution of psychoneuroendocrine controlling mechanisms'. The earlier article stresses uses of *our* knowledge, while the later manuscript suggests that a whole new approach has been overlooked by most workers in the sub-specialty.

From a rhetorical point of view, we might argue that the tone of these sentences is a strategic mistake, a miscalculation of his audience. But seeing the article in terms of a negotiation, we can see his tone as an assertion of the value of his knowledge claim. He is saying that this article differs from the views of most neuroendocrinologists, but is still important enough for the front section of *Science*, which contains the major papers. A more cautious article presented as a review of current knowledge, with a title like 'More on the Psychobiology of Reptilian Reproduction', would be more appropriate in tone, but less important to non-herpetologists, and thus less appropriate for *Science*.

Many of the reviewers' comments are concerned with the departures from standard organization and tone I have described. For instance, all Dr Bloch's reviewers comment on the length of his manuscript, even though he revised it, after the first rejection, to fall just within *Nature's* word limits, pointing out that it is 'about 2,930 words' (*Nature's* limit is 3000). As with most academic journals in which space is at a premium, appropriate length is determined not by the limits given in the 'Instructions to Authors' but by the importance granted to the article's claim; these reviewers do not think Dr Bloch has earned 3000 words yet. The most telling criticism of Dr Bloch's style comes from his most enthusiastic referee at the journal which finally published his paper, a reader who seems to worry that Dr Bloch's persona will endanger the reception of his work.

If the author is to have his observations seriously evaluated by others in the field, it is important that he not present himself as being overly speculative. Discussions of 'shards' and extremely speculative ideas

such as Figure 5 [his original model] and those beginning at the bottom
of page 7 [interspecies comparisons that form the basis for his current
work] will not improve the author's chances of being taken seriously at
this stage and would best be removed.

This response shows that Dr Bloch is perceived as a newcomer to the field, whose
use of personal metaphors, asides, and 'notions' is inappropriate 'at this stage'.
Perhaps a more personal and expansive style is permitted to those whose work
has already been recognized.

Some of the rather vague criticisms from reviewers of the form and style of
Dr Crews's paper also seem to be directed at his departures from convention, in
this case the format of the review article. For instance the more favourable referee
of the first version says 'the manuscript is not well-written'. It is always hard to
know exactly what this sort of comment means, but if we read on we find the
more definite criticism that 'a review paper of this nature which has pretensions
to generalization should not be based on a preliminary review of the literature!'
The meaning of 'preliminary' here is relative: the 1983 article has more references
than the review article published by the same journal in 1975, and probably has,
already, more than *Science* wants to print. The problem is, perhaps, that a review
article must not be so much a review of one's own work, in which the work of
others serves mainly as a background; the reviewer could be objecting not to the
number of citations but to the emphasis implied in the organization. A favourable
reviewer of the second manuscript submitted to *Science* shows, as we saw in Dr
Bloch's case, that criticisms of form—especially length—can sometimes be
interpreted as attempts to redefine the claim. 'A short paper will be read more
often—a point briefly made is often the point well made.' This may be good
advice for any academic author, but the specific passages the reviewer would like
to cut suggest that the reviewer is more concerned that Dr Crews will alienate
readers, than that he will bore them. The garter snake sections can be deleted
because 'anyone interested in reproductive biology must have noticed the article
by Dr Crews in *Scientific American* a few months back'. This change, like the
comment on the 'preliminary' review, can be read as an insistence that he move
the emphasis from his own work. The reviewer also suggests that the whiptail
lizard sections should be cut because they are controversial. 'Female mating in
the wild has never been observed (judging from a heated discussion by *Cnemidophorus* workers after a seminar by Dr Crews at a recent ASZ symposium)...Per-
sonally, I think Dr Crews is on shaky ground here, and there would be great danger
of a hostile reaction to an otherwise important contribution.' The reviewer does
not attack the *Cnemidophorus* work directly but can rely on this vague consensus;
it is inappropriate in a review article that claims to represent the work of the
specialty, not because it is wrong but because it is the author's own work and has
not yet been accepted.

The changes the authors make in various revisions in response to these
reviews show they take these apparently superficial matters of organization and

style as issues affecting the status of their claims: they make most of the changes suggested, but reluctantly. Dr Bloch describes his revision of the article for *Science*, after he read the *Nature* reviews, as 'cutting some of the speculation and adding some new data'. This he certainly does, extending his list of matches and including his recent reading in the reference list. He also changes his self-presentation radically, becoming the judge of, rather than the advocate for, the claim for the common origins of tRNA and rRNA. The title, 'An Argument for a Common Evolutionary Origin of tRNAs and rRNAs', becomes 'tRNA-rRNA Sequence Homologies: Evidence for a Common Evolutionary Origin?' The new title puts the data first, changes an *argument* to *evidence*, and strikes a note of scepticism with the question mark.

The structure of the new version moves much closer to the structure of a conventional research report. Comparison of the two abstracts shows a reorganization along inductive rather than deductive lines, moving *towards* the conclusion of common origins rather than *from* this assumption. The first four pages of the review of theory are cut, as are the last two pages on prospects for 'further evolution'. The article now begins with the homologies, lists some of them, and gives his methods. The formula for determining the significance of these homologies, his link with a recognizable line of previous research on sequences, is now on page 2 instead of page 6. In general the exposition is tightened, introductory sentences are added, a few asides cut, and some sections are moved from 'Results' to 'Methods', where they flesh out that previously rather skimpy section. The splitting of the section on whether the homologies result from convergence, to take into account both convergence and function, shows a new refinement in his argument and recognition of a body of literature on sequences. The section on convergence, his refutation of a counter-interpretation, is shorter, and ends cautiously, saying that a larger data base is needed. The model that was first is now last; it occupies only a paragraph, and comes with no elaborate explanation of the conditions it satisfies. The article is six pages long instead of twelve, with ten notes instead of sixty (*Science* discourages 'exhaustive' reference lists). But only notes indicating the sources of sequences are cut; all the substantive references to related work by others are retained.

Dr Bloch's revisions for the *Journal of Molecular Evolution* continue the reorganization into more conventional format, with an emphasis on the data, and into a less personal and less assertive style. The introduction emphasizing the significance of his findings is cut, and a short summary of his method is put in its place. In the first sentence, where before rRNAs were 'peppered with stretches' homologous to tRNAs, now they 'were found to contain stretches'. In a gesture towards consideration of both sides of the data, suggested by a referee at *Science*, he adds a new table showing the tRNAs that *do not* have homologies. The heading 'Why the Homologies?' is replaced by a more conventional 'Discussion'. He finds more arguments against the possibility of coincidence or horizontal transmission and supporting the concept of a multifunctional molecule. He makes the tone even more cautious: 'We propose that' becomes 'one interpretation would be that'. Finally, the model is deleted entirely from the text and relegated to the caption of Figure 5, and the sentence on how the model led to the finding of the homologies, the last relic of the narrative of his thought that Dr Bloch gave in his first version, is deleted.

After the review at the *Journal of Molecular Evolution*, Dr Bloch makes nearly all the changes the reviewers and editor suggest. He adds a numerical example, some figures on the quality of the matches, an example of his calculation of the possibility of coincidence, and references to possible DNA-level transfections. The tone becomes still more cautious; a level at which he excludes homologies as coincidental that was 'acceptable' is now 'provisionally acceptable', and where he had said that these coincidental matches 'will be revealed' by further comparisons, now he says they 'should be revealed'. He admits a possible weak point of his method of argument, that 'the evidence so far has supported homology only by eliminating or weakening arguments favouring alternative explanations'. And finally he deletes Figure 5, which was criticized by the reviewers, and with it all trace of his model.

In addition to the changes suggested by reviewers, Dr Bloch makes an apparently minor formal change suggested by the editor that is relevant to the position of the article in the literature. The editor comments on Dr Bloch's use of the term 'homology', a complex term that usually means a *common sequence* in molecular biology, but means a feature resulting from *common origins* in evolutionary biology.[18] The editor says he has long had a policy of trying to keep the word univocal, and argues that use of the word in its more general sense marks an unnecessary division within the discipline: 'molecular biologists have to be biologists too.' Dr Bloch was happy to agree and change his use of the term; otherwise he would be begging the question in arguing that the homologies showed common origins. However, he also adds a note saying, 'Their distributions suggest...that they represent true homologies.'

The concluding metaphor of the shards is, alas, gone. Instead Dr Bloch ends the published paper with another metaphor, that of 'filling in the map'. But this, he tells me, refers to 'a phrase used back in '55 by Benzer, describing filling in the genetic (linkage) map with mutants', so it is an allusion to a tradition in the field, not an assertion of a personal style. We may, however, see a personal style in another sort of figure added by Dr Bloch to the final version of the paper, a diagram related to his current work on species comparisons. The version in print is full of tables and graphs; as a colleague says, 'every day he thinks of some new way to illustrate it'. His graphic figures may in some way replace the figures of speech he has had to cut; in fact, his Figure 2a is the equivalent of his shards metaphor. Both kinds of figures provide visual images that represent selected features of highly complex data. In these terms, Dr Bloch gradually changes his figures to the kind more acceptable to the *Journal of Molecular Evolution*.

Dr Crews's revisions also show some concessions to the views of the referees and the conventional form, with its implied placement of the claim in the context of the literature, in order to get his claim in print. In his first revision for *Science*, the accounts of his own studies are shortened and subordinated to the work of others. This version is more readable: digressions are deleted, especially near the beginning, transitions are added, some supporting but complicating details are moved to the notes (he now has eleven explanatory notes instead of two), and a concluding restatement of the argument replaces the appendix-like anticlimax of the earlier draft. He says in his letter asking for a second review that these changes make this version more 'straightforward', but these changes affect the persona of the article as well as its readability, for the article now makes a sharper claim and makes fewer demands on the reader.

I can draw no line of demarcation dividing the changes Dr Crews is willing to make from those he is not. But in general he is acutely aware of how his tone defines his relation to the work of others, and he is willing to change this tone wherever necessary. He is unwilling to modify his inferences from his evidence, preferring even to cut sections and use them in other articles rather than to compromise his argument. The change in tone at this stage is suggested by the title: the assertive 'New Concepts in Behavioral Endocrinology' becomes the descriptive 'Functional Associations in Behavioral Endocrinology: Gamete Production, Sex Hormone Secretion, and Mating Behavior'. The provocative opening remark about other researchers' 'highly inbred stocks of rodents and birds' becomes a milder comment on 'laboratory and domestic species'. Where before, in the summary attacked by one reviewer, he said 'this survey makes several points', now it 'raises several questions'. He adds a cautious note in saying that the lack of dependence of mating behaviour on hormones 'may be more common in vertebrates' than previously thought. Other changes in tone are apparent in his changing 'my laboratory has been investigating' to 'the most thoroughly investigated species', and in his phrasing of an assertion in the form, 'it is important to restate the obvious'. He responds to a reviewer's criticism of his 'really very simple model' by pointing out that 'the four reproductive tactics...represent extremes'. He is careful to incorporate the 'existing body of knowledge' referred to by the other reviewer, reminding the reader that many species *do* follow the conventionally accepted pattern. An example of his avoidance of confrontations (and witty understatement) is his mention of his most controversial point, the relevance of all this to humans, only in a note. By softening the confrontational tone of the earlier version, Dr Crews includes his readers on his side of the argument, whether they belong there or not.

While Dr Crews backs off in matters of tone in this resubmission to *Science*, he mounts a counter-attack in the form of his argument, adding a great deal of material. First he establishes the paradigmatic status of the concept he is attacking in a new transition: 'the concept...has persisted despite an increasing number of studies revealing variations to this rule.' Here he adds a number of counter-examples and then asserts, cautiously, that 'It is possible that the rule...may be due to a bias in the species most studied'. Since his associated/dissociated dichotomy was considered too simple, he adds more examples to develop it in detail. The brief comment that was called 'a quantum leap from faulty premises' is expanded into four paragraphs. Another comment that had been criticized, on explosive or opportunistic breeders, is moved from the beginning, where it seemed an aside, to the end, where it is introduced as 'a classic example', well known to all.

The article can no longer be called 'a preliminary review of the literature', and if it is to become, in the words of the reviewer, 'a straw man', it is a well stuffed one. The revision is only one page longer, but while the earlier version had 57 references, the new one had 195, far more than is usual in a *Science* review article. The considerable changes show again how Dr Crews's place in a research network of zoologists, psychologists, and endocrinologists allows him to respond to critics in his revision. Where the earlier version had listed twelve readers, mostly colleagues in the zoology department, the second lists thirty-one, mostly from other departments and schools. And this list includes only the actual readers, not those who raised questions or made criticisms and suggestions after the many lectures he gave while he was revising, or those who talked to him in the halls, on the phone, and after work at a Mexican restaurant. Any writer can cut the parts of an article criticized by reviewers, but perhaps only a writer who argues his claim every day can rebuild the article on broader foundations of evidence in a period of a month.

This remarkable flexibility continues in the five major revisions Dr Crews made between the *Science* version and the one for *Nature*. At first, like most authors, he tried to make a minimum number of changes, using his word-processor to change all occurrences of *behavior* to *behaviour* for the British journal, cutting the subtitle, which exceeded *Nature's* limit of eighty characters, and adapting his references to its style-sheet. But in later drafts, after many more readings by colleagues, he made what he considers 'wholesale cuts'. What finally emerged is a version of a little more than three pages, with the new, catchy, headline-style title, 'Gamete Production, Sex Hormone Secretion and Mating Behaviour Uncoupled'. The introduction is gone, and the paper begins immediately with the argument. Following a favourable reviewer's suggestion, almost all examples are relegated to the table, only two sentences are left on Dr Crews's garter snakes and the controversial *Cnemidophorus* studies are deleted entirely. One important sentence is added, making the current view seem one-sided: 'Indeed, all of the data supporting this paradigm have been obtained from species in which both sexes exhibit an associated reproductive tactic.' Now there are just fifty-two notes; significantly, only five of them refer to work done in his laboratory, and the first of these is carefully placed far down the reference list. The evolutionary argument the *Science* reviewer had called 'a clever and reasonable hypothesis' is now apparent only to the reader who compares a statement on the second page (saying that the old view had supported phylogenetic conservatism of these relations) to the last sentence ('The possibility that similarities in the mating behaviour in different vertebrate species [are] the result of convergent, rather than divergent, evolution, adds another perspective to our understanding'). As the form of the article has approached the conventional format, and the tone has become more cautious, the article has changed subtly from an attack on a paradigm by one scientist to an outline of the logical implications from the collective work of all the researchers in the field.

How do '*Journal of Molecular Evolution* (1983) 19:420–28' and '*Hormones and Behavior* (1984) 18:22–28' differ from the authors' first manuscripts, 'The Evolution of Evolution' and 'New Concepts in Behavioral Endocrinology'? The claims in the published versions are at a lower level of the hierarchy, Dr Bloch claiming only that matching sequences *may* indicate common origins, Dr Crews claiming only that his comparisons show that gamete production, sex hormone secretion and mating behaviour may be dissociated in some species. In Latour and Woolgar's terms, they have had to add modalities and move their claims away from fact-like status. In Pinch's terms, the authors, in this evidential context, have to settle for claims of somewhat less externality than those they had first proposed. They have to leave out their models, and this could be a loss for them, because whatever words have been excluded at this point, as the article goes into print, cannot be part of the authors' claims. So if, for instance, molecular biologists not only accept Dr Bloch's claim of common origins for these two molecules, but follow this claim to something like his model as well, this article would give him no way to assert his priority. (For this reason, he described the model in an abstract published separately.) We see this limitation of claim as well in the more conventional personae and forms the authors use in the published versions. These are, as one might have guessed, not as much fun to read as the earlier drafts, and not as clear to a non-specialist, since they are highly compressed, are allusive in their references, and give none of the background or history of the claims.

Perhaps the most serious change in the articles, in practical terms, is that they now reach much more limited audiences than those the authors had hoped to address when they submitted their manuscripts to *Nature* and *Science*. This means not only that the articles miss whatever prestige an article acquires by appearing in those journals, but also that they are less likely to be seen, in Dr Bloch's case, by the molecular biologists doing sequencing, and in Dr Crews's case, by the wide range of zoologists. These are the researchers who, if they reoriented their research programmes to pursue these new interpretations of published data, might provide more data to strengthen these claims; more sequences to check for matching tRNAs and rRNAs, or more animals for which the patterns of hormone levels, gamete production, and mating behaviour are reliably known. But we should remember that Dr Bloch and Dr Crews are asking for a great deal. If we wanted to explain all aspects of the scientific community in functional terms, we could see in the relegation of their articles to more specialized journals an example of how the publication process works, protecting these researchers in other fields from just this kind of claim from outside their own research programmes, and thereby preventing the capricious redirection of goals, the proliferation of research programmes and the scattering of resources.

The process of publication of a claim does not stop with the acceptance of one article; both writers have other outlets for their ideas, at other levels of the hierarchy of journals. Here again we see a sharp difference between Dr Bloch's opportunities and those of Dr Crews, with their differing positions in their fields. Dr Bloch has tirelessly presented his papers on RNA sequence matches at conferences. For instance, in one poster session at a huge cell biology convention, he would repeat to anyone who was interested his whole case for common origins, drawing the listener along from figure to figure. If the listener seemed interested, Dr Bloch might go on to his larger ideas about surrogation. Thus, in this forum, he could have his own form and choose his own level of claims, according to the responses of the individuals who made up his audience. But his audience on this occasion consisted largely of friends and students, nearly all still working in his old field, and a few passers-by, perhaps attracted first by his lively illustrated bulletin board, many of them apparently graduate students with time for an intriguing, if odd, idea. Dr Bloch puts a great deal of preparation and energy into these presentations, and is happy with the chance to persuade anyone, but it seems unlikely that he will persuade in this way the powerful molecular biologists whose interest he needs.

Dr Bloch found another outlet for his model in a very short version of a paper delivered at a European conference on the Origins of Life, the proceedings of which are being published.[19] In this unrefereed outlet he is freer to speculate, as the less cautious title suggests: 'tRNA-rRNA Sequence Homologies: A Model for the Origin of a Common Ancestral Molecule, and Prospects for Its Reconstruction.' This gives him a citation he can use in proposals and in other manuscripts to refer to his model, and a priority claim for the idea of primitive multifunctional RNA, should the idea be widely accepted. But he finds it rather too compressed to be easy reading. And he discounts the authority this publication would have for the experimentalists he needs to reach; with some praise and self-irony he calls the origin-of-life people (among whom he counts himself), 'a bunch of nuts'. He says that if and when a more recent detailed paper that includes the model is published, the early paper will be superseded.

So Dr Bloch continues to try to find outlets for the parts of his work cut from the published article and for the data and theoretical refinements that have emerged since the final version of the *JME* manuscript. He has submitted to *JME* a second article arguing that interspecies comparisons would show evolutionary convergence, and a third article with the details of the model. More recently, his collaboration with the physicist at the Center for Statistical Mechanics led to a manuscript in which 'second order spectral analysis is used to depict rigorously and to characterize principal periodicities in the positions of conserved sequences common to tRNAs and rRNAs'. Note that this claim takes as proven not only the existence of matching sequences but also the explanation that they are due to common origins. But, as the reviewer's comment about publishing the data first suggests, articles like this may have to wait until his earlier finding is known well enough to serve as the basis for new problems. He has also written an article for a popular audience, entitled 'The First Chromosome', which places the *JME* article in the context of larger evolutionary questions. But this too may have to wait for publication until the earlier article is more widely accepted—the popularization does not usually precede the refereed publication. So the publication of the *JME* article has been helpful, but it does not become a breakthrough that would open further outlets for publication until it is cited and becomes a part of the literature.

For Dr Crews the question is not so much whether an article can be published as where. Even though, as we saw, he cut out the first half of his manuscript before sending it out, and finally published an article about three pages long, he has been able to publish most of what he wrote. The material on his own work, which he cut to place more emphasis on the field as a whole, appears in two articles in an issue of *BioScience*, a glossy but rather serious popular biology journal. He was guest editor of this issue, chose two other articles on similar comparative research, and used the forum to make his polemical methodological point about the importance of studying such atypical species.[20]

Though the theoretical implications of Dr Crews's claim are cut, or at least well hidden, in the *Hormones and Behavior* article, he presents them undiluted in a paper for an unrefereed but nevertheless prestigious forum, a symposium at the Kinsey Institute. For this audience of physicians, psychotherapists and other researchers interested in sexuality, an audience that does not need to determine the status of his claims or place him in the literature of neuroendocrinological research, he can be as assertive as he was in earlier drafts. The argument has become more cautious since then, and is supported by all the additional data gathered during his revisions. But the tone, even in the abstract, is like the tone of 'New Concepts in Behavioral Endocrinology': 'The great diversity in reproductive tactics...has been unappreciated by behavioral endocrinology', and 'the deterministic paradigms in behavioral and endocrinology are overly narrow'. Where in the *Hormones and Behavior* article the evolutionary ideas were held until the end, here they are emphasized from the beginning. From the out-takes of the *Science* version he gets a sentence on the evolution of regulatory mechanisms, lists of exceptions to the paradigm and of animals with associated or dissociated patterns, and descriptions of his own work. Two pages on the development and evolution of functional associations at the end of the *Science* article, which had been the focus of criticism from the more hostile referees, are here expanded into six pages. The added pages make explicit the way his claim applies to other levels of the study of reproduction, so the place of the *Cnemidophorus* in this programme is now clearer.

A particularly telling difference between the Kinsey talk, for a general scientific audience, and the *Hormones and Behavior* paper, for an audience of neuroendocrinologists, is in Dr Crews's use of the scientific literature. He begins the Kinsey talk with a motto (not a usual practice in a scientific paper) from an article dating to 1946, and he refers prominently in the introduction to insights from masters in the comparative field, often from texts twenty to forty years old. The quotations seem to be a part of the persona he is developing here; on the one hand he is an outspoken dissenter from the rigid paradigm of neuroendocrinology, but on the other hand he is the inheritor of a rich tradition of comparative work.

While such self-presentation is not encouraged by the review article format, it is appropriate in this oral presentation to an audience of non-biologists, for whom he must represent biology (he is the only biologist there) and also present something lively, new and relevant to their own work with humans.

In the end, almost all of Dr Crews's first article appeared in print. But it appeared in three separate texts, for three separate audiences, and was inserted into the structure of scientific facts in three different ways. For Dr Crews as for Dr Bloch, the way his text enters the literature is crucial in determining the eventual status of its claim. Whether his claim or Dr Bloch's becomes a fact depends on how the articles are used by other researchers. But the form of the claims has been set; we don't know what the response of the research community will be, but we know exactly what it will be responding to. What is not printed is seldom cited.

Conclusion

What level of claim can I persuade this audience to accept? At the lowest level, I am saying that scientists sometimes revise their manuscripts considerably to get them published. To support this claim, I need only show you the stacks of manuscripts. This claim, though on a very low level of externality, is significant in some evidential contexts—for instance, the context of technical-writing teachers trying to convince their students of the value of rewriting assignments. But for the audience I am addressing here, that is arguing on the level of Pinch's 'splodges on a graph'. I can put the claim on a higher level of externality, to continue using Pinch's terms, by using these manuscripts to show that a scientific claim is socially constructed. But this claim, though it is in terms familiar to the audience of this journal [*Written Communication*], tells them nothing new. A more specific claim, that is likelier to tell the readers something new, is that the comments on, and revisions of, these manuscripts show one of the ways in which claims are socially constructed—that is, through the negotiation of the form of the article and thus the status of the claim. It seems to me that this claim may have relevance in two different evidential contexts, telling us about science or telling us about texts. In terms of one context, these cases suggest that the process of writing and revision of articles has an important consensus-building function. We have seen how this process maintains the homogeneity of the scientific literature. We have also seen how it shapes the research itself—Dr Bloch, for instance, putting more and more emphasis on his data. In another context, that of literary criticism, these cases show the relations between texts, within the genre of the scientific article. The question in this context is not how reality is transformed in texts, but how it is made by texts.

Like Dr Bloch and Dr Crews, I need more data to support my claims, so I address an audience that does research on scientific texts, though perhaps in other evidential contexts. I have found, in other cases, that a number of claims are possible from one line of research, and that disagreements about the status of claims do tend to focus on matters of appropriateness to the journal, organization and length, persona, and use of the literature. But I haven't yet seen enough descriptions of the publication process to know how far this description is useful. I see from Pinch's cases that negotiation in physics and biology is rather different; the biological arguments seem to involve the usefulness of alternative concepts for organizing large bodies of data that were collected for other reasons, rather than the sort of crucial experiments that are usually held to characterize the history of physics.[21] For instance, it seems to be an acceptable response to Dr Crews's argument from the mating of garter snakes to say, 'that's just one species', while it is not an acceptable response to an argument from the perihelion advance of Mercury to say, 'that's just one planet'. I wonder if students of scientific texts find characteristic differences between disciplines in the processes of negotiation of the texts.

I have focused on a narrow part of the social construction of a knowledge claim; by implication, I am saying that the rest of the process can also be seen in terms of the forms of texts. The later part of this production, the social processes reflected in reading and citation, have begun to be studied in terms of the form of articles as well as the quantity of citations.[22] The earlier part, the transformation of observations into a text, has been studied by Latour and Woolgar and by such ethnographic researchers as Michael Lynch.[23] Another aspect of this process suggests itself as worth studying: the production of the producers. How does a researcher learn all these complex conventions of the scientific article? What part do such negotiations play in the education of a doctoral student, or in choices of problem or shifts of specialty? And how do these negotiations lead to change within the research community? I have told the stories of just two texts; similar evidence may help us tell the story of a career or of a discipline.

NOTES

Earlier versions of this paper were presented at the Conference on College Composition and Communications, New York, March 1984; at the Discourse Analysis Workshop, University of Surrey, September 1984; and at the Sociology of Science Study Group, British Sociological Association, London, September 1984. I would like to thank Lester Faigley, Edward Smith, Charles Bazerman, Carolyn Miller, Susan Cozzens and Kenneth Bruffee for their comments on an earlier version, and Trevor Pinch for letting me see the manuscript of the article cited in note 8, and for helpful discussions of the levels of claims. One of the anonymous reviewers for *Written Communication* also made helpful suggestions. This study would not have been possible without the help of David Bloch and David Crews.

1. B. Latour and S. Woolgar, *Laboratory Life: The Social Construction of Scientific Facts* (London: Sage, 1979)

2. For my use of these terms, see G.N. Gilbert and M. Mulkay, 'Contexts of Scientific Discourse: Social Accounting in Experimental Papers', in K. Knorr, R. Krohn and R. Whitley (eds), *The Social Process of Scientific Investigation, Sociology of the Sciences Yearbook*, Vol. 4 (Dordrecht, D. Reidel, 1980), 269–94. In another study (J. Law and R. Williams, 'Putting Facts Together: A Study of Scientific Persuasion', *Social Studies of Science*, 12 [1982], 535–58). i.e., Science 12. Law and Williams reach conclusions similar to mine in an analysis of discussions among the co-authors of articles, and relate these conclusions to the concept of a network.

3. D. Bloch, B. McArthur, R. Widdowson, D. Spector, R. Guimares and J. Smith, 'tRNA-rRNA Sequence Homologies: Evidence for a Common Evolutionary Origin?', *Journal of Molecular Evolution*, 19 (1983), 420–28.

4. D. Crews, 'Gamete Production, Sex Hormone Secretion, and Mating Behavior Uncoupled', *Hormones and Behavior*, 18 (1984), 22–28.

5. For my use of the term 'network', see H.M. Collins, 'The TEA Set: Tacit Knowledge and Scientific Networks', *Science Studies*, 4 (1974), 165–86.

6. G.N. Gilbert. 'The Transformation of Research Findings into Scientific Knowledge', *Social Studies of Science*, 6 (1976), 281–306.

7. Latour and Woolgar, 78–86.

8. T. Pinch, 'Towards an Analysis of Scientific Observation: The Externality and Evidential Significance of Observational Reports in Physics' *Social Studies of Science*, 15 (1985), 3–36. H.M. Collins, *Changing Order: Replication and Induction in Scientific Practice* (London: Sage, 1985), has a discussion of these approaches to presentation of claims in Chapter 6. Another edition of Collins's book was published by The University of Chicago Press, 1992.

9. D. Crews and W. Garstka, 'The Ecological Physiology of a Garter Snake', *Scientific American* (November 1982), 159–68.

10. M. Gordon, 'The Role of Referees in Scientific Communication', in J. Hartley (ed.), *The Psychology of Written Communication: Selected Readings* (London: Kogan Page, 1980), 263–75.

11. B. Latour and F. Bastide, 'Writing Science-Fact and Fiction', in M. Callon and J. Law (eds), *Qualitative Scientometrics* (London: Macmillan, forthcoming).

12. C. Bazerman, 'What Written Knowledge Does: Three Examples of Academic Discourse', *Philosophy of the Social Sciences*, Vol. 11 (1981), 361–87.

13. For instance, P. Medawar, 'Is the Scientific Paper Fraudulent?', *Saturday Review* (August 1964), 42–43.

14. E. Mayr, *The Growth of Biological Thought: Diversity, Evolution, and Inheritance* (Cambridge, Mass.: Harvard University Press, 1982).

15. D. Bloch, C.-T. Fu and P. Dean, 'DNA and Histone Synthesis Rate Change During the S. Period in Erlich Ascites Tumor Cells', *Chromosoma*, 82 (1981), 611–26.

16. See the letter on this term from M. Gossler, 'Numerology', *Nature*, 306 (8 December 1983), 530.

17. D. Crews, 'Psychobiology of Reptilian Reproduction', *Science*, 189 (26 September 1975), 1057–65.

18. See Mayr, op. cit. note 14, 465, and M. Norell, 'Homology Defined', *Nature*, 306 (8 December 1983), 530.

19. D. Bloch, 'tRNA-rRNA Sequence Homologies: A Model for the Origin of a Common Ancestral Molecule, and Prospects for Its Reconstruction', *Origins of Life and Evolution of the Biosphere* 15.1 (Dordrecht: Teidel, 1984).

20. D. Crews, 'Diversity of Hormone-Behavior Controlling Mechanisms', and 'Alternative Reproductive Tactics in Reptiles', *BioScience*, 33 (1983), 545, 562–67.

21. Mayr, 494.

22. See, for instance, C. Bazerman, 'Physicists Reading Physics: Schema-Laden Purposes and Purpose-Laden Schemas', *Written Communication*, 2, (January 1985), 3–23; and the articles surveyed in S. Cozzens, 'Taking the Measure of Science: A Review of Citation Theories', *International Society for the Sociology of Science Newsletter*, 7 (1981), 21–60.

23. See, for instance, C. Bazerman, 'Making Reference: Empirical Contextx, Choices, and Contraints in the Literary Creation of the Compton Effect', *PRE/TEXT*, (1984); K. Knorr-Cetina, *The Manufacture of Knowledge: An Essay on the Constructivist and Contextual Nature of Science* (Oxford: Pergamon, 1981); Latour and Woolgar, and M. Lynch, E. Livingston and H. Garfinkel, 'Temporal Order in Laboratory Work', in Knorr-Cetina and M. Mulkay (eds), *Science Observed: Perspectives on the Social Study of Science* (London: Sage, 1984), 205–38. Also G. Myers, 'The Social Construction of Two Biologists' Proposals', *Written Communication*, 2, (July 1985), 219–45, describes the process shaping Dr Bloch's and Dr Crews's applications for research grants.

Bibliography

Since the literature on rhetoric of science, broadly construed, is diverse, vast, and far flung, I have selected the entries for this bibliography on a narrow construal of the phrase: material concerning science published in rhetorical journals; material published about science by rhetoricians; and/or material which declares titularly its explicit interest in the rhetorical investigation of science. These criteria excluded many fine and interesting analyses and discussions, particularly from the Sociology of Scientific Knowledge literature, but also from literary, philosophical, and historical fields, and they included some questionable material from rhetoric; quality was not a selection criterion.

Articles

Alford, Elizabeth M. "Thucydides and the Plague of Athens: The Roots of Scientific Writing." *Written Comunication* 5 (1988): 131-53.

Allen, Jo. "Thematic Repetition as Rhetorical Technique [on William Harvey's De Motu Cordis]." *Journal of Technical Writing and Communication* 21 (1991): 29-40.

_____. "Commentary: A Response to J.T.H. Connor and Jennifer J. Connor's Analysis [Of Allen, 1991]." *Journal of Technical Writing and Communication* 22 (1992): 203-10.

Anderson, Paul V., R. John Brockman, and Carolyn R. Miller, Ed. *New Essays in Technical and Scientific Communication: Research, Theory, Practice.* Farmingdale, NY: Baywood, 1983.

Anderson, Ray Lynn. "Rhetoric and Science Journalism." *Quarterly Journal of Speech* 56 (1970): 358-68.

Anderson, Wilda. "The Rhetoric of Scientific Language: An Example From Lavoisier." *Modern Language Notes* 96 (1981): 746-70.

Anderson, Wilda. "Scientific Nomenclature and Revolutionary Rhetoric." *Rhetorica* 7 (1989): 45-53.

Antczak, Frederick. "Hearing our Cassandras: Ethical Criticism and Rhetorical Receptions of Paul Ehrlich." *Social Epistemology* 8 (1994): 281-88.

Batschelet, Margaret W. "Plain Style and Scientific Style: The Influence of the Puritan Plain Style on Early American Science Writers." *Journal of Technical Writing and Communication* 18 (1988): 287-95.

Bazerman, Charles. "What Written Knowledge Does: Three Examples of Academic Discourse." *Philosophy of the Social Sciences* 11 (1981): 361-88.

_____. "Making Reference: Empirical Contexts, Choices, and Constraints in the Literary Creation of the Compton Effect." *PRE/TEXT* 5 (1984): 39-66.

_____. "Theoretical Integration in Experimental Reports in Twentieth Century Physics: Spectroscopic Articles in Physical Review, 1893-1980." *Social Studies of Science* 14 (1984): 163-96.

_____. "Physicists Reading Physics: Schema-Laden Purposes and Purpose-Laden Schemas." *Written Comunication* 2 (1985): 3-23.

_____. "Studies of Scientific Writing: E Pluribus Unum?" *4S Review* 3 (1985): 13-20.

_____. "Codifying the Social Scientific Style: The APAPublication Manual as A Behaviorist Rhetoric." *The Rhetoric of the Human Sciences: Language and Argument in Scholarship and Public Affairs.* Ed. John S. Nelson Allan Megill, and Donald Mcloskey. Madison, WI: University of Wisconsin Press, 1987.

_____. "How Natural Philosopers Can Cooperate: The Literary Technology of Coordinated Investigation in Joseph Priestley's History and Present State of Electricity." *Textual Dynamics of the Professions: Historical and Contemporary Studies of Writing in Professional Communities.* Ed. Charles Bazerman and James Paradis. Madison, WI: University of Wisconsin Press, 1991. 13-44.

_____. "Intertextual Self-Fashioning: Gould and Lewontin's Representations of Literature." *Understanding Scientific Prose.* Ed. Jack Selzer. Madison, WI: University of Wisconsin Press, 1993. 20-41.

Beer, Gillian, and Herminio Martins. "Introduction." The History of the Human Sciences 3 (1990): 163-75.

Billig, Michael. "Celebrating Argument within Psychology: Dialogue, Negation, and Feminist Critique." *Argumentation* 8 (1994): 49-62.

Bokeno, R. Michael. "The Rhetorical Understanding of Science: An Explication and a Critical Commentary." *The Southern Speech Communications Journal* 52 (1987): 285-311.

Brown, Richard Harvey. "Logics of Discovery as Narratives of Conversion: Rhetorics of Invention in Ethnography, Philosophy, and Astronomy." *Philosophy and Rhetoric* 27 (1994): 1-34.

Cadden, Joan. "Science and Rhetoric in the Middle Ages: The Natural Philosophy of William of Conches." *Jounal of the History of Ideas* 56 (1995): 1–24.

Campbell, John Angus. "Darwin and *The Origin of Species*." *Speech Monographs* 37 (1970): 1-14.

_____. "Charles Darwin and the Crisis of Ecology." *Quarterly Journal of Speech* 60 (1974): 442-9.

_____."Nature, Religion, and Emotional Response: A Reconsideration of Darwin's Affective Decline." *Victorian Studies* 18 (1974): 159-74.

_____."The Polemical Mr. Darwin." *Quarterly Journal of Speech* 61 (1975): 375-90.

_____."Creationism: The Argument That Time Forgot." *Argument in Transition: Proceedings of the Third Summer SCA/AFA Conference on Argumentation*. Ed. Jack Rhodes and Others. Annandale, VA: Speech Communication Association, 1983. 423-40.

_____."Scientific Revolution and the Grammar of Culture." *Quarterly Journal of Speech* 72 (1986): 351-76.

_____."Charles Darwin: Rhetorician of Science." *The Rhetoric of the Human Sciences: Language and Argument in Scholarship and Public Affairs*. Ed. John S. Nelson, Allan Megill, and Donald McCloskey. Madison, WI: University of Wisconsin Press, 1987. 69-86. Reprinted in this volume.

_____. "Poetry, Science, and Argument: Erasmus Darwin as Baconian Subversive." *Argument and Critical Practices: Proceedings of the Fifth SCA/AFA Conference on Argumentation*. Ed. Joseph W. Wenzel. Annandale, VA: Speech Communication Association, 1987. 499-506.

_____."The Invisible Rhetorician: Charles Darwin's Third-Party Strategy." *Rhetorica* 7 (1989): 55-85.

_____."Scientific Discovery and Rhetorical Invention: The Path to Darwin's *Origin*." *The Rhetorical Turn: Invention and Persuasion in the Conduct of Inquiry*. Ed. Herbert W. Simons. London: Sage, 1990. 58-91.

_____."Reply to Gaonkar and Fuller." *The Southern Communication Journal* 58 (1993): 312-18.

_____."Of Orchids, Insects and Natural Theology," *Argumentation* 8 (1994): 63-80.

Campbell, Paul Newell. "Poetic-Rhetorical, Philosophical, and Scientific Discourse." *Philosophy and Rhetoric* 6 (1973): 1-29.

_____. "The Personae of Scientific Discourse." *Quarterly Journal of Speech* 61 (1975): 391-405.

Carlisle, Fred E. "Literature, Science, and Language: A Study of Similarity and Difference." *PRE/TEXT* 1 (1980): 39-72.

Ceccarelli, Leah. "A Masterpiece in a New Genre: The Rhetorical Negotiation of Two Audiences in Schrödinger's What Is Life." *Technical Communication Quarterly* 3 (1994): 7-17.

Charney, Davida. "A Study in Rhetorical Reading: How Evolutionists Read "The Spandrels of San Marco." *Understanding Scientific Prose*. Ed. Jack Selzer. Madison, WI: University of Wisconsin Press, 1993. 203-231.

Condit, Celeste-Michelle. "The Birth of Understanding: Chaste Science and the Harlot of the Arts." *Communication Monographs* 57 (1990): 323-7.

Coney, Mary B. "Terministic Screens: A Burkean Reading of the Experimental Article." *The Journal of Technical Writing and Communication* 22 (1992): 149-59.

Connor, J. T. H., and Jennifer J. Connor. "Commentary on Rhetorical Analysis of William Harvey's *De Motu Cordis*." *Journal of Technical Writing and Communication* 22 (1992): 195-202.

Cook, Tom, and Ron Seamon. "Ein Feyerabenteur." *PRE/TEXT* 1.1-2 (1980): 124-60.

Couture, Barbara. "Provocative Architecture: A Structural Analysis of Gould and Lewontin's 'The Spandrels of San Marco'. *Understanding Scientific Prose*. Ed. Jack Selzer. Madison, WI: University of Wisconsin Press, 1993. 276-309.

Cox, Barbara, and Charles G. Roland. "How Rhetoric Confuses Scientific Issues." *IEEE Transactions on Professional Communication* PC-16. No. 3 (1973): 140-2.

Craig, Robert T. "The Speech Tradition." *Communication Monographs* 57 (1990): 309-14.

Crismore, Avon, and Rodney Farnsworth. "Mr. Darwin and His Readers: Exploring Interpersonal Metadiscourse as a Dimension of Ethos." *Rhetoric Review* 8 (1989): 91-112.

_____. "Scientific Rhetoric, Metadiscourse, and Power: Darwin's Origin." Rhetoric and Ideology: Compositions and Criticism of Power. Ed. Charles W. Kneupper. Arlington: Rhetoric Society of America, 1989. 174-88.

Cushman, Donald P. "A Window of Opportunity Argument." Communication Monographs 57 (1990): 328-32.

_____, and Branislav Kovacic. "The Rhetoric of the Reasoned Social Scientific Fact." Argumentation 8 (1994): 33-48.

Czubaroff, Jeanine. "The Deliberative Character of Strategic Scientific Debates." *Rhetoric in the Human Sciences: Inquiries in Social Construction* . Ed. Herbert W. Simons. London: Sage, 1989. 28-47.

Dombrowski, Paul M. "Challenger Through the Eyes of Feyerabend." *Journal of Technical Writing and Communication* 24 (1994): 7-18.

Dowdy, Diane. "Rhetorical Techniques of Audience Adaption in Popular Science Writing." *Journal of Technical Writing and Communication* 17 (1987): 275-85.

Ehninger, Douglas. "Science, Philosophy-and Rhetoric: A Look Toward the Future." *The Rhetoric of Western Thought*. Ed. James L. Golden, Goodwin F. Berquist, and William E. Coleman. Fourth Ed. Dubuque, Io: Kendall/Hunt, 1989 [1978]. 623-36.

Ehrlich, Paul. "Perils of a Modern Cassandra: Some Personal Comments." *Social Epistemology* 8 (1994): 239-40.

Fahnestock, Jeanne. "Accommodating Science." *Written Comunication* 3 (1986): 275-96.

_____. "Arguing in Different Forums: The Bering Crossover Controversy." *Science, Technology, and Human Values* 14 (1989): 26-42. Reprinted in this volume.

_____. "Tactics of Evaluation in Gould and Lewontin's "The Spandrels of San Marco"." *Understanding Scientific Prose*. Ed. Jack Selzer. Madison, WI: University of Wisconsin Press, 1993. 158-79.

_____, and Marie Secor. "The Stases in Scientific and Literary Argument." *Written Comunication* 5 (1988): 427-440.

Farnsworth, Rodney, and Avon Crismore. "On the Reefs: The Verbal and Visual Rhetoric of Darwin's Other Big Theory." *Rhetoric Society Quarterly* 21 (1991): 11-25.

Findlen, Paula. "Controlling the Experiment: Rhetoric, Court Patronage, and the Experimental Method of Francesco Redi." *History of Science* 31 (1993): 35.

Finocchiaro, Maurice A. "Logic and Rhetoric in Lavoisier's Sealed Note: Toward A Rhetoric of Science." *Philosophy and Rhetoric* 10 (1977): 111-22.

_____. "Varieties of Rhetoric in Science." *The History of the Human Sciences* 3 (1990): 177-93.

Fisher, Walter R. "Narrative Rationality and the Logic of Scientific Discourse." *Argumentation* 8 (1994): 21-32.

Fuller, Steve. "'The Rhetoric of Science': A Doubly Vexed Expression." *The Speech Communication Journal* 58 (1993): 306-11.

Gaonkar, Dilip Parameshwar. "The Idea of Rhetoric in the Rhetoric of Science." *The Southern Communication Journal* 58 (1993): 258-95.

Gould, Stephen Jay. "Fulfilling the Spandrels of World and Mind." *Understanding Scientific Prose*. Ed. Jack Selzer. Madison, WI: University of Wisconsin Press, 1993. 310-336.

Gragson, Gay, and Jack Selzer. "The Reader in the Text of 'The Spandrels of San Marco'." *Understanding Scientific Prose*. Ed. Jack Selzer. Madison, WI: University of Wisconsin Press, 1993. 180-202.

Graves, Heather Brodie. "Rhetoric and Reality in the Process of Scientific Inquiry." *Rhetoric Review* 14 (1995): 106-23.

Gross, Alan G. "The Rhetorical Invention of Scientific Invention: The Emergence and Transformation of A Scientific Norm." *Rhetoric in The Human Science: Inquiries into Social Construction*. Ed. Herbert W. Simons. London: Sage, 1989. 89-108.

Gross, Alan. "The Origin of Species: Evolutionary Taxonomy as An Example of the Rhetoric of Science." *The Rhetorical Turn: Invention and Persuasion in The Conduct of Inquiry*. Ed. Herbert W. Simons. London: Sage, 1990. 91-115.

_____. "Analogy and Intersubjectivity: Political Oratory, Scholarly Argument, and Scientific Reports." *Quarterly Journal of Speech* 69 (1983):

_____. "Public Debates as Failed Social Dramas: The Recombinant DNA Controversy." *Quarterly Journal of Speech* 70 (1984): 397-409.

_____. "The Form of an Experimental Paper: A Realizaton of the Myth of Induction." *Journal of Technical Writing and Communication* 15 (1985): 15-26.

_____. "Discourse on Method: The Rhetorical Analysis of Scientific Texts." *PRE/TEXT* 9 (1988): 169-85. Reprinted in *Humanistic Aspects of Technical Communication*, Ed. Paul M. Dombrowski. Amityville, NY: Baywood, 1994. 63-79.

_____. "On the Shoulders of Giants: Seventeenth-Century Optics as an Argument Field." *Quarterly Journal of Speech* 74 (1988): 1-17. Reprinted in this volume.

_____. "Philosophy versus Science: The Species Debate and the Practice of Taxonomy." *PSA* 1988 [Philosophy of Science Association]. Ed. Arthur Fine and Mickey Forbes. East Lansing, MI: 1988. 223-30.

_____. "Persuasion and Peer Review in Science: Habermas's Ideal Speech Situation Applied." *The History of the Human Sciences* 3 (1990): 195-209.

_____. "Reinventing Certainty: The Significance of Ian Hacking's Realism." *PSA* 1990 [Philosophy of Science Association]. Ed. Mickey Forbes, Arthur Fine, and Linda Wessels. East Lansing, MI: 1990. 421-31.

_____. "Rhetoric of Science is Epistemic Rhetoric." *Quarterly Journal of Speech* 76 (1990): 304.

_____. "Does Rhetoric of Science Matter? The Case of the Floppy-Eared Rabbits." *College English* 53 (1991): 933-43.

_____. "Rhetoric of Science Without Constraints." *Rhetorica* 9 (1991): 283-99.

_____. "Experiment as Text: The Limits of Literary Analysis." *Rhetoric Review* 11 (1993): 290-300.

_____. "The Rhetoric of Science and the Science of Rhetoric". Canadian Society for the Study of Rhetoric 1991-1992. Ed. Albert W. Halsall. 1993. 4: 8-35.

_____. "Rhetorical Imperialism in Science." *College English* 55 (1993): 82-87.

_____. "What if We're not Producing Knowledge? Critical Reflections on the Rhetorical Criticism of Science." *Southern Journal of Speech Communication* 58 (1993): 301-5.

_____. "Guest Editor's Column." *Technical Communication Quarterly* 3 (1994): 5-6.

_____. "Ending it all: Closure in Science and its Philosophy." *Argumentation* 8 (1994): 9-20.

_____. "Is a Rhetoric of Science Policy Possible." *Social Epistemology* 8 (1994): 273-80.

_____. "A Tale Told Twice: The Rhetoric of Discovery in the Case of DNA." *Argument and Critical Practices: Proceedings of the Fifth SCA/AFA Conference on Argumentation.* Ed. Joseph W. Wenzel. Annandale, VA: Speech Communication Association, 1987. 491-7.

_____. "The Roles of Rhetoric in the Public Understanding of Science." *Public Understanding of Science* 3 (1994): 3-23.

_____. "Renewing Aristotelian Theory: The Cold Fusion Theory as a Test Case." *Quarterly Journal of Speech* 81 (1995): 48-62.

Halloran, S. Michael. "Technical Writing and the *Rhetoric of Science*." Journal of Technical Writing and Communication 8 (1978): 77-88.

_____. "The Birth of Molecular Biology." Rhetoric Review 3 (1984): 70-83. Reprinted in this volume.

_____, and Annette Norris Bradford. "Figures of Speech in the Rhetoric of Science and Technology." *Essays on Classical Rhetoric and Modern Discourse.* Ed. Robert J. Connors, Lisa S. Ede, and Andrea A. Lunsford. Carbondale, IL: Southern Illinois University Press, 1984. 179-92.

Harmon, Joseph E. "An Analysis of Fifty Citation Superstars From the Scientific Literature." *Journal of Technical Writing and Communication* 22 (1992): 17-38.

Harmon, Joseph E. "Current Contents of Theoretical Scientific Papers." *Journal of Technical Writing and Communication* 22.4 (1992): 357-76.

Harris, R. Allen. "Argumentation in Syntactic Structures." *Rhetoric Society Quarterly* 19 (1989): 103-24.

_____. "Power and the Ethos of Syntactic Structures." *Rhetoric and Ideology: Compositions and Criticisms of Power.* Ed. Charles W. Kneupper. Arlington, TX: Rhetoric Society of America, 1989. 189-96.

_____. "Assent, Dissent, and Rhetoric in Science." *Rhetoric Society Quarterly* 20 (1990): 13-37. Reprinted in Humanistic Aspects of Technical Communication, Ed. Paul M. Dombrowski. Amityville, NY: Baywood, 1994. 33-62.

_____. "Rhetoric of Science." *College English* 53 (1991): 282-307.

_____. "Generative Semantics: Secret Handshakes, Anarchy Notes, and the Implosion of Ethos." *Rhetoric Review* 12 (1993): 125-59.

_____. "The Chomskyan Revolution I: Syntax, Semantics, and Science." *Perspectives on Science* 2 (1994): 38-75.

_____. "The Chomskyan revolution II: Sturm und Drang." *Perspectives on Science* 2 (1994): 176-230.

Hayes, James T. "'Creation-Science' Is Not 'Science': Argument Fields and Public Argument." *Argument in Transition: Proceedings of the Third Summer SCA/AFA Conference on Argumentation.* Ed. Jack Rhodes and Others. Annandale, VA: Speech Communication Association, 1983. 416-22.

Hikins, James W., and Kenneth S. Zagacki. "Rhetoric, Philosophy, and Objectivism: An Attenuation of the Claims of Rhetoric of Inquiry." *The Quarterly Journal of Speech* 74 (1988): 201-28.

Hollander, Rachelle D. "Rhetoric and Science." *Social Epistemology* 8 (1994): 241-2.

Holmquest, Anne. "Rhetorical Argument in Science: The Function of Presumption." *Argument in Transition: Proceedings of the Third Summer SCA/AFA Conference on Argumentation.* Ed. Jack Rhodes and Others. Annandale, VA: Speech Communication Association, 1983. 251-71.

_____. "Rhetoric and Semiotic in Scientific Argumentation: The Function of Presumption in Charles Darwin's Origin of Species Essays." *Argument and Social Practice: Proceedings of the Fourth SCA/AFA*

Conference on Argumentation. Ed. J. Robert Cox, Malcolm O. Sillars, and Greg B. Walker. Annandale VA: Speech Communication Association, 1985. 376-402.

Holton, Gerald. "Quanta, Relativity, and Rhetoric." *Persuading Science: The Art of Scientific Rhetoric*. Ed. Marcello Pera and William R. Shea. Canton MA: Science History Publications, 1992. 173-204.

Howe, Henry F., and John Lyne. "Gene Talk in Sociobiology." *Social Epistemology* 6 (1992): 1-54.

Hyde, Michael J. "Storytelling and Public Moral Argument: The Case of Medicine." *Argument and Social Practice. Proceedings of the Fourth SCA/AFA Conference on Argumentation*. Ed. J. Robert Cox, Malcolm O. Sillars, and Greg B. Walker. Annandale VA: Speech Communication Association, 1985. 364-75.

―――――. "Medicine, Rhetoric, and Euthanasia: A Case Study in the Workings of A Postmodern Discourse." *Quarterly Journal of Rhetoric* 79 (1993): 201-24.

Hyland, Ken. "Talking to the Academy: Forms of Hedging in Scientific Research Articles." *Written Comunication* 13 (1996): 251-81.

Johnson-Sheehan, Richard D. "Scientific Communication and Metaphors: An Analysis of Einstein's 1905 Special Relativity Paper." *Journal of Technical Writing and Communication* 25 (1995): 71-84.

Journet, Debra. "Rhetoric and Sociobiology." *Journal of Technical Writing and Communication* 14 (1984): 339-50.

―――――. "Forms of Discourse and the Sciences of the Mind." *Written Comunication* 7 (1990): 171-90.

―――――. "Biological Explanation, Political Ideology, and "Blurred Genres": A Bakhtinian Reading of the Science Essays of J. B. S. Haldane." *Technical Communication Quarterly* (1993): 185-204.

―――――. "Deconstructing 'The Spandrels of San Marco'. *Understanding Scientific Prose*. Ed. Jack Selzer. Madison, WI: University of Wisconsin Press, 1993. 232-255.

Katriel, Tamar, and Robert E. Sanders. "The Meta-Communicative Role of Epigraphs in Scientific Text Construction." *Rhetoric in The Human Sciences*. Ed. Herbert W. Simons. London: Sage, 1989. 183-94.

Keith, William. "Cognitive Science on A Wing and A Prayer." *Social Epistemology* 4 (1990): 343-55.

―――――. "Rhetorical Criticism and the Rhetoric of Science: Introduction." *The Southern Communication Journal* 58 (1993): 255-57.

Kelso, J. A. "Science and the Rhetoric of Reality." *Central States Speech Journal* 31 (1980): 17-29.

Latour, Bruno, and P. Fabbri. "La rhetorique de la science." *Actes de la recherche* 13 (1981): 81-95.

Leff, Michael. "The Idea of Rhetoric as Interpretive Practice: A Humanist's Response to Gaonkar." *The Southern Communication Journal* 58 (1993): 296-300.

Lessl, Thomas M. "Science and the Sacred Cosmos: The Ideological Rhetoric of Carl Sagan." The Quarterly *Journal of Speech* 71 (1985): 175-87.

―――――. "Heresy, Orthodoxy, and the Politics of Science." *The Quarterly Journal of Speech* 74 (1988): 18-34.

―――――. "The Priestly Voice." The Quarterly Journal of Speech 75 (1989): 183-97.

Lewin, Philip. "Categorization and the Narrative Structure of Science." *Philosophy and Rhetoric* 27 (1994): 35.

Lyne, John. "Ways of Going Public: The Projection of Expertise in the Sociobiology Controversy." *Argument in Transition: Proceedings of the Third Summer SCA/AFA Conference on Argumentation*. Ed. Jack Rhodes and Others. Annandale VA: Speech Communication Association, 1983. 400-15.

―――――. "Critical Reflections on the Iowa Conference, March 28-31, 1984: The Rhetoric of Human Sciences-Rhetorics of Inquiry." *Quarterly Journal of Speech* 71 (1985): 65-73.

―――――. "Punctuated Equilibria: A Case Study in Scientific and Para-Scientific Argument." *Argument and Social Practice: Proceedings of the Fourth SCA/AFA Conference on Argumentation*. Ed. J. Robert Cox, Malcolm O. Sillars, and Greg B. Walker. Annandale VA: Speech Communication Association, 1985. 403-19.

―――――. "Learning the Lessons of Lysenko: Biology, Rhetoric, and Politics in Historical Controversy." *Argument and Critical Practice: The Fifth SCA/AFA Conference on Argumentation*. Ed. Joseph W. Wenzel. Annandale, VA: Speech Communication Association, 1987. 507-12.

―――――. "Argument in the Human Sciences." *Perspectives on Argumentation*. Ed. Robert Trapp and Janice Schuetz. Prospect Heights: Waveland Press, 1990. 178-89.

―――――. "Bio-Rhetorics: Moralizing the Life Sciences." *The Rhetorical Turn: Invention and Persuasion in the Conduct of Inquiry*. Ed. Herbert W. Simons. London: Sage, 1990. 35-57.

―――――. "Angels in the Architecture: A Burkean Inventional Perspective on "Spandrels"." Understanding Scientific Prose. Ed. Jack Selzer. Madison, WI: University of Wisconsin Press, 1993. 144-157.

―――――. "Punctuated Equilibria: Rhetorical Dynamics of a Scientific Controversy." *Quarterly Journal of Speech* 72 (1986): 132-47. Reprinted in this volume.

_____, and Henry F. Howe. "Rhetorics of Expertise: E. O. Wilson and Sociobioloby." *Quarterly Journal of Speech* 76 (1990): 134-51.

Markel, Mike. "Induction, Social Constructionism, and the Form of the Science Paper." *Journal of Technical Writing and Communication* 23 (1992): 7-22.

McCarthy, Lucille Parkinson. "A Psychiatrist Using DSM-III: The Influence of a Charter Document in Psychiatry." *Textual Dynamics of the Professions: Historical and Contemporary Studies of Writing in Professional Communities.* Ed. Charles Bazerman and James Paradis. Madison, WI: University of Wisconsin Press, 1991. 358-79.

McGuire, J. E., and Trevor Melia. "Some Cautionary Strictures on the Writing of the Rhetoric of Science." *Rhetorica* 7 (1989): 87-99.

_____. "The Rhetoric of the Radical Rhetoric of Science." *Rhetorica* 9 (1991): 301-16.

McShea, Daniel W. "Evolutionary Trends and the Salience Bias (With Apologies to Oil Tankers, Karl Marx, and Others)." *Technical Communication Quarterly* 3 (1994): 21-38.

Miller, Carolyn R. "Technology as a Form of Consciousness." *Central States Speech Journal* 29 (1978): 228-36.

_____. "Public Knowledge in Science and Society." *PRE/TEXT* 3 (1982): 31-49.

_____. "Invention in Technical and Scientific Discourse: A Prospective Survey." *Research in Technical Communication: A Bibliographic Source Book.* Ed. M. G. Moran and D. Journet. Westport, CT: Greenwood Press, 1985. 117-62.

_____. "The Rhetoric of Decision Science, Or Herbert A. Simons Says [Abstract] ." *Science, Technology, and Human Values* 14 (1989): 43-6.

_____. "The Rhetoric of Decision Science, Or Herbert A. Simons Says." *The Rhetorical Turn: Invention and Persuasion in the Conduct of Inquiry.* Ed. Herbert W. Simons. Chicago: University of Chicago Press, 1990. 162-84.

_____. "Kairos in the Rhetoric of Science." *A Rhetoric of Doing.* Ed. Stephen P. Witte, Neil Nakadate, and Roger Cherry. Carbondale, IL: Southern Illinois University Press, 1992.

_____. "Opportunity, Opportunism, and Progress: Kairos in the Rhetoric of Technology." *Argumentation* 8 (1994): 81-96.

_____, and S. Michael Halloran. "Reading Darwin, Reading Nature; Or, On the Ethos of Historical Science." *Understanding Scientific Prose.* Ed. Jack Selzer. Madison, WI: University of Wisconsin Press, 1993. 106-126.

Mirel, Barbara. "Debating Nuclear Energy: Theories or Iisk and Purposes of Communication." *Technical Communication Quarterly* 3 (1994): 41-66.

Moran, M. G. "Joseph Priestley, William Duncan and Analytic Arrangement in Eighteenth-Century Scientific Discourse." *Journal of Technical Writing and Communication* 14 (1984): 207-15.

_____. "A History of Technical and Scientific Writing." *Research in Technical Communication: A Bibliographic Source Book.* Ed. M. G. Moran and D. Journet. Westport, CT: Greenwood Press, 1985. 25-38.

Moss, Jean Dietz. "The Interplay of Science and Rhetoric in Seventeenth Century Italy." *Rhetorica* 7 (1989): 23-43.

_____. "Galileo's Letter to Christina: Some Rhetorical Considerations." *Renaissance Quarterly* 36 (1983): 547-76

_____. "Galileo's Rhetorical Strategies in Defence of Copernicanism." *Novitê celesti e crisis del sapere.* Ed. P. Galluzzi. Florence: Giunti Barbèra, 1984. 95-103.

Moss, Jean Dietz. "The Rhetoric of Proof in Galileo's Writings on the Copernican System." *Reinterpretting Galileo.* Ed. W. W. Wallace. Washington: Catholic University of America Press, 1986.

Munévar, Gonzalo. "Rhetorical Grounds for Determining What is Fundamental Science: The Case of Space Exploration." *Argument and Social Practice: Proceedings of the Fourth SCA/AFA Conference on Argumentation.* Ed. J. Robert Cox, Malcolm O. Sillars, and Greg B. Walker. Annandale VA: Speech Communication Association, 1985. 420-34.

Myers, Greg. "Nineteenth Century Popularizations of Thermodynamics and the Rhetoric of Social Prophecy." *Victorian Studies* 29 (1985): 35-66.

_____. "The Social Construction of Two Biologists' Proposals." *Written Comunication* 2 (1985): 219-45.

_____. "Texts as Knowledge Claims: The Social Construction of Two Biology Articles." *Social Studies of Science* 15 (1985): 593-630. Reprinted in this volume.

_____. "Every Picture Tells A Story: Illustrations in E. O. Wilson's Sociobiology." *Human Studies* 11 (1988): 235-69.

_____. "Writing, Readings, and the History of Science." *Studies in the History and Philosophy of Science* 20 (1989): 271-84.

_____. "Science for Women and Children: The Dialogue of Popular Science in the Nineteenth Century." *Nature Transfigured: Literature and Science 1700-1800*. Ed. Sally Shuttleworth and J. R. R. Christie. Manchester: Manchester University Press, 1989.

_____. "Stories and Styles in Two Molecular Biology Review Articles." *Textual Dynamics of the Professions: Historical and Contemporary Studies of Writing in Professional Communities*. Ed. Charles Bazerman and James Paradis. Madison, WI: University of Wisconsin Press, 1991. 45-75.

_____. "Making Enemies: How Gould and Lewontin Criticize." *Understanding Scientific Prose*. Ed. Jack Selzer. Madison, WI: University of Wisconsin Press, 1993. 256-275.

_____. "Narratives of Science and Nature in Popularizing Molecular Genetics". *Advances in Written Text Analysis*. Ed. Malcolm Coulthard. New York: Routledge, 1994. 179-90

Overington, Michael A. "The Scientific Community as Audience: Toward a Rhetorical Analysis of Science." *Philosophy and Rhetoric* 10 (1977): 143-64.

Paradis, James. "Bacon, Linnaeus, and Lavoisier: Early Language Reform in the Sciences." *New Essays in Technical and Scientific Communication: Research*, Theory, Practice. Ed. Paul V. Anderson, R. John Brockman, and Carolyn R. Miller. Farmingdale, NY: Baywood, 1983. 200-226.

_____. "Montaigne, Boyle, and the Essay of Experience." *One Culture*. Ed. George Levine. Madison, WI: University of Wisconsin Press, 1987. 59-91.

Paul, Danette, and Davida Charney. "Introducing Chaos (Theory) into Science and Engineering: Effects of Rhetorical Strategies on Scientific Readers." *Written Communication* 12 (1995): 396–438.

Pera, Marcello. "The Role and Value of Rhetoric in Science." *Persuading Science: The Art of Scientific Rhetoric*. Ed. Marcello Pera and William R. Shea. Canton MA: Science History Publications, 1992. 29-54.

Pinch, Trevor. "Cold Fusion and the Sociology of Scientific Knowledge." *Technical Communication Quarterly* 3 (1994): 85-100.

Prelli, Lawrence J. "The Rhetorical Construction of Scientific Ethos." *Rhetoric in The Human Sciences: Inquiries in Social Construction*. Ed. Herbert W. Simons. London: Sage, 1989. 48-68. Reprinted in this volume.

_____. "Rhetorical Logic and the Integration of Rhetoric and Science." *Communication Monographs* 57 (1990): 312-322.

_____. "Rhetorical Perspective and the Limits of Critique." *The Southern Communication Journal* 58 (1993): 319-27.

_____. "Introduction." Argumentation 8 (1994): 1-7.

Reeves, Carol. "Establishing a Phenomenon: The Rhetoric of Early Medical Reports on AIDS." *Written Communication* 7 (1990): 393-416.

_____. "Owning a Virus: The Rhetoric of Scientific Discovery Accounts." *Rhetoric Review* 10 (1992): 321-36. Reprinted in this volume.

_____. "Strategies of Key AIDS Medical Scientists and Physicians." *Written Communication* 13 (1996): 137–50.

Rosner, Mary, and Georgia Rhoades. "Science, Gender, and 'The Spandrels of San Marco and the Panglossian Paradigm'." *Understanding Scientific Prose*. Ed. Jack Selzer. Madison, WI: University of Wisconsin Press, 1993. 82-105.

Rowan, K. E. "Moving Beyond the What to the Why: Differences in Professional and Popular Science Writing." *Journal of Technical Writing and Communication* 19 (1989): 161-79.

Runquist, Mark. "The Rhetoric of Geology: Ethos in the Writing of North American Geologists, 1823-1988." *Journal of Technical Writing and Communication* 22 (1992): 387-404.

Rymer, Jone. "Scientific Composing Processes: How Eminent Scientists Write Journal Articles." *Writing in Academic Disciplines*. Ed. David A. Joliffe. Norwood: Ablex, 1988. 211-51.

Schiappa, Edward. "Burkean Tropes and Kuhnian Science: A Social Constructivist Perspective on Language and Reality". *Journal of Advanced Composition* 13 (1995): 401-22.

Schollmeier, Paul. "A Classical Rhetoric of Modern Science." Philosophy and Rhetoric 17 (1984): 209-20.

_____. "A Rhetorical Ontology for Modern Science." *Rhetorica* 12 (1994): 327-41.

Segal, Judy Z. "Writing and Medicine: Text and Context." *Writing in the Workplace: New Research. Ed. Rachel Spilka. Carbondale, IL: Southern Illinois University Press, 1993*.

Shapin, Steven. "Cordelia's Love: Credibility [=Ethos] and the Social Studies of Science." *Perspectives on Science* 3 (1995): 255–75.

Simons, Herbert W. "Are Scientists Rhetors in Disguise?" *Rhetoric in Transition*. Ed. E. E. White. University Park, PA: The Pennsylvania University Press, 1980. 115-30.

Sipiora, Phillip. "Ethical Argumentation in Darwin's Origin of Species." *Ethos: New Essays in Rhetorical and Critical Theory.* Ed. James S. Baumlin and Tita French Baumlin. Dallas: Southern Methodist University Press, 1994: 265-92.

Soper, Kate. "Feminism and Ecology: Realism and Rhetoric in the Discourses of Nature." *Science, Technology, and Human Values* 20 (1995): 311–330.

Stephens, James. "Rhetorical Problems in Renaissance Science." *Philosophy and Rhetoric* 8 (1975): 213-29.

———. "Style as Therapy in Renaissance Science." *New Essays in Technical and Scientific Communication: Research, Theory, Practice.* Ed. Paul V. Anderson, R. John Brockman, and Carolyn R. Miller. Farmingdale, NY: Baywood, 1983. 187-199.

Stevenson, Dwight W. "Toward a Rhetoric of Scientific and Technical Discourse." *The Technical Writing Teacher* 5 (1977): 4-10.

Sullivan, Dale. "The Epideictic Rhetoric of Science." *Journal of Business and Technical Communication* 5 (1991): 229-45.

———. "Nova's Narrative Excommunication of Fleischmann and Pons." *Science, Technology, and Human Values* 19 (1994): 283–306.

Taylor, Charles Alan. "Of Audience, Expertise, and Authority: The Evolving Creationism Debate." *Quarterly Journal of Speech* 78 (1992): 277-95.

———. "Science as Cultural Practics: A Rhetorical Perspective." *Technical Communication Quarterly* 3 (1994): 67-81.

———, and Celseste Condit. "Objectivity and Elites: A Creation-Science Trial." *Critical Studies in Mass Communication* 5 (1988): 293-312.

Thacker, Brad, and James F. Stratman. "Transmuting Common Substances: The Cold Fusion Controversy and the Rheotric of Science". *Journal of Business and Technical Communication* 9 (1995): 389–424.

Thompson, D. "Arguing for Experimental 'Facts' in Science." *Written Comunication* 10 (1993): 106-28.

Turner, Brian. "Giving Good Reasons: Environmental Appeals in the Nonfiction of John McPhee." *Rhetoric Review* 13 (1994): 164-82.

Valletta, Clement L., and Robert A. Paoletti. "'In-Determinacy' in Science and Discoures: A Rhetoric of Disciplinary Levels." *Journal of Technical Writing and Communication* 25 (1995): 27-42.

Vickers, Brian. "Epideictic Rhetoric in Galileo's Dialogo". *Annali dell'Istituto e Museo de Storia della Scienza di Firenz* 8 (1983): 69–102.

Waddell, Craig. "Reasonableness Versus Rationality in the Construction and Justification of Science Policy Decisions: The Case of the Cambridge Experimentation Review Board." *Science, Technology, and Human Values* 14 (1989): 7-25.

———. "The Role of Pathos in the Decision-Making Process: A Study in the Rhetoric of Science Policy." Quarterly Journal of Speech 76 (1990): 381-400. Reprinted in this volume.

———. "Perils of a Modern Cassandra: Rhetorical Aspects of Public Indifference to the Population Explosion." *Social Epistemology* 8 (1994): 221-37.

———. "Rhetoric of Environmental Policy: From Critical Practice to the Social Construction of Theory." *Social Epistemology* 8 (1994): 289-310.

Wallace, William A. "Aristotelian Science and Rhetoric in Transition: The Middle Ages and the Renaissance." *Rhetorica* 7 (1989): 7-21.

Walzer, Arthur E., and Alan G. Gross. :"Positivists, Postmodernists, Aristotelians, and the Challenger Disaster." *College English* 56 (1994): 420-433.

Wander, Philip C. "The Rhetoric of Science." Western Speech Communication 50 (1976): 226-35.

———, and Denis Jaehne. "From Cassandra to Gaia: The Limits of Civic Humanism in a Post-Ecological World." *Social Epistemology* 8 (1994): 243-59.

Warnick, Barbara. "A Rhetorical Analysis of Episteme Shift: Darwin's Origin of the Species." *The Southern Speech Communication Journal* 49 (1983): 26-42.

Weaver, Richard M. "Dialectic and Rhetoric at Dayton, Tennessee." The Ethics of Rhetoric. Davis, CA: Hermagoras Press, 1985 [1953]. 27-54. Reprinted in this volume.

Weimer, Walter B. "Science as a Rhetorical Transaction: Toward a Nonjustificational Conception of Rhetoric." *Philosophy and Rhetoric* 10 (1977): 1-29.

———. "For and Against Method." *PRE/TEXT* 1 (1980): 161-203.

Wells, Susan. "'Spandrels,' Narration, and Modernity." *Understanding Scientific Prose.* Ed. Jack Selzer. Madison, WI: University of Wisconsin Press, 1993. 42-60.

Winkler, Victoria. "The Role of Models in Technical and Scientific Writing." *New Essays in Technical and Scientific Communication: Research, Theory, Practice*. Ed. Paul V. Anderson, R. John Brockman, and Carolyn R. Miller. Farmingdale, NY: Baywood, 1983. 111-22.

Winsor, Dorothy A. "'Constructing Scientific Knowledge in Gould and Lewontin's 'The Spandrels of San Marco.'" *Understanding Scientific Prose*. Ed. Jack Selzer. Madison, WI: University of Wisconsin Press, 1993. 127-143.

Woolgar, Steve. "What is the Analysis of Scientific Rhetoric for?" *Science, Technolgoy and Human Values* 14 (1989): 47–49.

Xiao, Xiaosui. "China Encounters Darwinism: A Case of Intercultural Rhetoric." *Quarterly Journal of Speech* 81 (1995): 83-99.

Zagacki, Kenneth S., and William Keith. "Rhetoric, Topoi, and Scientific Revolutions." *Philosophy and Rhetoric* 25 (1992): 59-78.

Zappen, James P. "Francis Bacon and the Rhetoric of Science." *College Composition and Communication* 26 (1975): 244-7.

_____. "A Rhetoric for Research in Science and Technologies." *New Essays in Technical and Scientific Communication: Research, Theory, Practice*. Ed. Paul V. Anderson, R. John Brockman, and Carolyn R. Miller. Farmingdale, NY: Baywood, 1983. 123-38.

_____. "Historical Perspectives on the Philosophy and the Rhetoric of Science." *PRE/TEXT* 6 (1985): 9-29.

_____. "Historical Studies in the Rhetoric of Science and Technology." *The Technical Writing Teacher* 14 (1987): 285-98.

_____. "The Discourse Community in Scientific and Technical Communication." *Journal of Technical Writing and Communication* 19 (1989): 74-88.

_____. "Francis Bacon and the Historiography of Scientific Rhetoric." *Rhetoric Review* 8 (1989): 74-90.

_____. "Scientific Rhetoric in the Nineteenth and Early Twentieth Centuries: Herbert Spencer, Thomas H. Huxley, and John Dewey." *Textual Dynamics of the Professions: Historical and Contemporary Studies of Writing in Professional Communities*. Ed. Charles Bazerman and James Paradis. Madison, WI: University of Wisconsin Press, 1991. 145-69.

_____. "The Rhetoric of Science and the Challenge of Post-Liberal Democracy." *Social Epistemology* 8 (1994): 261-71.

Books and Special Journal Issues

Bazerman, Charles. *Shaping Written Knowledge: The Genre and Activity of the Experimental Article in Science*. Madison, WI: University of Wisconsin Press, 1988.

_____, Guest ed. *Science, Technology, and Human Values: Rhetoricians on the Rhetoric of Science* 14 .1 (1989).

Beer, Gillian, and Herminio Martins, Guest eds. *The History of the Human Sciences: Rhetoric and Science* 3.2 (1990).

Fuller, Steve, Ed. Social Epistemology: *The Rhetoric of Sociobiology* 6.2 (1992).

_____. *Social Epistemology: Public Indifference to Population Issues* 8.3 (1994).

Gross, Alan G. *The Rhetoric of Science*. Cambridge: Harvard University Press, 1990.

_____.Guest ed. *Technical Communication Quarterly: Rhetoric and Science* 3.1 (1994).

Harrington, E. W. *Rhetoric and The Scientific Method of Inquiry: A Study of Invention*. Vol. 1 of University of Colorado Studies, Series in Language and Literature. Boulder: University of Colorado Press, 1948.

Holloway, Rachel L. *In The Matter of J. Robert Oppenheimer: Politics, Rhetoric, and Self-Defense*.Westport, Ct: Praeger, 1993.

Keith, William, Guest ed. *The Southern Communication Journal: The Rhetoric of the Rhetoric of Science* 58.4 (1993).

Leff, Michael, and John Angus Campbell, Guest eds. *Rhetorica: Rhetoric of Science*. 7.1 (1989).

Moss, Jean Dietz. *Novelties in The Heavens: Rhetoric and Science in The Copernican Controversy*. Chicago: University of Chicago Press, 1993.

Nelson, John S., Allan Megill, and Donald McCloskey, ed. *The Rhetoric of the Human Sciences: Language and Argument In Scholarship and Public Affairs*. Madison, WI: University of Wisconsin Press, 1987.

Myers, Greg. *Writing Biology*. Madison, WI: University of Wisconsin Press, 1991.

Prelli, Lawrence. *A Rhetoric of Science*. Carbondale, IL: University of South Carolina Press, 1990.

_____, Guest ed. *Argumentation*: Rhetoric of Science. 8.1 (1994).

Selzer, Jack, Ed. *Understanding Scientific Prose*. Madison, WI: University of Wisconsin, 1993.

Simons, Herbert W., Ed. *Rhetoric in The Human Sciences: Inquiries in Social Construction*. London: Sage, 1989.

_____, Ed. *The Rhetorical Turn: Invention and Persuasion in The Conduct of Inquiry*. London: Sage, 1990.

Reviews and Review Articles

Bazerman, Charles. "Scientific Writing as a Social Act." *New Essays in Technical and Scientific Communication: Research, Theory, Practice*. Ed. Paul V. Anderson, R. John Brockman, and Carolyn R. Miller. Farmingdale, NY: Baywood, 1983. 156-81.

Campbell, John Angus. "Review of Thomas Henry Huxley." *Quarterly Journal of Speech* 80 (1994): 245-7.

Condit, Celeste-Michelle. "The New Science of Human Reproduction: A Reflection on the Inadequacy of 'Disciplines' for the Understanding of Human Life." *Quarterly Journal of Speech* 79 (1993): 232-47.

Finocchiaro, Maurice A. "Review of Moss (1994)." *Philosophy and Rhetoric* 29 (1996): 206-9.

Gross, Alan G. "Forum: Response to Harris." *Rhetoric Society Quarterly* 22 (1992): 35-37.

_____. "Review of Jean D. Moss's Novelties in The Heavens." *Rhetorica* 11 (1993): 205-7.

_____. "Review of Reconstructing Scientific Revolutions and World Changes" *Quarterly Journal of Speech* 80 (1994): 225-7.

Harris, R. Allen. "Rhetoric of Science Meets the Nazis: Review of Robert Proctor's *Racial Hygiene*." *Rhetoric Society Quarterly* 19 (1989): 66-7.

_____. "Forum: Review of Alan G. Gross, *The Rhetoric of Science*." *Rhetoric Society Quarterly* 22 (1992): 32-35.

_____. "Review of Marcello Pera's Discourses of Science." *Rhetoric Review* 14 (1995): 207-13.

Houk, Davis W. "Review of Donald McCloskey's *Knowledge and Persuasion in Economics*." *Quarterly Journal of Speech* 81 (1995): 533-5.

Jenkins, Liz. "Review of M. Billig's *Arguing and Thinking*." *Philosophy and Rhetoric* 28 (1995): 83–86.

Melia, Trevor. "And Lo the Footprint ... Selected Literature in Rhetoric and Science." *Quarterly Journal of Speech* 70 (1984): 303-334.

_____. "Review of Dear (1991), Gross (1990), Myers (1990), Prelli (1989)." *Isis* 83 (1992): 100-6.

Miller, Carolyn R. "Some Perspectives on Rhetoric, Science, and History." *Rhetorica* 7 (1989): 101-14.

Myers, Greg. "Persuasion, Power, and the Conversational Model." *Economy and Society* 18 (1989): 221-44.

_____. "Writing Research and the Sociology of Scientific Knowledge: A Review of Three New Books." *College English* 48 (1986): 595-610.

Paradis, James. "Review of Selzer (1994)." *IEEE Transactions on Professional Communication* 37 (1994): 110-11.

Sillars, Malcolm O. "When Science Comes to Rhetoric's House." *Text and Performance Quarterly* 9 (1989): 229-42.

Dissertations

Ackerman-Ross, Floria Susan. "An Epsitemological Analysis of the Rhetorical Nature of Science as Manifested in the Academic Disciplines of Speech Communication and Psychology." The Pennsylvania State University, 1976.

Altimore, Michael. "The Rhetoric of Scientific Controversy: Recombinant DNA." The University of Iowa, 1990.

Anderson, Charles M. "Richard Selzer and the Rhetoric of Surgery." The University of Iowa, 1985.

Anderson, Ray Lynn. "Persuasive Functions of Science-Fiction: A Study in the Rhetoric of Science." University of Minnesota, 1968.

Berhardt, Stephen Arthur. "Text Structure and Rhetoric in Scientific Prose." The University of Michigan, 1981.

Blakeslee, Anne Madeline. "Inventing Scientific Discourse: Dimensions of Rhetorical Knowledge in Physics." Carnegie-Mellon University, 1993.

Carr, Diane Rose. "Science Writing and AIDS: The Effectiveness of Popular Information." University of South Carolina, 1990.

Cifoletti, Giovanna Cleonice. "Mathematics and Rhetoric: Peletier and Gosselin and the Making of the French Algebraic Tradition." Princeton University, 1993.

Ciolli, Russ Thomas. "A Rhetorical Analysis of Two Public Addresses by C. P. Snow: 'The Two Cultures and the Scientific Revolution' and 'The Moral Un-Neutrality of Science'." University of Pittsburgh, 1991.

Dowdey, Diane. "Literary Science: A Rhetorical Analysis of An Essay Genre and its Tradition." The University of Wisconsin-Madison, 1984.

Engnell, Richard Arthur. "The Rhetorical Criticism of Communication in Science: An Analysis of Early Publications in Transformational-Generative Grammar." University of California-Los Angeles, 1973.

Futrell, Wiley Michael. "Jeremy Rifkin Challenges Recombinant Research: A Rhetoric of Heresy." The Louisiana State University, 1993.

Graves, Heather Ann Brody. "The Rhetoric of Physics: An Ethnography of the Writing Processes in a Physics Laboratory." The Ohio State University, 1993.

Hansen, Kristine. "Rhetoric and Epistemology in Texts from the Social Sciences: An Analysis of Three Disciplines' Discourse about Modern American Blacks." The University of Texas at Austin.

Harris, Randy Allen. "The Life and Death of Generative Semantics." Rensselaer Polytechnic Institute, 1990.

Hatfield, David Lee. "The Rhetoric of Science: A Case Study of the Cold Fusion Controversy." The Louisiana State University, 1993.

Hearell, W. Dale. "Clarity and Rhetorical Intent in English Sixteenth Century Scientific Prose." Washington State University, 1990.

Holmquest, Anne. "Rhetorical Argument in Science: The Function of Presumption in the Origin of Species 'Revolution'." University of Iowa, 1989.

Jensen, John Vernon. "The Rhetoric of Thomas H. Huxley and Robert G. Ingersoll In Relation to the Conflict Between Science and Theology." University of Minnesota, 1966.

Johnstone, Anne Coffin. "Uses for Journal Keeping: An Ethnography of Writing in a University Science Class." State University of New York at Albany.

Karanikas, Marianthe Vaia. "Biology, Metaphor and Invention: Case Studies in the Writing of Biology." University of Illinois at Chicago, 1992.

Lee, Donald Paul. "A Rhetorical Analysis of the Scientific-Romantic Synthesis in the Popular Scientific Writings of Lewis Thomas." The Louisiana State University, 1991.

Lessl, Thomas Mark. "The Public Scientist: Rhetoric and the American Space Movement." The University of Texas at Austin, 1985.

Loveland, Jeff. "Rhetoric and Science in Buffon's Natural History." Duke University, 1994

Marston, Peter J. "Rhetorical Forms and Functions of Cosmological Argument." University of Southern California, 1987.

Mcomber, James Brant. "Philosophy in the Service of Rhetoric: Rhetoric and Antirhetoric in the Creation Science Controversy." The University of Iowa, 1992.

Miller, Carolyn Rae. "Environmental Impact Statements and Rhetorical Genres: An Application of Rhetorical Theory to Technical Communication." Rensselaer Polytechnic Institute, 1980.

Peters, Thomas Nathan. "Rhetorical Project for Understanding Scientific Change." University of Illinois at Urbana-Champaign, 1992.

Prelli, Lawrence John. "A Rhetorical Perspective for the Study of Scientific Discourse." The Pennsylvania State University, 1984.

Pickett, James Randolph. "Rhetorical Movements in the Recombinant DNA Debates: Containment, Pollution, and the Politics of Science." University of Pittsburgh, 1994.

Ragonnet, James Lawrence. "The Relationship between Rhetorical and Scientific Invention." Rensselaer Polytechnic Institute, 1982.

Reeves, Carol Ann. "The Characterization of a Medical Problem: An Analysis of the Writing on AIDS in Medical Science." Texas Christian University, 1989.

Rowan, Katherine Ellen. "Producing Written Explanations of Scientific Concepts for Lay Readers: Theory and a Study of Individual Differences Among Collegiate Writers." Purdue University, 1986.

Scott, Darrell William. "Thomas Reid and Eighteenth-Century Science: A Re-Evaluation of the Philosophical and Rhetorical Significance of his Philosophy of Common Sense." Wayne State University, 1977.

Segal, Judith Zelda. "Reading Medical Prose as Rhetoric: A Study in the Rhetoric of Science." University of British Columbia, 1988.

Swoboda, Merrily Kodis. "The American Rhetorical Career of Louis Agassiz: A Case Study of Transformations in American Science." University of Pittsburgh, 1978.

Taylor, Charles Alan. "The Rhetorical Construction of Science: Demarcation as Rhetorical Practice." University of Illinois, 1990.

Waddell, Craig. "The Role of Pathos in the Decision-Making Process: A Study in the Rhetoric of Science Policy." Rennselaer Polytechnic Institute, 1989.

Weimer, Donna Schimeneck. "A Rhetorical Analysis of a Scientific Controversy: Margaret Mead versus Derek Freeman in Cultural Anthropology." The Pennsylvania State University, 1991.

Welsh, Susan Booker. "Edgar Allan Poe and the Rhetoric of Science." Drew University, 1987.

Wenzel, Joseph Wilfred. "A Rhetorical Perspective for the Study of Scientific Discourse." University of Illinois at Urbana-Champaign, 1963.

Yanos, Susan Brenda. "A Rhetorical Analysis of the Current Challenges to the Evolutionary Paradigm." Ball State University, 1990.

Index

A

AARST, see: American Association for the Rhetoric of Science and Technology, The
abstract rhetors, see: rhetors, abstract
accommodation, see: audience (accommodation of)
accountability 170, 171
acquired immunodeficiency syndrome, see: AIDS
action xxvii, 107, 128
active voice, see: voice (active)
adherence 30, 43, 46, 90
Adler, Mortimer, J. 122
  quoted 107, 109
Adovasio, J. M.
  quoted 55
AFA, see: American Forensic Association, The
agreement xi, 98
  vs. conviction 128
AIDS xxxiv, 128, 151, 153, 155, 156, 165
Alhazen, Ibn al- 30
allegiance
  and research xxvi–xxviii
Allison, Paul D. 103
allopatric speciation 73, 80, 85
  see also: Mayr, Ernst
American Association for the Rhetoric of Science and Technology, The (AARST) xlii
American Forensic Association, The (AFA) xxi, xlii
American Psychology Association, The (APA) 8
American Sign Language (ASL) xxxiii, 92
amplification 108
Anaxagoras xxxvii
Anderson, Ray Lynn xl, xli
Anderson, Wilda xli, xliii
anecdote, representative, see: representative anecdote
anemia, sickle cell 135
Annals of Internal Medicine (journal) 152, 153, 156, 159, 162, 163
anomaly 28, 59, 64, 95, 101
Antczak, Frederick
  quoted 144
anthropology xx, xxxvii, xlii, 60, 92
  of science xxvii, xliii
anthropomorphism 11, 13, 102
APA, see: American Psychology Association, The
apologia 61
appeal, rhetorical 107–108, 128
  Kuhn as a xxxii, 54–56, 95
  Mark-Antony a 5, 6
  see also: ethos, logos, pathos
appropriateness 165

and Quintilian 142
and social construction 140–143
of article for journal 189, 199, 200
Aquinas, Thomas 37
archeology xii, xxix, xxxii, 53–67, 202
  evidence in 53–54
argument field xvi, xix, xxi–xxiii, xlii, 19–38
  optics as af 21, 26–27
  scientific af 22
  see also: community (argumentative)
argument (argumentation) xi, xx, xxvii, xxxv, xliv, 43, 53, 124, 128, 141, 153, 161, 165, 170, 184, 196, 200, 202, 203, 206, 207, 208, 209, 212
  a fortiori a 12, 33, 57
  and biology 200
  and experimentation 176–177, 179
  and forum xxxii
  and topics 99
  counter-a xxviii, 59
  dialectical vs. rhetorical a 107–109
  Kuhn on a xxviii, 152
  lesser-of-two-evils a 57
  public a 85
  reconstruction of a 138–139, 146–147
  technical a 85, 128
  et passim
  see also: debate; deduction; induction; syllogism
Aristarchus 3
Aristotle xii, xxiv, xxviii, xxix, xxxiv, xxxviii, xlv, 26, 30, 37, 54, 120–121, 143
  and the wall of certainty xiv
  and the wall of expertise xiv, xl
  A's definition of rhetoric 20–21
  on appeals 54
  on contingency 140
  on dialectic xii
  on explanation 19
  on ethos 87, 89
  on light 36
  on science 36
  quoted 89, 138, 142
Armstrong, Herbert W. 80
Arnold, Matthew 48
arrangement 20, 29, 30, 31, 32, 33, 34, 49, 153, 182–183, 189, 200, 201, 203, 205, 209, 213, 214
articles, journal xviii, xxxv–xxxvi, 169, 172, 187, 188, 189, 200, 213
  r a 152, 189, 200, 201, 203, 205, 213
  see also: experimental report, peer review
ASL, see: American Sign Language
assent 107, 128, 131, 188